AF581781

Hemorheology and Metabolism

Hemorheology and Metabolism

Editor

Norbert Nemeth

MDPI • Basel • Beijing • Wuhan • Barcelona • Belgrade • Manchester • Tokyo • Cluj • Tianjin

Editor
Norbert Nemeth
Department of Operative
Techniques and Surgical
Research, Faculty of Medicine
University of Debrecen
Debrecen
Hungary

Editorial Office
MDPI
St. Alban-Anlage 66
4052 Basel, Switzerland

This is a reprint of articles from the Special Issue published online in the open access journal *Metabolites* (ISSN 2218-1989) (available at: www.mdpi.com/journal/metabolites/special_issues/Hemorheology_Metabolism).

For citation purposes, cite each article independently as indicated on the article page online and as indicated below:

LastName, A.A.; LastName, B.B.; LastName, C.C. Article Title. *Journal Name* **Year**, *Volume Number*, Page Range.

ISBN 978-3-0365-2895-3 (Hbk)
ISBN 978-3-0365-2894-6 (PDF)

Contents

About the Editor vii

Preface to "Hemorheology and Metabolism" ix

Jean-Frédéric Brun, Emmanuelle Varlet-Marie, Justine Myzia, Eric Raynaud de Mauverger and Etheresia Pretorius
Metabolic Influences Modulating Erythrocyte Deformability and Eryptosis
Reprinted from: *Metabolites* **2021**, *12*, 4, doi:10.3390/metabo12010004 1

Alicja Szołna-Chodór and Bronisław Grzegorzewski
The Effect of Glucose and Poloxamer 188 on Red-Blood-Cell Aggregation
Reprinted from: *Metabolites* **2021**, *11*, 886, doi:10.3390/metabo11120886 31

Katalin Biro, Gergely Feher, Judit Vekasi, Peter Kenyeres, Kalman Toth and Katalin Koltai
Hemorheological Parameters in Diabetic Patients: Role of Glucose Lowering Therapies
Reprinted from: *Metabolites* **2021**, *11*, 806, doi:10.3390/metabo11120806 43

Bence Tanczos, Viktoria Somogyi, Mariann Bombicz, Bela Juhasz, Norbert Nemeth and Adam Deak
Changes of Hematological and Hemorheological Parameters in Rabbits with Hypercholesterolemia
Reprinted from: *Metabolites* **2021**, *11*, 249, doi:10.3390/metabo11040249 53

Sandor Szanto, Tobias Mody, Zsuzsanna Gyurcsik, Laszlo Balint Babjak, Viktoria Somogyi and Barbara Barath et al.
Alterations of Selected Hemorheological and Metabolic Parameters Induced by Physical Activity in Untrained Men and Sportsmen
Reprinted from: *Metabolites* **2021**, *11*, 870, doi:10.3390/metabo11120870 65

Takeshi Hashimoto, Hayato Tsukamoto, Soichi Ando and Shigehiko Ogoh
Effect of Exercise on Brain Health: The Potential Role of Lactate as a Myokine
Reprinted from: *Metabolites* **2021**, *11*, 813, doi:10.3390/metabo11120813 79

Korbinian Sebastian Hermann Ksoll, Alexander Mühlberger and Fabian Stöcker
Central and Peripheral Oxygen Distribution in Two Different Modes of Interval Training
Reprinted from: *Metabolites* **2021**, *11*, 790, doi:10.3390/metabo11110790 91

Diana Schrick, Margit Tőkés-Füzesi, Barbara Réger and Tihamér Molnár
Plasma Fibrinogen Independently Predicts Hypofibrinolysis in Severe COVID-19
Reprinted from: *Metabolites* **2021**, *11*, 826, doi:10.3390/metabo11120826 103

Gustavo Sampaio de Holanda, Samuel dos Santos Valença, Amabile Maran Carra, Renata Cristina Lopes Lichtenberger, Bianca de Castilho and Olavo Borges Franco et al.
Translational Application of Fluorescent Molecular Probes for the Detection of Reactive Oxygen and Nitrogen Species Associated with Intestinal Reperfusion Injury
Reprinted from: *Metabolites* **2021**, *11*, 802, doi:10.3390/metabo11120802 117

Mihai Oltean, Jasmine Bagge, George Dindelegan, Diarmuid Kenny, Antonio Molinaro and Mats Hellström et al.
The Proteomic Signature of Intestinal Acute Rejection in the Mouse
Reprinted from: *Metabolites* **2021**, *12*, 23, doi:10.3390/metabo12010023 131

Ibitamuno Caleb, Luca Erlitz, Vivien Telek, Mónika Vecsernyés, György Sétáló and Péter Hardi et al.
Characterizing Autophagy in the Cold Ischemic Injury of Small Bowel Grafts: Evidence from Rat Jejunum
Reprinted from: *Metabolites* **2021**, *11*, 396, doi:10.3390/metabo11060396 . . . . **147**

Csaba Korei, Balazs Szabo, Adam Varga, Barbara Barath, Adam Deak and Erzsebet Vanyolos et al.
Hematological, Micro-Rheological, and Metabolic Changes Modulated by Local Ischemic Pre- and Post-Conditioning in Rat Limb Ischemia-Reperfusion
Reprinted from: *Metabolites* **2021**, *11*, 776, doi:10.3390/metabo11110776 . . . . **163**

About the Editor

Norbert Nemeth

Norbert Nemeth, MD, PhD, DSc is a full professor, head of the Department of Operative Techniques and Surgical Research, Faculty of Medicine, University of Debrecen, Hungary (since 2013), and vice-dean for educational affairs of the Faculty of Medicine, University of Debrecen (since 2017). He graduated from the Faculty of Medicine, University of Debrecen in 2000 as MD, received his PhD degree in 2004, completed habilitation in 2011, and obtained the doctor of science (DSc) degree at the Hungarian Academy of Sciences in 2017. His main research focus: ischemia-reperfusion, sepsis, microcirculation and hemorheology, experimental surgery and microsurgery. Former president of the International Society for Experimental Microsurgery (ISEM, 2014–2016), president of the Hungarian Society of Hemorheology (2016–2021). Honorary member of the Sociedade Brasileira para o Desenvolvimento da Pesquisa em Cirurgia (SOBRADPEC, 2012). Board member of the International Society for Clinical Hemorheology (since 2008), and of the European Society for Surgical Research (ESSR, Award Committee, 2008–2013). President of the Experimental Surgical Session of the Hungarian Surgical Society (2013–2015, 2021–), vice-president of the Hungarian Society of Laboratory Animal Science (2012-2021). Member of the Editorial Board of the journals Microsurgery (since 2006), the Annals of Plastic Surgery (since 2008), the Clinical Hemorheology and Microcirculation (since 2014), the Acta Cirurgica Brasileira (associate editor since 2015), the Frontiers in Physiology (Red Blood Cell Physiology section, review editor since 2018), the Series on Biomechanics (since 2019), and the Journal of Cellular Biotechnology (since 2021). Author and co-author of more than 130 scientific full papers.

Preface to "Hemorheology and Metabolism"

By A.L. Copley's definition, hemorheology deals with macro- and micro-dimensions of blood flow properties, and the vessel wall with which the flowing blood comes into direct contact. In this sense, hemorheology plays an important role in determining the characteristics of blood flow, the tissue perfusion, microcirculation and the shear-stress-related endothelial functions in health and disease. In the past decades, clinical and experimental studies revealed that micro-rheological parameters, such as red blood cell deformability and aggregation, are influenced by numerous factors, including metabolic ones. However, several questions remained unanswered related to the magnitude and dynamics of changes, border of reversibility/irreversibility, pathophysiology, as well as preventive and therapeutic issues. Thanking to the modern investigative methods new insights have been provided into the relation of blood rheology and metabolites. This relation has a significance in cardiovascular and metabolic disorders, in inflammatory processes, including sepsis and ischemia-reperfusion, in intensive therapy, sports medicine, as well as in exercise and comparative physiology.

This e-book version of the Special Issue "Hemorheology and Metabolites" has been dedicated to the novel findings and recent advances in this field, presenting clinical or clinically oriented experimental research and review articles in the context of metabolites, metabolic alterations and blood macro- and micro-rheology. Many thanks to all the authors of these papers for their valuable contributions to the new results and concepts!

Norbert Nemeth
Editor

 metabolites

Review

Metabolic Influences Modulating Erythrocyte Deformability and Eryptosis

Jean-Frédéric Brun [1,*], Emmanuelle Varlet-Marie [2], Justine Myzia [1], Eric Raynaud de Mauverger [1] and Etheresia Pretorius [3]

[1] UMR CNRS 9214-Inserm U1046 Physiologie et Médecine Expérimentale du Cœur et des Muscles-PHYMEDEXP, Unité D'explorations Métaboliques (CERAMM), Département de Physiologie Clinique, Université de Montpellier, Hôpital Lapeyronie-CHRU de Montpellier, 34295 Montpellier, France; j-myzia@chu-montpellier.fr (J.M.); eric.raynaud-de-mauverger@chu-montpellier.fr (E.R.d.M.)

[2] UMR CNRS 5247-Institut des Biomolécules Max Mousseron (IBMM), Laboratoire du Département de Physicochimie et Biophysique, UFR des Sciences Pharmaceutiques et Biologiques, Université de Montpellier, 34090 Montpellier, France; emmanuelle.varlet@umontpellier.fr

[3] Department of Physiological Sciences, Stellenbosch University, Stellenbosch, Private Bag X1 MATIELAND, Stellenbosch 7602, South Africa; resiap@sun.ac.za

* Correspondence: j-brun@chu-montpellier.fr; Tel.: +04-67-33-82-84

Abstract: Many factors in the surrounding environment have been reported to influence erythrocyte deformability. It is likely that some influences represent reversible changes in erythrocyte rigidity that may be involved in physiological regulation, while others represent the early stages of eryptosis, i.e., the red cell self-programmed death. For example, erythrocyte rigidification during exercise is probably a reversible physiological mechanism, while the alterations of red blood cells (RBCs) observed in pathological conditions (inflammation, type 2 diabetes, and sickle-cell disease) are more likely to lead to eryptosis. The splenic clearance of rigid erythrocytes is the major regulator of RBC deformability. The physicochemical characteristics of the surrounding environment (thermal injury, pH, osmolality, oxidative stress, and plasma protein profile) also play a major role. However, there are many other factors that influence RBC deformability and eryptosis. In this comprehensive review, we discuss the various elements and circulating molecules that might influence RBCs and modify their deformability: purinergic signaling, gasotransmitters such as nitric oxide (NO), divalent cations (magnesium, zinc, and Fe^{++}), lactate, ketone bodies, blood lipids, and several circulating hormones. Meal composition (caloric and carbohydrate intake) also modifies RBC deformability. Therefore, RBC deformability appears to be under the influence of many factors. This suggests that several homeostatic regulatory loops adapt the red cell rigidity to the physiological conditions in order to cope with the need for oxygen or fuel delivery to tissues. Furthermore, many conditions appear to irreversibly damage red cells, resulting in their destruction and removal from the blood. These two categories of modifications to erythrocyte deformability should thus be differentiated.

Keywords: erythrocyte deformability; metabolism; hormones; homeostasis; eryptosis; stress; COVID-19; sleep apnea

check for updates

Citation: Brun, J.-F.; Varlet-Marie, E.; Myzia, J.; de Mauverger, E.R.; Pretorius, E. Metabolic Influences Modulating Erythrocyte Deformability and Eryptosis. *Metabolites* **2022**, *12*, 4. https://doi.org/10.3390/metabo12010004

Academic Editor: Norbert Nemeth

Received: 29 November 2021
Accepted: 18 December 2021
Published: 21 December 2021

1. Introduction

Red blood cells (RBCs) are known to markedly modify their shape in order to transit into small capillary vessels, whose radius is smaller than their own [1]. This ability to deform also results in RBC elongation in flow. This property plays an important role in the blood viscosity at high shear rates, so that in this situation blood can be modeled as a Newtonian fluid [2,3], whose viscosity reflects RBC deformability.

In fact, the term 'red cell deformability' is not so easy to define, because it appears more and more obvious that RBCs can undergo many varieties of deformation in narrow channels or in flow according to the experimental or physiological situation. Classical studies using microscopic flow visualization led to the idea that the deformation of RBCs resulted from continuous viscous deformation, which was called "fluid drop-like adaptation" by H Schmid-Schönbein [4]. This kind of deformation was determined by the cytoplasm fluidity and the surface-area-to-volume ratio of the red cells.

Over recent years, these classical concepts have been reassessed with new sophisticated experimental approaches, resulting in a more complex picture, mostly in the context of a new emphasis given to the rheological behavior of RBCs in sickle-cell disease [5]. Studying the stiffness of RBCs from individuals with sickle-cell trait, Zheng [6] developed a microsystem able to measure the individual mechanical properties (i.e., shear modulus and viscosity) of a single red cell submitted to a shear stress. After the RBCs were deformed under the influence of this shear stress, the dynamic RBC recovery was monitored and analyzed according to the Kelvin–Voigt model, allowing the measurement of an elastic shear modulus of RBCs submitted to different shear rates. Even more recently, another group [7–9] developed a microfluidic impedance red cell assay (MIRCA) in order to measure RBC transition through narrow openings and also challenged to some extent the concept of 'fluid drop-like' RBCs whose deformation was assumed to be mostly related to viscosity with little or no elastic component. The authors defined new parameters such as an RBC occlusion index (ROI) and an RBC electrical impedance index (REI), which measure the cumulative percentage of vessel occlusion and the impedance change, respectively.

These recent experiments suggest that the process of RBC deformability was until now misinterpreted by previous microfluidic measurements. Furthermore, Lanotte and coworkers [10] recently performed experiments on and simulations of microcirculatory flow in various conditions of volume fractions and shear rates. They showed that RBCs undergo a large variety of morphological modifications during their deformation. With an increasing shear rate, the RBCs were first shown to tumble, then they were shown to roll, then they adopted the form of a tumbling stomatocyte, and finally they exhibited various polylobed shapes that were only observed above a threshold value of the viscosity contrast between the plasma and cytosol. In another paper, the same team further described the complexity of the mechanisms involved in these transition processes from one shape to another under the influence of an increase in shear stress [11].

All this recent literature emphasizes the complexity of red cell deformation, which is far from a simple phenomenon. In light of this quite recent literature, it becomes clear that the experiments performed in recent decades on so-called 'red cell deformability' explored only a limited aspect of this physiological mechanism. Until now, our knowledge on the regulation of RBC deformability is mostly based on the measurement of the deformation of RBCs entering a narrow channel and the deformation of RBCs submitted to a shear stress in flow. These two approaches are likely to rely on different cellular mechanisms. It is likely that most of this information needs to be investigated again with newer experimental approaches.

Some experiments have been conducted with artificially stiffened erythrocytes, showing that impaired deformability dramatically decreases perfusion, with a quite different effect in various tissues [12]. It is clear that such experiments do not reflect typical in vivo situations, but they are not meaningless. Examples of extremely rigid erythrocytes can be found in situations such as sickle-cell disease [5,13]. In this case, consistent with these experiments, stiffened RBCs can clearly be responsible for vessel occlusion. In fact, in the majority of cases, the modification of the erythrocyte rigidity is not so dramatic and the RBCs remain able to deform and transit through the microcirculation. However, such moderately rigidified erythrocytes transit mostly in the largest microvessels, a situation that has been termed capillary maldistribution [14,15].

Research has also focused on studying the physiological and pathological changes that happen to RBCs during either disease or when RBCs are artificially rigidified. Over the last 30 years of the 20th century, an impressive body of literature has been published on the factors that modify erythrocyte deformability. This early research did not, however, clearly separate reversible and irreversible RBC rigidification, and most of this early research was carried out before the emergence of the concept of eryptosis (or programmed RBC death, which is similar to apoptosis but specific to the anuclear RBC) [16–23].

Conditions where eryptosis have been noted, have all been reported to impair erythrocyte deformability. This is the case for hypoxia, iron deficiency, cancers, dehydration, metabolic syndrome, phosphate depletion, hemolytic anemia, heart failure, diabetes mellitus, chronic kidney disease, mycoplasma infection, malaria, hemolytic uremic syndrome, sepsis, sickle-cell disease, etc. Additionally, factors known to inhibit eryptosis such as catecholamines, erythropoietin, adenosine, resveratrol, urea, vitamin E, and caffeine have also been reported to modify erythrocyte rheology [24]. It is clear that the most important regulator of erythrocyte deformability is the clearance of rigid erythrocytes within the spleen. A mechanical checking of the deformability of circulating erythrocytes is regularly performed in the splenic microcirculation, so that the RBCs that are not able to correctly squeeze through the narrow splenic slits are trapped and removed from circulation [25–27].

Therefore, we suggest that all the literature dealing with the factors that modify erythrocyte deformability should be analyzed in light of the new concept of eryptosis. For example, the pathologic alterations of erythrocytes that occur in metabolic diseases such as diabetes [28,29] should probably be considered as a completely different process to the physiological reversible erythrocyte stiffening observed during muscular exercise [30]. The exercise-induced decrease in red cell deformability is a very interesting example of this difference. In healthy athletes, exercise transiently modifies the blood rheology without evidence of increased eryptosis [31,32], while in sickle-cell disease patients, it induces a long-lasting stiffening of the red cells that seems to be explained by irreversible damage and probable further eryptosis [5]. Sickle-cell disease is clearly a condition associated with increased eryptosis [33].

Additionally, there are many influences that can modify RBC deformability. This paper is an attempt to summarize this large body of literature and to integrate our knowledge with regards to the classical definitions of deformability and eryptosis.

2. The Main Classical Physicochemical Modifiers of RBC Deformability

Classically, the most important modifiers of RBC deformability were the physicochemical characteristics of the surrounding environment [34]. A biphasic influence of pH and osmolality on RBC deformability displaying a "u-shaped curve" has been described. The deformability of red cells appears to be optimal within the physiological range and is markedly impaired above and below these narrow physiological boundaries. This stiffening effect of pH and osmolality changes has been assumed to increase erythrocyte trapping in the spleen and thus decrease RBC lifespan [35]. It has also been shown that an environment containing proteins such as albumin is mandatory for preventing alterations of the erythrocyte shape, since albumin has the ability to prevent and even reverse echinocytosis [36].

Recently, the physiological relevance of such osmotic changes to red cell water content has been emphasized by studies showing that aquaporin-1 (AQP1), which is expressed in red cell membranes, may drive rapid water exchange and that this exchange results in an important volume change (up to 39%) [37]. This effect is almost suppressed in AQP1-knockout (KO) erythrocytes [37]. Such alterations of the erythrocyte volume in microvessels result in an increase in the osmotic gradient between the plasma and interstitial fluid. Red cells thus appear to be "micropumps" that regulate in situ local osmolarity [37]. Accordingly, red cells are likely to be major regulators of water exchange in the body, and thus, to contribute to body water homeostasis [37].

Some studies have also been devoted to divalent cations. For example, magnesium has been shown to protect RBCs from in vitro experimental rigidification by several procedures [38,39].

3. A Brief Overview of Eryptosis

The mechanisms of eryptosis are described in the classical publications by the Lang group [16–19,21,24,40–46]. Eryptosis occurs in conditions such as heart failure, uremia, haemolytic uremic syndrome, sepsis, fever, dehydration, mycoplasma infection, anemia, metabolic syndrome, cancer, diabetes, hepatic failure, Wilson's disease, malaria, sickle-cell anaemia, iron deficiency, thalassemia, glucose-6-phosphate dehydrogenase deficiency, Parkinson's disease, type 2 diabetes, Alzheimer's disease, and rheumatoid arthritis [20]. Figure 1 shows a brief overview of eryptosis, which is triggered by various signaling pathways, including the presence of circulating inflammatory molecules that results in oxidative stress. Eryptosis is therefore the culminative term for the end-stage of the process, resulting in cell death; even so, as RBCs are known to be exceptionally resilient, they do have the ability to recover if the stressor molecule or environment is changed. However, there is, as with all physiological and molecular pathways, a point of no return, after which recovery is not possible.

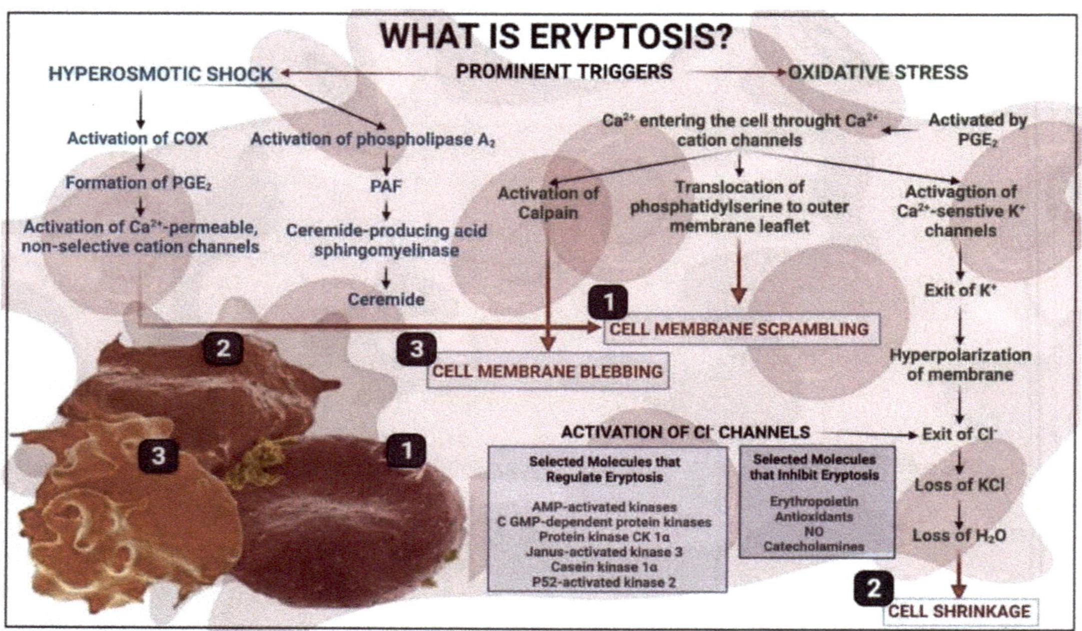

Figure 1. A brief overview of eryptosis (adapted from [47]) showing the pathways that initiate it, under the influence of osmotic shock or oxidative stress, resulting in activation of intracellular pathways, leading in turn to phospholipid membrane scrambling (1); cell shrinkage (2); and membrane blebbing (3). Figure created using www.biorender.com accessed on 17 December 2021.

4. RBC Receptors

The receptors expressed on the red cell membrane play an important role in its optimal functioning. The following paragraphs will provide a brief summary of these receptors and their role in deformability and eryptosis. Several receptors for various ligands are present on the red cell membrane [1]. Some of these are mentioned later in this review. Among them, we should mention the N-methyl D-aspartate (NMDA) receptors, which are expressed on the red cell membrane. These receptors are major targets of divalent cations and mediate most of their effects in various conditions. They contribute to the regulation of intracellular calcium in erythrocytes [48]. However, it has been shown that the experimental activation of NMDA receptors has no measurable effect on the rheological properties of erythrocytes [49], suggesting that the effects of divalent cations on red cell deformability are not mediated by NMDA receptors. Notwithstanding, an abnormally high abundance of N-methyl D-aspartate receptors on the erythrocyte membranes of sickle-cell disease patients has been reported and is associated with an excessive calcium uptake. Presumably, this process can trigger the cascade of red-cell-damaging events that include RBC rigidification [50]. Moreover, Unal and coworkers recently reported that memantine, an antagonist of NMDA receptors, impairs RBC deformability in rats [51]. All this suggests that, despite the lack of a measurable effect of NMDA receptor activation on the deformability of normal erythrocytes, the inactivation of these receptors does have such an effect, whose physiological significance needs to be more precisely established.

There is a large body of literature about the purinergic receptors in RBCs, and despite a relative paucity of hemorheological studies dealing with this issue, this literature suggests that they may be important modulators of red cell deformability. P1 and P2 purinergic receptors are expressed on the red cell membrane and are able to bind extracellular nucleosides and nucleotides [52]. P1 receptors are stimulated by adenosine and P2 receptors are stimulated by adenosine triphosphate (ATP). P2 receptors comprise P2X and P2Y receptors. Their activation triggers the intracellular signaling pathways in progenitor erythrocytes, resulting in reactive oxygen species formation, microparticle release, and apoptosis. In mature erythrocytes, P2 receptor stimulation is involved in cell volume regulation, phosphatidylserine exposure, eicosanoid release, hemolysis, impaired ATP release, and susceptibility or resistance to infection [52]. Furthermore, the P1 receptor agonist adenosine protects against eryptosis via the activation of a pathway which most probably acts downstream of PKC. Purinergic signaling in erythrocytes is probably involved in the maintenance of microcirculation in ischemic tissue [45]. Erythrocytes, in fact, are not only targets of purinergic stimulation but are also able to release ATP and ADP [53]. It has been shown that ATP and ADP are continuously released by RBCs and are later converted outside the cell into adenosine, which then re-enters the red cell. ATP release by erythrocytes is triggered by hypoxia, hypercapnia, mechanical deformation, reduced O_2 tension, acidosis, cell swelling, prostacyclin analogues, and β-adrenoceptor agonists. According to Sprague and coworkers [54], erythrocyte transit through narrow capillaries, where the shear stress makes them deform, results in ATP release. This ATP then binds to purinergic P2 receptors expressed on endothelial cell membranes, resulting in a release of NO and PGI2 [55]. Caffeine increases ATP release from RBCs, probably via its effect on the intracellular cAMP levels. Intracellular ATP is essential for maintaining the function and structural integrity of erythrocytes. By contrast, ATP depletion has been shown to sensitize RBCs to the eryptotic effects of Ca^{2+} [20].

The A3 adenosine receptor is also expressed on red cells. Its antagonist reversine (2-(4-morpholinoanilino)-6-cyclohexylaminopurine) has important effects in nucleated cells, since it is known to induce cell cycle arrest, inhibit cell proliferation, influence cellular differentiation, induce cell swelling, and trigger apoptosis. In fact, since erythrocytes lack mitochondria, they exhibit a different response to reversine. In this case, reversine powerfully inhibits cell membrane scrambling after energy depletion, Ca^{2+} loading, and oxidative stress, and therefore prevents the occurrence of eryptosis [41].

Purinergic signaling is involved in the response to low blood O_2 which triggers ATP release by erythrocytes, leading to the stimulation of P2 × 2/3 receptors in the aortic body [56]. The breakdown of ATP into ADP inhibits ATP release via a negative feedback, which involves the P2Y13 receptors in human erythrocytes [57].

Undoubtedly, all these purinergic effects are likely to modulate RBC rheology, but, surprisingly, there is very little published information on this issue. I. Juhan-Vague reported forty years ago that the ADP released by rigid RBCs impaired the deformability of normal erythrocytes [28]. More recently, it was reported that the ADP release from RBCs in healthy human volunteers was lower in middle-aged than in young healthy individuals, and that fish oil intake improved the erythrocyte deformability, parallel to a 50% decrease in the ADP release [58].

Thus, erythrocyte deformability is included in a regulatory loop involving purinergic signaling. Erythrocyte stiffening inhibits the release of ATP, which is in turn increased when red cells become more deformable, e.g., when treated by hydroxyurea or the HMG-CoA reductase inhibitor simvastatin [59]. Subsequently, the stiffened erythrocytes release

ADP, which inhibits the release of ATP. This mechanism probably aims at maintaining RBC energy stores, but it is also likely to induce a self-potentiating loop resulting in RBC rigidification. Moreover, when erythrocytes are well-deformable and release ATP, they induce more NO production by the vessel wall, and this NO results in vasodilation and increased RBC deformability [60].

Acetylcholine (Ach) can also bind to RBCs that express both muscarinic [61] and nicotinic cholinergic receptors [62]. The hemorheologic effects of Ach are both an increase in RBC deformability and a decrease in RBC aggregation [63]. Further studies by the team of A. Muravyov have helped to describe the signaling pathways involved in the effects of Ach [64].

The RBC membrane also expresses receptors for the endogenous ligands of benzodiazepines [65], corticotropin-releasing factor (CRF) [66], and prolactin [67]. Most of these chemical messengers have been reported to modify in vitro or in vivo erythrocyte rheologic properties, but the data on these effects remain to some extent conflicting.

It is logical to hypothesize that the receptor-mediated alterations in erythrocyte deformability induced by chemical messengers that physiologically circulate in the blood are reversible adaptative processes that do not involve eryptosis. However, physiological factors such as NO, anandamide, iron, adenosine, retinoic acid, and zinc appear in the list of eryptosis-inducing substances that can be found in the recent review on this topic published by Pretorius and coworkers [20]. Therefore, it can be assumed that eryptosis (which involves an irreversible modification of erythrocytes leading to their premature death) can actually be triggered outside of any pathologic context, as an adaptation to a fully physiological situation.

The following paragraphs will revisit our previous knowledge on the effect of various circulating molecules on RBC deformability and integrate our new knowledge regarding eryptosis.

5. Iron and Oxidative Stress as Drivers of RBC Deformability

Among trace elements, iron is surely the most studied and the best-known. The frequency of iron deficient states and the possibility of treating them with iron supplementations is likely to explain this. It is known that roughly 30 to 40% of the total body iron is stored in the form of ferritin and hemosiderin. Additionally, a lower amount of iron is stored as transferrin [68]. Circulating ferritin is the classical quantitative marker of iron stores. It is also a marker of inflammation, since it appears in blood as a leakage product from damaged cells [69,70]. It is known that oxidant stress caused by increased serum ferritin levels and dysregulated iron rigidifies RBCs and induces eryptosis [71]. In the presence of high serum ferritin, e.g., in conditions such as hemochromatosis, the RBC structure is compromised [72–74]; see also Figure 2. Therefore, erythrocyte deformability may be moderately and reversibly impaired in some physiological situations but can also be irreversibly damaged, according to the severity of the applied stress. Another example is the early postburn period. In this situation, Bekyarova and coworkers have evidenced a decrease in RBC deformability that is related to the activation of lipid peroxidation [75] and is very likely to reflect an increase in programmed cell death in response to a major oxidant stress [76]. The susceptibility to spontaneous eryptosis which increases with erythrocyte age and oxidative stress is abrogated by antioxidants such as N-acetyl-L-cysteine [24]. An increased rate of eryptosis in such situations of erythrocyte damaging is perhaps another kind of homeostatic adaptation, protecting the body against exposure to older erythrocytes, which is now evidenced as independently associated with an increased risk of fatality [77,78].

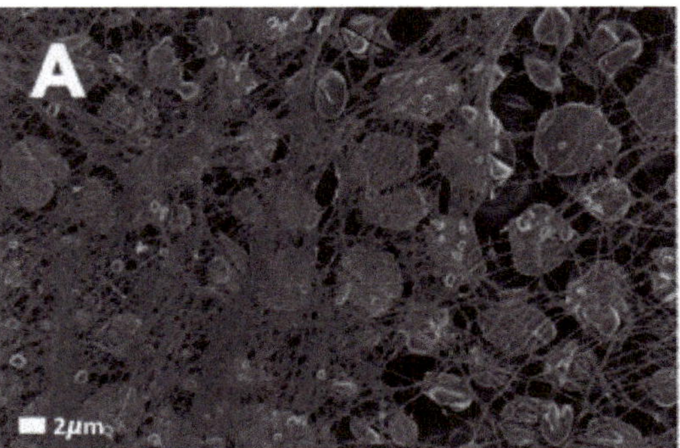

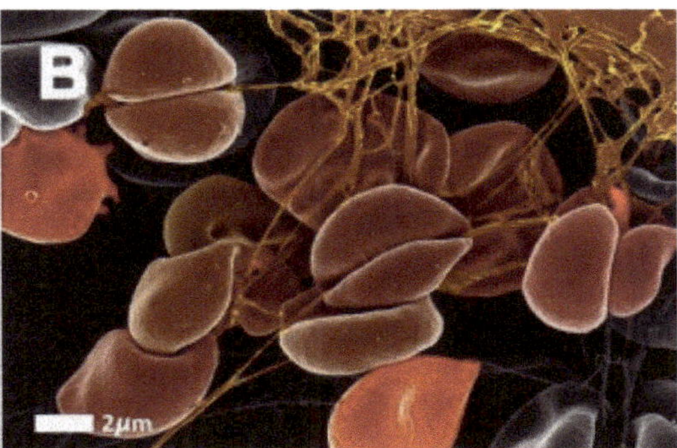

Figure 2. Morphology of RBCs in the presence of high circulating serum ferritin. This picture shows that in conditions like hemochromatosis, RBC structure is markedly compromised. (**A**) Individual with hereditary hemochromatosis (H63D/C2882Y), serum ferritin level 374 ng/mL^{-1}; (**B**) individual with hereditary hemochromatosis (H63D/wild type), serum ferritin level 1500 ng/mL^{-1}. Raw data from [72].

The high levels of serum ferritin found in those with conditions such as Alzheimer's disease and Parkinson's disease may also modify erythrocyte function and structure [47,79–83]. It is now well established that Fe^{++} triggers eryptosis [20]. This eryptotic effect of iron may explain why erythrocyte deformability is significantly impaired in hemochromatosis and hyperferritinemia [74,84,85].

Although a high value of serum ferritin cannot rule out the existence of iron deficiency, a low ferritin value is well recognized as being highly specific to iron deficiency. Experimental studies in iron-deficient rats have evidenced a lower erythrocyte deformability that appeared to be related at least in part to the lower hemoglobin content of the erythrocytes [86,87]. Athletes with low plasma ferritin also exhibit a higher value of blood viscosity, a higher plasma viscosity, and a higher RBC aggregation index when compared to athletes exhibiting normal values of plasma ferritin. By contrast, no difference in hematocrit or RBC deformability could be evidenced between these two subgroups [88]. In fact, iron is known to damage the structure of RBC membranes [89], resulting in more rouleaux networks.

6. Antioxidants

Erythrocyte deformability is improved by the antioxidants vitamin E [90,91], alpha-tocopherol [92], alpha-tocotrienol [93], fish oil, and dietary tea catechins [94]. Muscular activity is a situation known to be associated with an increase in oxidant stress, which in this case should of course be considered as a purely physiological event. However, this rise in oxidative stress can be very important and become harmful. Trained athletes appear to be protected against the potential deleterious effects of this oxidative stress. This is evidenced by the fact that in both rats [95] and humans [96], exercise-induced oxidative stress decreases RBC deformability in sedentary individuals but not in exercise-trained ones.

7. Zinc

Zinc is also known to increase the deformability of artificially hardened RBCs in vitro [97] and is frequently low in the serum of athletes, reflecting some degree of deficiency. It is known that athletes who exhibit low serum zinc values have a higher blood viscosity and impaired erythrocyte deformability [98], and this hemorheologic profile is associated with a decrease in exercise performance. Experimentally, a double-blind randomized trial with oral zinc gluconate in healthy volunteers was found to decrease the blood viscosity [99], while no significant effect on performance could be evidenced. Zinc was also shown to decrease erythrocyte aggregation both in vitro and in vivo [100]. More recently, in contrast with these findings, it was shown that zinc is also able to promote eryptosis [42]. It is likely that, as discussed above for iron, the impact of this mineral on RBCs can be different according to the severity of the applied stress, which explains this apparent paradox.

8. RBCs and Their Energy Needs

Since erythrocytes need energy to undergo deformation, the depletion of their energetic stores has a marked effect on their deformability. This is regularly observed when RBCs are stored in vitro. In this situation, there is a gradual temperature- and time-dependent decrease in the glucose and ATP levels, with a simultaneous rise in the intracellular levels of lactate and LDH. Parallel to this process, there is a time- and temperature-dependent swelling and an echinocytic transformation of RBCs. At the same time, a gradual increase in the RBC rigidity can be evidenced with the measurement of blood viscosity at a high shear rate. This process of echinocytosis can be partially reversed if the erythrocytes are resuspended in a buffer containing 0.2% albumin [101]. The literature on eryptosis shows that the glucose depletion of RBCs (and more generally, an energy crisis, see Figure 1 and Table 1) triggers the process of programmed cell death in RBCs [46].

Table 1. Factors influencing red cell deformability and eryptosis.

	Increases RBC Deformability	Decreases RBC Deformability	Increases Eryptosis (After [24] and [20])	Decreases Eryptosis (After [24] and [20])
Biologically active molecules and metabolites	ATP NO H2S Carbon monoxide Zn^{++} Lactate (in trained athletes)	Ketone bodies Cholesterol Glucose > 200 dg/mL Lactate (in sedentary subjects)	Aluminium Arsenic Cadmium Carbon monoxide Ceramide (acylsphingosine) Chromium Copper Fe^{++}, Energy depletion Glucose (via glycation) Osmotic shock Zn^{++}	NO Erythropoietin Catecholamines β and α
Hormones and chemical messengers	Acetylcholine Epinephrine Endothelin 1 Apelin Leptin Progesterone Erythropoietin Somatostatin Prostaglandin E1 DHEA	Glucagon Melatonin ADP PGE2 Norepinephrine (?) Leukotriene B4 Thyroxin IGF-I Estradiol	Anandamide Estradiol Leukotriene C(4) Lithium Lysophosphatidic acid Mercury PAF Phosphate Progesterone Prostaglandin E2 Silver ions Sphingosine	Adenosine Chloride Erythropoietin Nitroprusside (NO-donor) Urea
Nutritional factors	Tea catechins Vitamin E α-tocopherol, α tocoterol	Carbohydrate intake	Curcurmin Gossypol Oxysterol Phytic acid Retinoic acid Retinoic acid Selenium (sodium selenite) Tannic acid Vitamin K	Caffeine Glutathione Monohydroxyethylrutoside N-acetylcysteine Naringin Vitamin E

On the other hand, hyperglycemia also has an effect on erythrocytes that has been extensively investigated. In older studies, it was reported that short-term hyperglycemia does not markedly impair the blood rheology unless extremely high concentrations (hundreds of millimoles per liter) were reached. The relevance of these experiments is unclear, since such concentrations can never be found in human diseases [102]. However, the chronic exposure of RBCs to high concentrations of glucose is known to increase intracellular sorbitol, and high values of sorbitol within RBCs were shown to be associated with impaired erythrocyte deformability [103]. In fact, the concentrations of sorbitol used in the abovementioned experiments were extremely high and probably irrelevant to what can be found in human diseases.

Obviously, it is important to study this issue of high glucose levels because it can be relevant to the pathophysiology of the vascular complications of diabetes mellitus. In diabetes, it is well known that the blood rheology is altered [104], but these alterations are rather moderate when the disease is correctly equilibrated [105,106]. Thus, one cannot expect in the case of diabetes the conditions of rheologic occlusions as have been observed in classical experiments with hardened RBCs [12] or sickle-cell disease [5,13]. In contrast with these conditions of "overtly abnormal blood rheology", diabetes represents an example of "covertly abnormal" blood rheology [107], which has a different pathophysiological relevance but is also likely to induce some microcirculatory disturbances.

A recent study on 300 patients showed that there is a threshold value for the effect of chronic hyperglycemia on red cell rigidity at a value of 9.05% glycated hemoglobin. This means that the average blood glucose levels need to chronically remain above roughly 200 mg/dL to result in a measurable decrease in red cell deformability [108]. This is likely to reflect nonreversible alterations of red cell structure and properties that do not have the same significance and may be involved in a pathologic process. Not surprisingly, in a recent prospective study on 247 diabetics, it was shown that RBCs' characteristics are predictors of the development of diabetic retinopathy [109]. In this context, eryptosis has been reported to occur [110]. In fact, in the metabolic syndrome which represents a disorder that combines hyperglycemia, dyslipidemia, hypertension, and obesity and leads to diabetes and atherosclerosis, increased eryptosis is already observed [111]. In vivo, an acute hyperglycemic "spike", which is a situation associated with a rise in oxidant stress in diabetes, has been shown to impair blood rheology [112]. Recently, Babu and Singh [113] reported an effect of glucose added in vitro to a medium containing erythrocyte from diabetic patients. Increasing the glucose concentrations resulted in an increase in RBC aggregation and a decrease in RBC deformability. This decrease in deformability was associated with a change in the shape of the erythrocytes. The RBCs' perimeter-to-area ratio was increased, and this effect likely explained, at least in part, the alteration in deformability. This effect was observed in the RBCs from diabetic patients but not in the RBCs from healthy subjects. This question was also investigated by Shin [114], who observed significant hemorheological changes in red cells incubated with glucose. Both the deformability and aggregation of the erythrocytes decreased in a dose- and time-dependent manner. These authors interpreted these hemorheological modifications as a consequence of the glucose-induced (auto)oxidation and glycation of the erythrocytes.

In fact, in normal red cells, glucose deprivation (energy crisis) [20,24], rather than hyperglycemia, induces irreversible red cell damage and eryptosis [115,116]. Interestingly, in physiological conditions, RBC rigidity is positively correlated with carbohydrate intake in trained athletes [117]. Presumably, this physiological effect does not have the same significance as the red cell stiffening induced by chronic hyperglycemia. Physiological changes in blood glucose concentration may be associated with some hemorheological alterations, but when these changes reach a pathological range, they may be associated with red cell damage due to free radicals or other factors and thus trigger irreversible alterations and eryptosis. Figure 3 shows an RBC from an individual with diabetes.

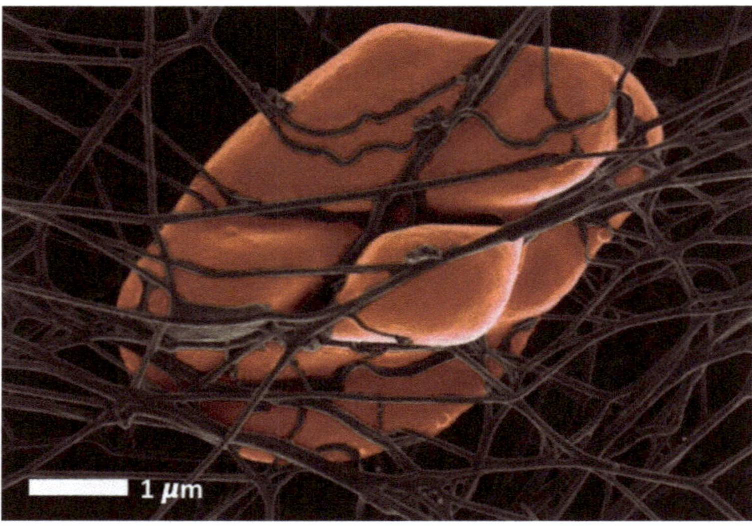

Figure 3. A representative RBC from a type 2 diabetes patient (Raw data from [83]. Erythrocyte deformability is often found to be moderately impaired in diabetes mellitus patients, due to several metabolic and hormonal disturbances (glycation and oxidative stress) that may also promote eryptosis. This is an example of "covertly abnormal blood rheology", which is supposed to induce microcirculatory disturbances. The main glucose-regulating hormones insulin and glucagon have also been reported to exert an influence on red cell deformability, whose pathophysiological relevance remains unclear.

9. RBCs and Circulating Lipids

Among the factors that are the strongest statistical determinants of blood viscosity, blood lipid concentrations deserve a special emphasis. All the studies investigating the relationships between serum cholesterol and erythrocyte deformability have evidenced a strong positive correlation between cholesterol levels and RBC rigidity [118]. This was interpreted as a reflection of the changes in the membrane lipid composition which modified the cell membrane fluidity and thus the whole deformability of erythrocytes. If RBC membrane cholesterol content is decreased under the effects of treatment by the lipid-lowering drug simvastatin, there is an increase in red cell deformability and an increase in ATP release by the erythrocytes [119]. Polyunsaturated fatty acids of the omega 3 family (3PUFA), on the other hand, improve RBC deformability in both healthy volunteers [58,120,121] and patients with disease [122–124].

Postprandial lipemia is a physiological condition involving increased lipid concentrations in the blood. In this condition, of course, the abovementioned changes in red cell deformability are observed, and it is also observed that lipids and fibrinogen may act synergistically, suggesting that the effect of large triglyceride-rich lipoproteins can be potentiated by fibrinogen [125].

10. The Effect of Lactate and Ketones on RBCs

Lactate is an important metabolite generated by carbohydrate breakdown upstream in the Krebs cycle and released into the blood in situations of hypoxia or simply during exercise, and it has been shown to exert hemorheological effects [126]. If erythrocytes are submitted to increased concentrations of lactate in vitro, there is a decrease in erythrocyte deformability. In vivo, a rigidification of erythrocytes during muscular exercise is only found when the circulating lactate concentrations rise above 4 mmol/L, i.e., the level of the onset of acidosis [127]. Interestingly, in highly trained endurance athletes, this stiffening effect of lactate on RBCs is no longer found. Conversely, lactate appears in this case to improve erythrocyte deformability [128]. This specific training-induced pattern of response to lactate may be one of the explanations for the exercise-induced arterial hypoxemia that occurs in extreme athletes.

Ketone bodies are another metabolite that can be used by tissues as an alternative fuel in some physiological situations such as starvation. Situations such as a short-term ketogenetic diet [129] have been experimentally shown to impair red cell flexibility.

11. Nitric Oxide and RBC Function

Nitric oxide (NO) is surely one of the most important substances known to interact with erythrocytes, which are in turn able to release it [130,131]. The major source of nitric oxide is the endothelial cell, but nitric oxide can also be produced by the erythrocyte, which possesses functional NO-synthesizing mechanisms [132]. NO synthesis in RBCs via nitric oxide synthase (NOS) and NO release into the blood stream can be induced by mechanical stress, so that NO is released by the red cell in close proximity to the vessel wall [133]. One of the effects of NO on RBCs is to protect them from subhemolytic mechanical damage [134], but NO also increases RBC deformability, as demonstrated by M. Bor-Kuçukatay and coworkers [135]. These investigators also reported that, in contrast to NO donors which improved erythrocyte deformability, NOS inhibitors above a threshold concentration value decreased erythrocyte deformability. Nitric oxide donors, as well as the NO precursor L-arginine and the potassium blocker TEA, have been shown to reverse the effects of NOS inhibitors [135]. Therefore, NO is not only a potent regulator of vascular tone but is also a major physiological regulator of blood rheology via its direct effect on RBC deformability. Furthermore, NO release by polymorphonuclear leukocytes increases RBC deformability [136]. In fact, the effect of NO on erythrocyte rigidity depends on the NO concentration, as studied by Mesquita and coworkers [137], who reported a biphasic effect. At a NO concentration of $10^{(-7)}$ M, the erythrocyte deformability improved, while at $10^{(-5)}$ M, the membrane lipid fluidity decreased. At a NO concentration of $10^{(-3)}$ M, there was an increase in the methemoglobin concentration and the RBC deformability decreased, although the membrane fluidity and lipid peroxidation were not changed compared to the control. We should also mention the experiments with spermine NONOate that resulted in an increase in the RBC deformability, due to an effect on the RBC membrane associated with an improvement in its oxygen carrying properties [63]. Older erythrocytes exhibit a decrease in both internal NO synthesis and sensitivity to external NO, which likely explains, at least in part, why older erythrocytes are less deformable [138]. It is important to point out that nitric oxide is also a protector of the red cell against eryptosis [44]. Presumably, this antieryptotic effect is likely to prevent a further decrease in RBC deformability.

In human diseases, the effects of nitric oxide on RBC deformability have some potentially interesting applications. In *Plasmodium falciparum* malaria, hypoargininemia has been reported to impair nitric oxide production and decrease erythrocyte deformability, even more so at febrile temperatures [139]. In sickle-cell anemia, oxidative stress impairs the effectiveness of RBC NOS for producing NO, so that the stimulating effect of NO on erythrocyte deformability is blunted [140,141]. In experimental hypertension, the effect of NO on RBC deformability is also impaired [142]. NO donors such as nitroglycerine help to maintain red cell deformability in conditions such as cardiopulmonary bypass. High-dose nitroglycerin has been shown in this case to improve erythrocyte deformability through activating the phosphorylation of aquaporin 1 [143].

In fact, NO is not the only representative of the novel family of gasotransmitters, which are signaling molecules that easily diffuse across lipid membranes and exert their effect only in the area of their biosynthesis. Another gasotransmitter that is generating an increasing interest in hemorheology is hydrogen sulfide (H_2S). Hydrogen sulfide is produced from L-cysteine and D-cysteine under the influence of enzymes such as cystathionine β-synthase, cystathionine γ-lyase, 3-mercaptopyruvate sulfurtransferase, and cysteine aminotransferase [144]. This gasotransmitter has been shown to exert a cardioprotective effect and to regulate vascular tone via an effect on the contractility of vascular smooth muscle cells. H_2S has been reported to play a role in angiogenesis, the functional properties of platelets, thrombus stability, and erythrogenesis. Its involvement in the pathogenesis of atherosclerosis and arterial hypertension is a matter of current research. On the whole, all three known gaseous mediators, NO, CO, and H_2S, improve RBC deformability and decrease RBC aggregation. [145]. H_2S, like the other gasotransmitters, is assumed to act as an oxygen sensor and to be in close synergistic interaction with NO and CO to perform this function.

12. Hormones and Circulating Chemical Messengers

As explained above, the number of chemical messengers and hormones exhibiting specific receptors on RBCs is regularly increasing. We propose a tentative list of these in Table 1. Among the chemical messengers, we should mention immunoglobulins (IgG), complements [146], and lectins [147].

12.1. Insulin and IGF-I

The fuel-regulating hormone **insulin** deserves a special mention in regard to this. Insulin binds on the red cell membrane and activates intracellular pathways, with an effect on erythrocyte deformability that was evidenced in several prior studies [28,148,149] and has been confirmed by more recent investigations [141]. The influence of insulin on erythrocyte rheology seems to be mediated by an effect on the cell membrane [148] that includes changes in the molecular composition of the lipid membrane bilayer and thus in its microviscosity, which is associated with alterations to the function of membrane Na/K ATPase [150]. Similarly to the other factors presented in this review, the effects of insulin may be an improvement or a decrease in erythrocyte deformability. When very high, supraphysiological levels of insulin are applied in vitro, there is a decrease in RBC deformability, as was recently reconfirmed during insulin clamp experiments in hypertensives [151]. Interestingly, the ATP concentrations in RBCs are closely correlated with the free insulin levels in plasma [152]. Since, as reported above, intracellular ATP is an important determinant of RBC deformability, this positive relationship between the insulin and ATP content of the red cell may be involved in the regulation of red cell deformability.

C-peptide is a pancreatic hormone co-secreted with insulin. It is mostly a useful index of endogenous insulin secretion but is also supposed to exert some hormonal effects. Among these effects, C-peptide appears to increase eNOS in diabetics, resulting in a fluidification of the RBC membranes [153]. This effect is associated with some blood flow redistributions, a reduction in the NaK ATPase pump function, and an improvement in the renal function [153].

Another important hormone closely related to insulin is insulin-like growth factor I (IGF-I). IGF-I can bind on insulin receptors and thus, via this binding, can exert some insulinlike effects. It also has its own receptors that mediate important anabolic effects throughout the entire body. There are also IGF-I receptors on the red cell membrane [154]. Some clinical reports suggest that IGF-I may be a regulator of blood viscosity. In trained athletes, values of IGF-I within the upper quintile of distribution are associated with an impairment of blood fluidity. Since in this study IGF-I is correlated with a lower RBC deformability, measured with viscometry at high shear rate [155], this observation raises the possibility of a direct effect of IGF-I on RBC deformability via the binding of this hormone on its receptors on the red cell membrane.

12.2. Glucagon and RBCs

Glucagon is another important hormone involved in the regulation of fuel metabolism, in which it exerts an action opposite to that of insulin. Glucagon also exerts catabolic effects on the body's carbohydrate and protein stores. It was reported to decrease RBC deformability by P. Valensi and coworkers in 1986 [156]. However, more recently, the opposite effect has been reported by R. Komatsu and coworkers. These investigators showed that intravenously injected glucagon improved the erythrocyte deformability (as measured with a technique of filterability), resulting in a significant decrease in whole blood viscosity, which was associated with an increase in the blood flow [157]. Both of these reports are in disagreement with another publication by W. Reinhart, who reported that neither C-peptide, insulin, or glucagon had any measurable influence on RBC deformability, as assessed by blood viscometry [158].

12.3. *Thyroid Hormones*

Erythrocytes also exhibit receptors for the thyroid hormone L-triodothyronine [159]. Whether thyroid hormones are regulators of blood rheology remains unclear, but a decrease in RBC deformability has been reported to exist in hyperthyroidism and to be reversible after the successful treatment of the disease [160].

12.4. *Leptin*

In this context, leptin, a hormone released by adipocytes and thus belonging to the family of adipokines [161], is involved in a feedback loop linking the size of the body fat stores and the energy intake, but it has also been shown to improve red cell deformability via a NO- and cGMP-dependent mechanism [162]. In a preliminary work, we reported that leptin was correlated with plasma viscosity and erythrocyte disaggregation [163], and more recently, we confirmed that it was closely associated with increased red cell deformability and aggregation [164]. This hormone also regulates the body water stores via a direct effect on the adrenal production of aldosterone [165,166]. Therefore, leptin, which has been until now barely investigated in hemorheological research, may be one of the important physiological regulators of red cell rheology, involving it in regulatory loops that link energy stores and circulation.

12.5. *Erythropoietin*

Erythropoietin (EPO) is undoubtedly a major regulator of blood viscosity. This hormone released by the kidney is stimulated by hypoxia and inhibited by increases in plasma viscosity at the level of the juxtaglomerular apparatus in the nephron. It stimulates erythrocyte development in the bone marrow. The team of W. Reinhardt has elegantly demonstrated in a seminal paper that EPO mediates the homeostatic regulation of viscosity ("viscoregulation") that follows a rise in plasma viscosity [167] and results in a decrease in RBC mass. In the early 1990s, several teams closely analyzed the evolution of blood viscosity factors during the natural history of chronic renal failure. The studies by our Portuguese colleagues J. Martins e Silva and C. Saldanha, and by M. Delamaire, are more thoroughly reviewed in our preceding review [168]. CKD-associated hemorheological disturbances (less deformable RBCs, increased plasma viscosity) were corrected by a treatment with recombinant human erythropoietin (rhEPO). In cancer patients, rhEPO increases red cell deformability and decreases red cell aggregation [169]. It is important to notice that EPO, in addition to being a hormone that improves red cell deformability, also has antieryptotic properties, although it remains unable to completely counteract the triggering of eryptosis induced by an intracellular calcium influx in RBCs [170]. Young erythrocytes appear to be particularly prone to eryptosis following a decline of EPO, a phenomenon which has been termed neocytolysis [24].

12.6. *Somatostatin*

Somatostatin is another important hormone synthesized in the pancreas and in other tissues which circulates in blood [171]. Somatostatin can induce dramatic circulatory changes and increase the peripheral blood flow in humans [172]. In addition, somatostatin interferes with platelet functions [173]. Consistent with this circulatory effect, somatostatin seems to improve erythrocyte deformability, as suggested by its beneficial effect when assessed in vivo by our group using several techniques [174].

12.7. Melatonin

Melatonin is a hormone released by the pineal gland, which plays an important role in the circadian rhythms in the body. Additionally, melatonin has been reported to decrease RBC deformability in experiments. However, pinealectomy by itself did not induce any statistically significant change in erythrocyte deformability [175]. Erythrocyte deformability, measured by a method of filterability, was not modified by in vitro incubation of blood samples with melatonin [176]. Therefore, this issue remains controversial and requires more study.

12.8. Leukotrienes and Prostaglandins

Leukotrienes belong to the family of arachidonic acid derivatives, and some of them are known to exert powerful biological effects. Some leukotrienes, but not all [177,178], have been reported to impair erythrocyte deformability. Leukotriene B4 decreases the filterability of washed resuspended erythrocytes measured with the hemorheometre [178]. Consistent with what is observed for the other arachidonic derivatives, some prostaglandins have an opposite effect. Prostaglandin E1 and the prostacyclin analogue iloprost improve red and white cell filterability in vitro and in vivo [179,180]. PGE2 decreases the deformability of RBCs and increases their aggregability [181].

12.9. Sex Hormones

The literature about the hemorheological effects of sex hormones has primarily been driven by the issue of the thrombogenic effects of oral contraceptives (OCs). We previously reviewed this classical literature more thoroughly [168]. Briefly, it was shown that OC users in the late 1970s exhibited a lower erythrocyte deformability that was associated with a moderate increase in the whole blood viscosity, while the values of the plasma viscosity remained in the normal range. Hematocrit was also found to be increased under OCs in some studies but not in others. It was assumed that the progestin component of the OC was responsible for a rise in the circulating fibrinogen that explained most of this pattern. In fact, this picture observed in old OC pills (whose progestin compounds were mostly 19-nortestosterone derivatives) is no longer observed. More recent low-dose compounds appear to be almost devoid of hemorheological side-effects, in contrast to the older compounds, although they still induce moderately higher RBC aggregation. In addition, more recent physiological investigations have been conducted in women with normal menstrual cycles, showing that their estradiol levels were positively correlated with their whole blood viscosity, plasma viscosity, and fibrinogen, and that their RBC deformability was impaired correlatively to their estradiol levels. Thus, in physiological conditions, estrogens appear to decrease blood fluidity. On the other hand, progesterone had the opposite effect in physiological conditions, decreasing both the fibrinogen and the blood viscosity, and increasing the RBC deformability. Therefore, the RBC deformability was lower in the follicular than in the luteal phase. Interestingly, the recent literature on estrogens and red cell rheology has pointed out the role of nitric oxide. Estrogens are known to physiologically increase both NO synthesis and release by the endothelial cells. This effect is likely to be mediated by the estrogen-responsive elements located in the promoting region of the gene coding for endothelial NOS.

Beside the classical mode of action of estrogens, which involves intracellular receptors, recent emphasis has been placed on the membrane receptors, on which estrogens can specifically bind and thus act more rapidly, inducing an increase in cytosolic Ca^{++} in some cells. In addition, estrogens also exhibit antioxidant properties which can delay NO clearing from blood [168]. All these mechanisms are therefore likely to improve RBC deformability via NO-mediated mechanisms. This seems to be in disagreement with the reports showing that in vivo hormonal treatment by either transcutaneous or oral estrogens impairs RBC deformability by increasing membrane rigidity [168].

In vitro, 17 β-estradiol at a concentration of 10^{-5} M also decreased the RBC aggregation in the blood samples of postmenopausal women undergoing hormone therapy [182].

A recent paper by M. Grau [183] suggests that the gender differences in hematological parameters in males compared to females (with higher RBC deformability in females) might be related to the higher estradiol levels that are associated with higher RBC NO levels. However, Pretorius and coworkers also reported on the effects of estrogen and progesterone on the RBC structure, and they noted increased eryptosis [184,185].

12.10. Dehydroepiandrosterone

The incorporation of dehydroepiandrosterone sulfate (DHEAS) into the human red cell membrane increases the acyl chain motion in the middle portion of the membrane and induces the echinocytosis of red cells. It is suggested that the increase in the viscosity of a red cell suspension, the decreased deformability, and the decrease in the deoxygenation rate of hemoglobin in the presence of DHEAS probably reflect the presence of echinocytes. In the presence of plasma proteins, the incorporation of DHEAs into red cells was remarkably suppressed [186].

12.11. Apelin

Apelin is a recently discovered hormone belonging to the family of adipokines, which can bind on the G-protein-coupled APJ receptor which is expressed over the surface of many cells in the body. It is widely expressed in various organs such as the heart, lungs, and kidney and exerts a hypotensive effect via the activation of the APJ receptors on the surface of endothelial cells, thus inducing the release of NO. In addition, it decreases the hypothalamic secretion of vasopressin and increases water intake, thus exerting important effects on the homeostatic regulation of the body's fluid stores. In rats that exhibited a decrease in erythrocyte deformability due to the experimental induction of diabetes and ischemia-reperfusion injury of the heart, apelin-13 has been shown to reverse this loss of deformability [187].

12.12. Catecholamines

The catecholamines norepinephrine and epinephrine are the two circulation messengers of the sympathetic nervous system that can be released into the blood by the adrenal medulla and act on RBCs via specific α- and β- receptor agonists. It is well known that norepinephrine is mostly an α-adrenergic agonist, while epinephrine is mostly a potent β-adrenergic agonist. Over recent years, evidence has been accumulated that both of these catecholamines are important regulators of RBC rheology. The α1-adrenergic receptor can be detected on human erythrocyte membranes [188]. There are also β-adrenergic receptors on the red cell membrane that have been shown in animal studies to mediate the regulating effects of catecholamines on cell volume and ion transport [189]. The first studies on catecholamines and red cell rheology were conducted in the late 1980s by Pfafferott and Volger, who reported that in vitro, *both* norepinephrine (α-adrenergic agonist) and isoprenaline (β-adrenergic agonist) decreased erythrocyte deformability [190]. More recently, however, Hilario demonstrated that epinephrine (β-adrenergic agonist) actually *improves* RBC deformability when measured with a more accurate technique but is also able to induce the transformation of human RBCs into echinocyte [191]. Catecholamines (both β-agonists and α-agonists) also have an antieryptotic effect [43], consistent with the short-term beneficial role of these hormones regarding the body's adaptation to unusual stresses. In fact, in experiments on species whose RBCs are nucleated, catecholamines induced a dramatic increase in the cell volume as a result of an accumulation of sodium and chloride due to the activation of an amiloride-sensitive, cyclic, AMP-dependent Na^+-H^+ exchanger which allowed Na^+ to enter in exchange for internal H^+. At the same time, the RBC deformability was improved (despite the increase in the cell volume). Both the RBC fluidification and the activation of this ionic exchange were likely to be an adaptive response to hypoxia which resulted in the increased oxygen-carrying capacity of the RBCs [192]. The literature, however, suggests that this epinephrine may improve RBC deformability [193–195], presumably via β-adrenergic receptors, while there is apparently no effect of either α1- and

α2-receptor agonists. It was also shown that RBCs incubated with epinephrine and isoproterenol exhibited significant changes of deformability, by 10% and 30%, respectively. This is consistent with the other classical effects of catecholamines mediated by β-adrenergic receptors (vasodilation, increased cardiac output, etc.) that all lead to an increased blood flow. The team of A. Muravyov extensively studied the effect of catecholamines on the rheological properties of the human RBC, showing that the effect of these hormones on RBC deformability is mostly under the control of intracellular Ca^{2+}-regulating pathways [196]. In contrast to this positive effect of catecholamines on RBC deformability in physiological conditions, an increased viscosity and decreased RBC deformability were observed in untreated pheochromocytoma [197].

12.13. Cortisol

Beside catecholamines, cortisol is a major hormone involved in the body's adaptation to stress. Cortisol has been shown to bind to the erythrocyte membrane, impairing the epinephrine binding at this level and resulting in an increase in the microviscosity of the membranes and a rise in $Na(^{+})$,$K(^{+})$-ATPase activity [198]. This is in line with the other effects of corticosteroids that moderate the effects of stress hormones in order to cope with them. Windberger has also described the hemorheological profile of Cushing syndrome in dogs, i.e., hypercortisolism due to increased and sustained cortisol release by the adrenal cortex, showing that this situation is associated with increased plasma viscosity and RBC aggregation [199]. This area surely deserves more investigation.

12.14. Endocannabinoids

The Endocannabinoid System (ECS) is an important regulatory system aiming at maintaining homeostasis in a variety of body functions such as temperature, mood, the immune system, and energy input and output. It exerts a wide variety of effects on emotional behavior, feeding behavior, appetite, nervous functions, fertility, and pre-and postnatal development. This system involves endogenous lipid mediators called endocannabinoids, of which the most well-known are 2-arachidonoylglycerol and anandamide, which are synthesized from the pool of arachidonic acid in cell membrane phospholipids. The modulation of this system with cannabinergic, cannabimimetic, and cannabinoid-based therapeutic drugs that are currently under development offers some potential for treating a number of diseases [200]. The endocannabinoid system (ECS), via the activation of the type-1 cannabinoid receptor (CB1), is involved in the process of exostasis, i.e., overeating and storage due to the anticipation of future energy needs, and thus adds some adaptatory regulations to homeostasis [201]. Endocannabinoids and their receptors are almost ubiquitous in the body and regulate intermediary metabolism and energy expenditure. They interfere with glucose and lipid metabolism so as to promote energy storage and reduce energy expenditure. This system is overactive in obesity and appears to play a role in the maintenance of fat mass [202]. All these functions modulated by endocannabinoids are also known to be associated with hemorheological processes. However, little is known about endocannabinoid involvement in the regulation of blood rheology. It is thus interesting to point out that the endocannabinoid anandamide has been reported to induce apoptosis in several varieties of nucleated cells, and to increase the activity of RBC cytosolic Ca^{2+}, resulting in the cell shrinkage and cell membrane scrambling of mature RBCs and, subsequently, inducing eryptosis [40].

12.15. Other Hormones

Erythrocytes are also known to be able to release endothelin-1 (ET1), a potent vasoconstrictor which is also produced by endothelial and smooth muscle cells. ET1 has been reported to improve the deformability (as assessed by filterability) of stiffened RBCs via an activation of the protein kinase C [203]. However, another study was unable to evidence this effect when the RBC deformability was evaluated with viscometry [204].

Finally, we should mention the surprising lack of a hemorheologic effect of calcium regulating hormones (parathormone, calcitonin, and vitamin D), which contrasts to the major importance of intracellular calcium in regard to erythrocyte rheology [205].

13. RBCs in Various Pathophysiological Situations

In the previous paragraphs, we referred to a pathological structure of the RBCs in conditions such as type 2 diabetes and hereditary hemochromatosis. We have published extensively on the RBC structural changes in other inflammatory conditions such as rheumatoid arthritis, Alzheimer's disease, and Parkinson's disease and also on the effect of selected dysregulated inflammatory markers including cytokines on RBC structure [47,79,82,206–209]. In all inflammatory conditions, the circulating dysregulated inflammatory markers have a profound effect on RBCs, platelets, and plasma protein structure. Both deformability patterns and eryptosis are visible in these conditions. Similar changes have been noted in septic shock and sleep apnea patients.

13.1. Stress

Stress is a neuroendocrine complex reaction ("general adaptation syndrome") which aims at maintaining homeostasis in the body when it is submitted to an unusual situation. The stress can be sufficient to cope with this unusual situation (eustress) or result in a prolonged disturbance of the body's functions (distress) [210]. All stress hormones (as indicated above in this review) are likely to exert hemorheologic effects, although the situation is not very clear concerning cortisol. Catecholamines exert complementary effects: epinephrine increases red cell deformability (thus favoring microcirculatory perfusion), while norepinephrine increases red cell aggregation and induces a fluid shift reduction of plasma volume with a parallel rise in hematocrit due to α1-receptor stimulation [211]. Therefore, unsurprisingly, stress induces hemorheological modifications. A pioneering study was presented by A. Ehrly [212], evidencing a rise in blood and plasma viscosity after video-film-induced emotional stress. More recently, a hyperviscosity syndrome was evidenced among evacuees who survived the earthquake and tsunami of Fukushima in Japan [213]. In this case, there was polycythemia.

A study on American veterans with Gulf War Illness (GWI) also deals with this issue. These patients experience chronic symptoms that include fatigue, pain, and cognitive impairment. Assuming that this symptom cluster may be related to impaired tissue oxygen delivery, the investigators measured the red cell deformability and found that it was increased in veterans suffering from GWI. [214]. Further studies on GWI showed that the increased deformability of the red cells was not affected during maximal exercise [215]. No clear explanation for this finding was given, but the effect of epinephrine on red cell deformability may be a logic explanation.

The issue of stress and hemorheology is more thoroughly developed in our previous review paper [216] but requires, of course, further investigation. On the whole, we can assume that "eustress" induces reversible adaptative alterations in the blood viscosity factors, while "distress" may induce more important damage that may involve eryptosis.

13.2. Chronic Fatigue Syndrome

By contrast, in people suffering from chronic fatigue syndrome, the red blood cells were found to be significantly stiffer than those in healthy controls [217]. A previous report on the same disease did not evidence this decrease in deformability [218], probably due to methodological concerns. Since oxidant stress is one of the mechanisms underlying chronic fatigue syndrome [219] and this oxidant stress damages red cells and makes them stiffer, this finding is logical. In this situation, the increased apoptosis of various blood cell lines was described [220], and, although we are not aware of a report on increased eryptosis, it is logical to assume that red cells are stiffened in chronic fatigue syndrome as a result of oxidant stress, which damages the cells and may lead to their programmed death.

13.3. Septic Shock

There are conditions that combine many disturbances to various functions, including RBC properties. The most impressive example of this is perhaps septic shock. Sepsis has been shown to be associated with impaired blood rheology [221–223]. Furthermore, in this context, decreased red blood cell deformability has been reported to be associated with a poor outcome in septic patients [224,225]. The hemorheologic disturbances observed during sepsis are associated with an altered metabolism; decreased 2,3-bisphosphoglycerate; the redistribution of membrane phospholipids; changes in the RBC volume, the affinity between hemoglobin and oxygen, the morphology, the antioxidant status, the intracellular Ca^{2+} homeostasis, and the membrane proteins; membrane phospholipid redistribution; and RBC O_2–dependent adenosine triphosphate efflux. During septic shock, a phenomenon of RBC autoxidation is observed and has been supposed to worsen the disorder [226,227]. Unsurprisingly, marked alterations in RBC rheology, including reduced deformability and increased aggregation, occur early on in septic patients, and reductions in RBC deformability over time are associated with a poor prognosis [224,228]. The sepsis-associated impairment of erythrocyte deformability is associated with a decrease in erythrocyte NO-releasing activity, both abnormalities being apparently a consequence of the inflammatory reaction [229]. In experimentally induced septic shock in rats, prostacyclin (iloprost) and nitric oxide prevented the sepsis-induced loss of red blood cell deformability via a complex mechanism [230].

13.4. Sleep Apnea

Another situation associated with a decrease in RBC deformability which is probably multifactorial is obstructive sleep apnea syndrome (OSAS). This syndrome has emerged over recent decades as an important risk factor for atherosclerosis and cardiovascular disorders [231–233]. In OSAS, endothelial dysfunction is markedly disturbed due to recurrent hypoxemic events, and there is a decrease in erythrocyte deformability [231]. Additionally, a decrease in NO bioavailability is also observed and may be both a consequence and a worsening factor of the defect in RBC deformability [232]. In this disease, there is also an increase in plasma viscosity and an increase in RBC aggregation [231]. OSAS is also known to be associated with apoptosis in various tissues and thus, presumably, eryptosis [233]. In contrast with hypoxemia, it is interesting to point out that the opposite condition, hyperoxia, seems to have no measurable effect on blood rheology, so that its use for donor-organ preservation before graft is not known to induce any hemorheologic disturbance [234].

13.5. COVID-19

The recent COVID-19 pandemic has generated a host of studies in all areas of biomedical research [235], and, unsurprisingly, disturbances in microcirculation [236] and the rheologic properties of blood [237] have been described. It is clear that endotheliopathies are important clinical features of acute COVID-19 [238,239]. Various circulating and dysregulated inflammatory coagulation biomarkers, including fibrin(ogen), D-dimer, P-selectin, the von Willebrand Factor (VWF), C-reactive protein (CRP), and various cytokines directly bind to endothelial receptors and are likely to be indicative of a poor prognosis [240–243]. This poor prognosis is further worsened by a substantial deposition of microclots in the lungs [244–246]. The plasma of COVID-19 patients also carries a massive load of preformed amyloid clots and there are numerous reports of damage to erythrocytes [247–249] and platelets and the dysregulation of inflammatory biomarkers [240–243,250]. Recently, we also determined whether the spike protein may interfere with blood flow by comparing naïve healthy plasma samples with and without added spike protein to plasma samples from COVID-19-positive patients (before treatment). We concluded that the spike protein may have direct pathological effects on blood rheology [251]. Significant pathological changes in microcirculation and the presence of persistent microclots have also been noted in Long COVID/PASC [250].

14. Concluding Remarks

Since the review on this topic that we published 20 years ago [168], our knowledge of the effects of various factors on erythrocyte deformability has expanded significantly. However, many issues remain incompletely understood at this time. It is clear, however, that this entire body of information we have put together suggests that there is an exquisitely regulated system involved in the homeostasis of blood rheology and flow distribution. The viscoregulatory loops hypothesized in the middle of the 20th century are now well described [167]. Others will probably be evidenced over the coming years.

The best known is probably the endothelium–leukocyte–liver axis, which involves polymorphonuclear neutrophils and monocytes that are able to release many biologically active substances that may in turn modify the blood viscosity factors. The release of cytokines such as IL-6 induces a rise in fibrinogen, which increases plasma viscosity and RBC aggregation. Circulating free radicals and arachidonic acid derivatives modify erythrocyte membrane properties and thus RBC deformability. Free radicals increase the rigidity and aggregation of erythrocytes. These interactions between white cell activation and erythrocyte rheologic properties have been remarkably well-investigated by the team of O. Başkurt [252]. This study showed that most of the hemorheological profiles of inflammatory diseases (and vascular diseases) are explained by these leukocyte–erythrocyte interactions and are likely to play an important role in the body's response to an inflammatory stimulation [253].

As indicated above, some other integrated responses of the body such as stress, the regulation of energy stores, and growth involve a hemorheologic response whose relevance is not completely understood. The relationship between the size of energy stores and the blood rheology even in normal conditions is also an issue that deserves more research, since some reports suggest that the hormones released by the adipose tissue (leptin and adiponectin) have circulatory effects [254]. Our recent report showing that leptin is more closely correlated with red cell aggregation than other determinants quantifying adipose stores may suggest that this hormone is involved in a regulatory loop linking the body's energy status and the hemorheological modulation of microcirculation [163]. Another interesting recent finding is that red cell rigidity in obesity is associated with higher values of angiogenesis [255]. Such a relationship needs to be more closely investigated, but it may lead to speculation that obesity-associated RBC stiffness may result in an adaptation of the microcirculation network. Presumably, NO, which is known to modulate angiogenesis in vitro and in vivo [256,257], may be involved in this effect.

In addition, according to several teams, dietary habits appear to be associated with the modification of blood rheology [258]. Very little is known about this issue, which requires further investigation. In order to propose a hypothesis, taking into account the abovementioned effects of circulating metabolites, exercise, and nutrition on blood rheology, we recently proposed the "healthy primitive lifestyle paradigm". This hypothesis assumes that evolution has selected in *Homo sapiens* genetic polymorphisms leading to insulin resistance as an adaptative strategy to cope with the lifestyle of our Paleolithic ancestors. According to the current scientific data, our Paleolithic ancestors performed a lot of prolonged low-intensity physical activity and their food was mostly based on lean meat and wild herbs (i.e., poor in saturated fat, rich in low-glycemic-index carbohydrates, and moderately high in protein). According to this hypothesis, an individual whose exercise and nutritional habits are close to this lifestyle represents the mainstream phenotype, and highly trained athletes represent a minority group. Sedentary subjects undergo a host of metabolic (and hemorheologic) modifications that aim at coping with the lack of physiological activity, which represents a pathologic situation in these populations [259]. We previously published data in agreement with this theory, thus integrating them into a logical picture of hemorheologic homeostasis [260].

Therefore, although this issue remains incompletely studied, we think that the body of information summarized in this review supports the concept that among the numerous situations where RBC rigidification has been reported to occur, some involve reversible changes in a fully physiological context, while others reflect quite irreversible damage to the red cells that will lead to eryptosis (Figure 4). In some situations, however, there may be a continuum between the two processes if the stimulus is too strong. Exercise, for example, generally involves reversible changes in RBC rheology but may also induce irreversible damage and programmed erythrocyte death. Such a triggering of eryptosis in this case may represent a protective mechanism against the deleterious effect of old RBCs, which is known to be associated with an increased risk of death [77,78]. However, further research is required to more precisely delineate the respective importance of reversible (physiological) and almost-irreversible (eryptosis related) RBC stiffening in various situations.

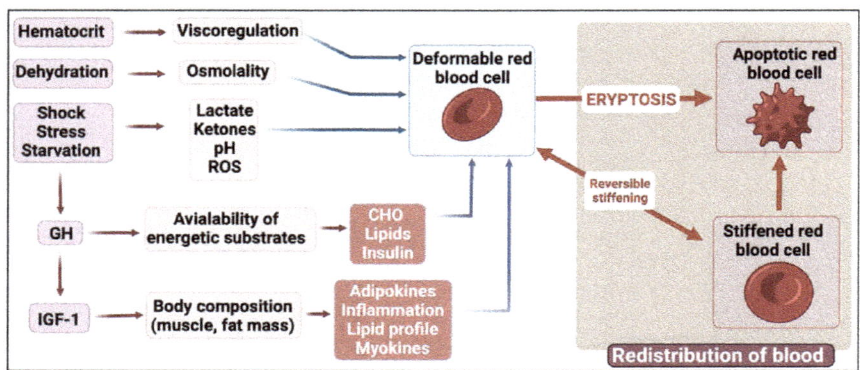

Figure 4. Regulatory loops involved in the modulation of red cell deformability. According to the physiological or pathological context, the factors thoroughly enumerated in this review increase or decrease red cell deformability, thus contributing to the adaptation of microcirculatory blood flow to this context. Erythrocyte stiffening may be a reversible event or one of the components of the cascade of events leading to programmed red cell death (eryptosis). Figure created using www.biorender.com accessed on 17 December 2021.

Author Contributions: Conceptualization and writing, J.-F.B., E.P., E.V.-M., J.M. and E.R.d.M.; Review, editing and visualization, E.P. All authors have read and agreed to the published version of the manuscript.

Funding: This research received no external funding.

Institutional Review Board Statement: Not applicable.

Informed Consent Statement: Not applicable.

Conflicts of Interest: The authors declare no conflict of interest.

References

1. Huisjes, R.; Bogdanova, A.; Van Solinge, W.W.; Schiffelers, R.M.; Kaestner, L.; Van Wijk, R. Squeezing for Life—Properties of Red Blood Cell Deformability. *Front. Physiol.* **2018**, *9*, 656. [CrossRef] [PubMed]
2. Munter, W.A.; Stein, P.D. Newtonian behavior of blood at high rates of shear. *Biorheology* **1973**, *10*, 501–508. [CrossRef]
3. Skalak, R.; Chien, S. Rheology of blood cells as soft tissues. *Biorheology* **1982**, *19*, 453–461. [CrossRef] [PubMed]
4. Schmid-Schönbein, H.; Gaehtgens, P. What is red cell deformability? *Scand. J. Clin. Lab. Investig.* **1981**, *156*, 13–26. [CrossRef]
5. Nader, E.; Skinner, S.; Romana, M.; Fort, R.; Lemonne, N.; Guillot, N.; Gauthier, A.; Antoine-Jonville, S.; Renoux, C.; Hardy-Dessources, M.D.; et al. Blood Rheology: Key Parameters, Impact on Blood Flow, Role in Sickle Cell Disease and Effects of Exercise. *Front. Physiol.* **2019**, *10*, 1329. [CrossRef]
6. Zheng, Y.; Cachia, M.A.; Ge, J.; Xu, Z.; Wang, C.; Sun, Y. Mechanical differences of sickle cell trait (SCT) and normal red blood cells. *Lab Chip* **2015**, *15*, 3138–3146. [CrossRef] [PubMed]
7. Man, Y.; Kucukal, E.; An, R.; Watson, Q.D.; Bosch, J.; Zimmerman, P.A.; Little, J.A.; Gurkan, U.A. Microfluidic assessment of red blood cell mediated microvascular occlusion. *Lab Chip* **2020**, *20*, 2086–2099. [CrossRef] [PubMed]

8. Man, Y.; Kucukal, E.; An, R.; Bode, A.; Little, J.A.; Gurkan, U.A. Standardized microfluidic assessment of red blood cell-mediated microcapillary oc-clusion: Association with clinical phenotype and hydroxyurea responsiveness in sickle cell disease. *Microcirculation* **2021**, *28*, e12662. [CrossRef]
9. Man, Y.; Maji, D.; An, R.; Ahuja, S.P.; Little, J.A.; Suster, M.A.; Mohseni, P.; Gurkan, U.A. Microfluidic electrical impedance assessment of red blood cell-mediated microvascular occlusion. *Lab Chip* **2021**, *21*, 1036–1048. [CrossRef] [PubMed]
10. Lanotte, L.; Mauer, J.; Mendez, S.; Fedosov, D.A.; Fromental, J.M.; Claveria, V.; Nicoud, F.; Gompper, G.; Abkarian, M. Red cells' dynamic morphologies govern blood shear thinning under microcirculatory flow conditions. *Proc. Natl. Acad. Sci. USA* **2016**, *113*, 13289–13294. [CrossRef]
11. Mauer, J.; Mendez, S.; Lanotte, L.; Nicoud, F.; Abkarian, M.; Gompper, G.; Fedosov, D.A. Flow-Induced Transitions of Red Blood Cell Shapes under Shear. *Phys. Rev. Lett.* **2018**, *121*, 118103. [CrossRef]
12. Chien, S. Red cell deformability and its relevance to blood flow. *Annu. Rev. Physiol.* **1987**, *49*, 177–192. [CrossRef]
13. Connes, P.; Alexy, T.; Detterich, J.; Romana, M.; Hardy-Dessources, M.D.; Ballas, S.K. The role of blood rheology in sickle cell disease. *Blood Rev.* **2016**, *30*, 111–118. [CrossRef]
14. Barshtein, G.; Ben-Ami, R.; Yedgar, S. Role of red blood cell flow behavior in hemodynamics and hemostasis. *Expert Rev. Cardiovasc. Ther.* **2007**, *5*, 743–752. [CrossRef]
15. Le Devehat, C.; Khodabandehlou, T.; Vimeux, M. Relationship between hemorheological and microcirculatory abnormalities in diabetes mellitus. *Diabete Metab.* **1994**, *20*, 401–404.
16. Lang, F.; Abed, M.; Lang, E.; Föller, M. Oxidative stress and suicidal erythrocyte death. *Antioxid. Redox Signal.* **2014**, *21*, 138–153. [CrossRef]
17. Lang, F.; Gulbins, E.; Lang, P.A.; Zappulla, D.; Föller, M. Ceramide in suicidal death of erythrocytes. *Cell. Physiol. Biochem.* **2010**, *26*, 21–28. [CrossRef]
18. Lang, F.; Lang, E.; Foller, M. Physiology and pathophysiology of eryptosis. *Transfus. Med. Hemother.* **2012**, *39*, 308–314. [CrossRef]
19. Lang, E.; Qadri, S.M.; Lang, F. Killing me softly–suicidal erythrocyte death. *Int. J. Biochem. Cell Biol.* **2012**, *44*, 1236–1243. [CrossRef]
20. Pretorius, E.; Du Plooy, J.N.; Bester, J. A Comprehensive Review on Eryptosis. *Cell. Physiol. Biochem.* **2016**, *39*, 1977–2000. [CrossRef] [PubMed]
21. Qadri, S.M.; Bauer, J.; Zelenak, C.; Mahmud, H.; Kucherenko, Y.; Lee, S.H.; Ferlinz, K.; Lang, F. Sphingosine but not sphingosine-1-phosphate stimulates suicidal erythrocyte death. *Cell. Physiol. Biochem.* **2011**, *28*, 339–346. [CrossRef] [PubMed]
22. Qadri, S.M.; Bissinger, R.; Solh, Z.; Oldenborg, P.A. Eryptosis in health and disease: A paradigm shift towards understanding the (patho)physiological implications of programmed cell death of erythrocytes. *Blood Rev.* **2017**, *31*, 349–361. [CrossRef] [PubMed]
23. Qadri, S.M.; Donkor, D.A.; Bhakta, V.; Eltringham-Smith, L.J.; Dwivedi, D.J.; Moore, J.C.; Pepler, L.; Ivetic, N.; Nazi, I.; Fox-Robichaud, A.E.; et al. Phosphatidylserine externalization and procoagulant activation of erythrocytes induced by *Pseudomonas aeruginosa* virulence factor pyocyanin. *J. Cell. Mol. Med.* **2016**, *20*, 710–720. [CrossRef]
24. Lang, E.; Lang, F. Triggers, inhibitors, mechanisms, and significance of eryptosis: The suicidal erythrocyte death. *BioMed Res. Int.* **2015**, *2015*, 513518. [CrossRef] [PubMed]
25. Duez, J.; Holleran, J.P.; Ndour, P.A.; Pionneau, C.; Diakite, S.; Roussel, C.; Dussiot, M.; Amireault, P.; Avery, V.M.; Buffet, P.A. Mechanical clearance of red blood cells by the human spleen: Potential therapeutic applications of a biomimetic RBC filtration method. *Transfus. Clin. Biol.* **2015**, *22*, 151–157. [CrossRef]
26. Klei, T.R.; Dalimot, J.; Nota, B.; Veldthuis, M.; Mul, F.P.; Rademakers, T.; Hoogenboezem, M.; Nagelkerke, S.Q.; van Ijcken, W.F.; Oole, E.; et al. Hemolysis in the spleen drives erythrocyte turnover. *Blood* **2020**, *136*, 1579–1589. [CrossRef] [PubMed]
27. Miko, I.; Nemeth, N.; Sogor, V.; Kiss, F.; Toth, E.; Peto, K.; Furka, A.; Vanyolos, E.; Toth, L.; Varga, J.; et al. Comparative erythrocyte deformability investigations by filtrometry, slit-flow and rotational ektacytometry in a long-term follow-up animal study on splenectomy and different spleen preserving operative techniques: Partial or subtotal spleen resection and spleen autotransplantation. *Clin. Hemorheol. Microcirc.* **2017**, *66*, 83–96. [CrossRef]
28. Juhan, I.; Vague, P.; Buonocore, M.; Moulin, J.P.; Jouve, R.; Vialettes, B. Abnormalities of erythrocyte deformability and platelet aggregation in insulin-dependent diabetics corrected by insulin in vivo and in vitro. *Lancet* **1982**, *1*, 535–537. [CrossRef]
29. Vague, P.; Juhan, I. Red cell deformability, platelet aggregation, and insulin action. *Diabetes* **1983**, *32* (Suppl. 2), 88–91. [CrossRef] [PubMed]
30. Brun, J.F.; Varlet-Marie, E.; Connes, P.; Aloulou, I. Hemorheological alterations related to training and overtraining. *Biorheology* **2010**, *47*, 95–115. [CrossRef] [PubMed]
31. Nader, E.; Guillot, N.; Lavorel, L.; Hancco, I.; Fort, R.; Stauffer, E.; Renoux, C.; Joly, P.; Germain, M.; Connes, P. Eryptosis and hemorheological responses to maximal exercise in athletes: Comparison between running and cycling. *Scand. J. Med. Sci. Sports* **2018**, *28*, 1532–1540. [CrossRef] [PubMed]
32. Nader, E.; Monedero, D.; Robert, M.; Skinner, S.; Stauffer, E.; Cibiel, A.; Germain, M.; Hugonnet, J.; Scheer, A.; Joly, P.; et al. Impact of a 10 km running trial on eryptosis, red blood cell rheology, and electrophysiology in endurance trained athletes: A pilot study. *Eur. J. Appl. Physiol.* **2020**, *120*, 255–266. [CrossRef] [PubMed]
33. Nader, E.; Romana, M.; Guillot, N.; Fort, R.; Stauffer, E.; Lemonne, N.; Garnier, Y.; Skinner, S.C.; Etienne-Julan, M.; Robert, M.; et al. Association Between Nitric Oxide, Oxidative Stress, Eryptosis, Red Blood Cell Microparticles, and Vascular Function in Sickle Cell Anemia. *Front. Immunol.* **2020**, *11*, 551441. [CrossRef]

34. Koutsouris, D.; Delatour-Hanss, E.; Hanss, M. Physico-chemical factors of erythrocyte deformability. *Biorheology* **1985**, *22*, 119–132. [CrossRef]
35. Engström, K.G.; Meiselman, H.J. Analysis of red blood cell membrane area and volume regulation using micropipette aspiration and perfusion. *Biorheology* **1995**, *32*, 115–116. [CrossRef]
36. Reinhart, W.H.; Piety, N.Z.; Deuel, J.W.; Makhro, A.; Schulzki, T.; Bogdanov, N.; Goede, J.S.; Bogdanova, A.; Abidi, R.; Shevkoplyas, S.S. Washing stored red blood cells in an albumin solution improves their morphologic and hemorheologic properties. *Transfusion* **2015**, *55*, 1872–1881. [CrossRef] [PubMed]
37. Sugie, J.; Intaglietta, M.; Sung, L.A. Water transport and homeostasis as a major function of erythrocytes. *Am. J. Physiol. Heart Circ. Physiol.* **2018**, *314*, H1098–H1107. [CrossRef]
38. Dupuy-Fons, C.; Brun, J.F.; Mallart, C.; Carvajal, J.; Fussellier, M.; Bardet, L.; Orsetti, A. In vitro influence of zinc and magnesium on the deformability of red blood cells artificially hardened by heating. *Biol. Trace Elem. Res.* **1995**, *47*, 247–255. [CrossRef]
39. Semenov, A.N.; Shirshin, E.A.; Muravyov, A.V.; Priezzhev, A.V. The Effects of Different Signaling Pathways in Adenylyl Cyclase Stimulation on Red Blood Cells Deformability. *Front. Physiol.* **2019**, *10*, 923. [CrossRef]
40. Bentzen, P.J.; Lang, F. Effect of anandamide on erythrocyte survival. *Cell. Physiol. Biochem.* **2007**, *20*, 1033–1042. [CrossRef]
41. Jemaà, M.; Fezai, M.; Lang, F. Inhibition of Suicidal Erythrocyte Death by Reversine. *Cell. Physiol. Biochem.* **2017**, *41*, 2363–2373. [CrossRef] [PubMed]
42. Kiedaisch, V.; Akel, A.; Niemoeller, O.M.; Wieder, T.; Lang, F. Zinc-induced suicidal erythrocyte death. *Am. J. Clin. Nutr.* **2008**, *87*, 1530–1534. [CrossRef] [PubMed]
43. Lang, P.A.; Kempe, D.S.; Akel, A.; Klarl, B.A.; Eisele, K.; Podolski, M.; Hermle, T.; Niemoeller, O.M.; Attanasio, P.; Huber, S.M.; et al. Inhibition of erythrocyte "apoptosis" by catecholamines. *Naunyn. Schmiedeberg's Arch. Pharmacol.* **2005**, *372*, 228–235. [CrossRef]
44. Nicolay, J.P.; Liebig, G.; Niemoeller, O.M.; Koka, S.; Ghashghaeinia, M.; Wieder, T.; Haendeler, J.; Busse, R.; Lang, F. Inhibition of suicidal erythrocyte death by nitric oxide. *Pflugers Arch.* **2008**, *456*, 293–305. [CrossRef] [PubMed]
45. Niemoeller, O.M.; Bentzen, P.J.; Lang, E.; Lang, F. Adenosine protects against suicidal erythrocyte death. *Pflugers Arch.* **2007**, *454*, 427–439. [CrossRef]
46. Qadri, S.M.; Mahmud, H.; Lang, E.; Gu, S.; Bobbala, D.; Zelenak, C.; Jilani, K.; Siegfried, A.; Foller, M.; Lang, F. Enhanced suicidal erythrocyte death in mice carrying a loss-of-function mutation of the adenomatous polyposis coli gene. *J. Cell. Mol. Med.* **2012**, *16*, 1085–1093. [CrossRef]
47. Pretorius, E.; Swanepoel, A.C.; Buys, A.V.; Vermeulen, N.; Duim, W.; Kell, D.B. Eryptosis as a marker of Parkinson's disease. *Aging* **2014**, *6*, 788–819. [CrossRef]
48. Makhro, A.; Hänggi, P.; Goede, J.S.; Wang, J.; Brüggemann, A.; Gassmann, M.; Schmugge, M.; Kaestner, L.; Speer, O.; Bogdanova, A. N-methyl-D-aspartate receptors in human erythroid precursor cells and in circulating red blood cells contribute to the intracellular calcium regulation. *Am. J. Physiol. Cell Physiol.* **2013**, *305*, C1123–C1138. [CrossRef]
49. Reinhart, W.H.; Geissmann-Ott, C.; Bogdanova, A. Activation of N-methyl D-aspartate (NMDA) receptors has no influence on rheological properties of erythrocytes. *Clin. Hemorheol. Microcirc.* **2011**, *49*, 307–313. [CrossRef]
50. Hänggi, P.; Makhro, A.; Gassmann, M.; Schmugge, M.; Goede, J.S.; Speer, O.; Bogdanova, A. Red blood cells of sickle cell disease patients exhibit abnormally high abundance of N-methyl D-aspartate receptors mediating excessive calcium uptake. *Br. J. Haematol.* **2014**, *167*, 252–264. [CrossRef] [PubMed]
51. Unal, Y.; Comu, F.M.; Emik, U.; Alkan, M.; Pampal, H.K.; Arslan, M. The effects of propofol and memantine on erythrocyte deformability. *Bratisl. Lek. Listy* **2014**, *115*, 253–255. [CrossRef]
52. Sluyter, R. P2X and P2Y receptor signaling in red blood cells. *Front. Mol. Biosci.* **2015**, *2*, 60. [CrossRef]
53. Aursnes, I.; Gjesdal, K.; Abildgaard, U. Platelet aggregation induced by ADP from unsheared erythrocytes at physiological Ca^{++}-concentration. *Br. J. Haematol.* **1981**, *47*, 149–152. [CrossRef]
54. Sprague, R.S.; Ellsworth, M.L.; Stephenson, A.H.; Lonigro, A.J. ATP: The red blood cell link to NO and local control of the pulmonary circulation. *Am. J. Physiol.* **1996**, *271*, H2717–H2722. [CrossRef]
55. Vulpis, V. Endothelin, microcirculation and hemorheology. *Clin. Hemorheol. Microcirc.* **1999**, *21*, 273–276. [CrossRef] [PubMed]
56. Piskuric, N.A.; Zhang, M.; Vollmer, C.; Nurse, C.A. Potential roles of ATP and local neurons in the monitoring of blood O_2 content by rat aortic bodies. *Exp. Physiol.* **2014**, *99*, 248–261. [CrossRef]
57. Wang, L.; Olivecrona, G.; Gotberg, M.; Olsson, M.L.; Winzell, M.S.; Erlinge, D. ADP acting on $P2Y_{13}$ receptors is a negative feedback pathway for ATP release from human red blood cells. *Circ. Res.* **2005**, *96*, 189–196. [CrossRef]
58. Kobayashi, S.; Hamazaki, T.; Sawazaki, S.; Nakamura, H. Reduction in the ADP release from shear-stressed red blood cells by fish oil administration. *Thromb. Res.* **1992**, *65*, 353–364. [CrossRef]
59. Burnstock, G. Blood cells: An historical account of the roles of purinergic signalling. *Purinergic Signal.* **2015**, *11*, 411–434. [CrossRef]
60. Lockwood, S.Y.; Erkal, J.L.; Spence, D.M. Endothelium-derived nitric oxide production is increased by ATP released from red blood cells incubated with hydroxyurea. *Nitric Oxide* **2014**, *38*, 1–7. [CrossRef] [PubMed]
61. Tang, L.C. Identification and characterization of human erythrocyte muscarinic receptors. *Gen. Pharmacol.* **1986**, *17*, 281–285. [CrossRef]

62. Kersh, G.J.; Tomich, J.M.; Montal, M. The M2 delta transmembrane domain of the nicotinic cholinergic receptor forms ion channels in human erythrocyte membranes. *Biochem. Biophys. Res. Commun.* **1989**, *162*, 352–356. [CrossRef]
63. Mesquita, R.; Pires, I.; Saldanha, C.; Martins-Silva, J. Effects of acetylcholine and spermineNONOate on erythrocyte hemorheologic and oxygen carrying properties. *Clin. Hemorheol. Microcirc.* **2001**, *25*, 153–163.
64. Muravyov, A.; Tikhomirova, I. Signaling pathways regulating red blood cell aggregation. *Biorheology* **2014**, *51*, 135–145. [CrossRef]
65. Olson, J.M.; Ciliax, B.J.; Mancini, W.R.; Young, A.B. Presence of peripheral-type benzodiazepine binding sites on human erythrocyte membranes. *Eur. J. Pharmacol.* **1988**, *152*, 47–53. [CrossRef]
66. Dave, J.R.; Eskay, R.L. Demonstration of corticotropin-releasing factor binding sites on human and rat erythrocyte membranes and their modulation by chronic ethanol treatment in rats. *Biochem. Biophys. Res. Commun.* **1986**, *136*, 137–144. [CrossRef]
67. Gopalakrishnan, V.; Ramaswamy, S.; Pillai, N.P.; Ranganathan, S.; Ghosh, M.N. Effect of prolactin on human red cell sodium transport. *Experientia* **1980**, *36*, 1423–1425. [CrossRef] [PubMed]
68. Cook, J.D.; Finch, C.A. Assessing iron status of a population. *Am. J. Clin. Nutr.* **1979**, *32*, 2115–2119. [CrossRef]
69. Kell, D.B. Iron behaving badly: Inappropriate iron chelation as a major contributor to the aetiology of vascular and other progressive inflammatory and degenerative diseases. *BMC Med. Genom.* **2009**, *2*, 2. [CrossRef]
70. Kell, D.B.; Pretorius, E. Serum ferritin is an important inflammatory disease marker, as it is mainly a leakage product from damaged cells. *Metallomics* **2014**, *6*, 748–773. [CrossRef]
71. Powell, R.J.; Machiedo, G.W.; Rush, B.F.; Dikdan, G. Oxygen free radicals: Effect on red cell deformability in sepsis. *Crit. Care Med.* **1991**, *19*, 732–735. [CrossRef] [PubMed]
72. Pretorius, E.; Bester, J.; Vermeulen, N.; Lipinski, B.; Gericke, G.S.; Kell, D.B. Profound morphological changes in the erythrocytes and fibrin networks of patients with hemochromatosis or with hyperferritinemia, and their normalization by iron chelators and other agents. *PLoS ONE* **2014**, *9*, e8527. [CrossRef] [PubMed]
73. Pretorius, E.; Kell, D.B. Diagnostic morphology: Biophysical indicators for iron-driven inflammatory diseases. *Integr. Biol.* **2014**, *6*, 486–510. [CrossRef]
74. Du Plooy, J.N.; Bester, J.; Pretorius, E. Eryptosis in Haemochromatosis: Implications for rheology. *Clin. Hemorheol. Microcirc.* **2018**, *69*, 457–469. [CrossRef] [PubMed]
75. Bekyarova, G.; Yankova, T.; Kozarev, I.; Yankov, D. Reduced erythrocyte deformability related to activated lipid peroxidation during the early postburn period. *Burns* **1996**, *22*, 291–294. [CrossRef]
76. Ham, T.H.; Shen, S.C. Studies on the destruction of red blood cells; thermal injury; action of heat in causing increased spheroidicity, osmotic and mechanical fragilities and hemolysis of erythrocytes; observations on the mechanisms of destruction of such erythrocytes in dogs and in a patient with a fatal thermal burn. *Blood* **1948**, *3*, 373–403. [PubMed]
77. Kaukonen, K.M.; Vaara, S.T.; Pettila, V.; Bellomo, R.; Tuimala, J.; Cooper, D.J.; Krusius, T.; Kuitunen, A.; Reinikainen, M.; Koskenkari, J.; et al. Age of red blood cells and outcome in acute kidney injury. *Crit. Care* **2013**, *17*, R222. [CrossRef]
78. Pettilä, V.; Westbrook, A.J.; Nichol, A.D.; Bailey, M.J.; Wood, E.M.; Syres, G.; Phillips, L.E.; Street, A.; French, C.; Murray, L.; et al. Age of red blood cells and mortality in the critically ill. *Crit. Care* **2011**, *15*, R116. [CrossRef]
79. Bester, J.; Buys, A.V.; Lipinski, B.; Kell, D.B.; Pretorius, E. High ferritin levels have major effects on the morphology of erythrocytes in Alzheimer's disease. *Front. Aging Neurosci.* **2013**, *5*, 88. [CrossRef]
80. Lipinski, B.; Pretorius, E.; Oberholzer, H.M.; Van Der Spuy, W.J. Interaction of fibrin with red blood cells: The role of iron. *Ultrastruct. Pathol.* **2012**, *36*, 79–84. [CrossRef]
81. Pretorius, E. The adaptability of red blood cells. *Cardiovasc. Diabetol.* **2013**, *12*, 63. [CrossRef]
82. Pretorius, E. Erythrocyte deformability and eryptosis during inflammation, and impaired blood rheology. *Clin. Hemorheol. Microcirc.* **2018**, *69*, 545–550. [CrossRef] [PubMed]
83. Pretorius, E.; Bester, J.; Vermeulen, N.; Alummoottil, S.; Soma, P.; Buys, A.V.; Kell, D.B. Poorly controlled type 2 diabetes is accompanied by significant morphological and ultrastructural changes in both erythrocytes and in thrombin-generated fibrin: Implications for diagnostics. *Cardiovasc. Diabetol.* **2015**, *14*, 30. [CrossRef]
84. Mcnamee, A.P.; Sabapathy, S.; Singh, I.; Horobin, J.; Guerrero, J.; Simmonds, M.J. Acute Free-Iron Exposure Does Not Explain the Impaired Haemorheology Associated with Haemochromatosis. *PLoS ONE* **2016**, *11*, e0146448. [CrossRef]
85. Pretorius, E.; Vermeulen, N.; Bester, J.; Du Plooy, J.L.; Gericke, G.S. The effect of iron overload on red blood cell morphology. *Lancet* **2014**, *383*, 722. [CrossRef]
86. Gelmini, G.; Quaini, F.; Ricci, R.; Schianchi, L.; Zanichelli, P.; Delsignore, R. Effects of iron-copper free diet on whole blood filtrability in rats. *Clin. Hemorheol. Microcirc.* **1989**, *9*, 469.
87. Wen, Z.Y.; Ma, W.Y.; Sun, D.G.; Chen, J.D. The study on RBC deformability of iron deficiency anemia. *Clin. Hemorheol. Microcirc.* **1995**, *15*, 81–87. [CrossRef]
88. Khaled, S.; Brun, J.F.; Wagner, A.; Mercier, J.; Bringer, J.; Prefaut, C. Increased blood viscosity in iron-depleted elite athletes. *Clin. Hemorheol. Microcirc.* **1998**, *18*, 309–318.
89. Cicha, I.; Suzuki, Y.; Tateishi, N.; Maeda, N. Iron-induced oxidative damage in human red blood cells and the effect of thiol-containing antioxidants. *Biorheology* **1999**, *36*, 48.
90. Kuçukatay, V.; Bor-Kucukatay, M.; Gundogdu, G.; Erken, G.; Ozcan, T.O.; Miloglu, F.D.; Kadioglu, Y. Vitamin E treatment enhances erythrocyte deformability in aged rats. *Folia Biol.* **2012**, *58*, 157–165.

91. Oostenbrug, G.S.; Mensink, R.P.; Hardeman, M.R.; De Vries, T.; Brouns, F.; Hornstra, G. Exercise performance, red blood cell deformability, and lipid peroxidation: Effects of fish oil and vitamin E. *J. Appl. Physiol.* **1997**, *83*, 746–752. [CrossRef]
92. Sentürk, U.K.; Gündüz, F.; Kuru, O.; Aktekin, M.R.; Kipmen, D.; Yalcin, O.; Bor-Kucukatay, M.; Yesilkaya, A.; Baskurt, O.K. Exercise-induced oxidative stress affects erythrocytes in sedentary rats but not exercise-trained rats. *J. Appl. Physiol.* **2001**, *91*, 1999–2004. [CrossRef] [PubMed]
93. Yerer, M.B.; Aydogan, S. The in vivo antioxidant effectiveness of alpha-tocopherol in oxidative 8stress induced by sodium nitroprusside in rat red blood cells. *Clin. Hemorheol. Microcirc.* **2004**, *30*, 323–329. [PubMed]
94. Begum, A.N.; Terao, J. Protective effect of alpha-tocotrienol against free radical-induced impairment of erythrocyte deformability. *Biosci. Biotechnol. Biochem.* **2002**, *66*, 398–403. [CrossRef]
95. Nanjo, F.; Honda, M.; Okushio, K.; Matsumoto, N.; Ishigaki, F.; Ishigami, T.; Hara, Y. Effects of dietary tea catechins on alpha-tocopherol levels, lipid peroxidation, and erythrocyte deformability in rats fed on high palm oil and perilla oil diets. *Biol. Pharm. Bull.* **1993**, *16*, 1156–1159. [CrossRef] [PubMed]
96. Sentürk, U.K.; Gündüz, F.; Kuru, O.; Koçer, G.; Ozkaya, Y.G.; Yesilkaya, A.; Bor-Küçükatay, M.; Uyüklü, M.; Yalcin, O.; Baskurt, O.K. Exercise-induced oxidative stress leads hemolysis in sedentary but not trained humans. *J. Appl. Physiol.* **2005**, *99*, 1434–1441. [CrossRef]
97. Brun, J.F.; Fons, C.; Fussellier, M.; Bardet, L.; Orsetti, A. Zinc salts improve in vitro erythrocyte flexibility. *Rev. Port. Hemorreol.* **1991**, *5*, 231–238.
98. Khaled, S.; Brun, J.F.; Micallef, J.P.; Bardet, L.; Cassanas, G.; Monnier, J.F.; Orsetti, A. Serum zinc and blood rheology in sportsmen (football players). *Clin. Hemorheol. Microcirc.* **1997**, *17*, 47–58.
99. Khaled, S.; Brun, J.F.; Cassanas, G.; Bardet, L.; Orsetti, A. Effects of zinc supplementation on blood rheology during exercise. *Clin. Hemorheol. Microcirc.* **1999**, *20*, 1–10.
100. Khaled, S.; Brun, J.F.; Cassanas, G.; Bardet, L.; Mercier, J.; Préfaut, C. In vitro effects of zinc gluconate and acetate on red cell aggregability. *Clin. Hemorheol. Microcirc.* **2000**, *22*, 325–329. [PubMed]
101. Reinhart, S.A.; Schulzki, T.; Bonetti, P.O.; Reinhart, W.H. Studies on metabolically depleted erythrocytes. *Clin. Hemorheol. Microcirc.* **2014**, *56*, 161–173. [CrossRef]
102. Jain, R.K.; Traykov, T.T. Effect of glucose and galactose on RBC deformability. *Biorheology* **1986**, *23*, 292.
103. Robey, C.; Dasmahapatra, A.; Cohen, M.P.; Suarez, S. Sorbinil partially prevents decreased erythrocyte deformability in experimental diabetes mellitus. *Diabetes* **1987**, *36*, 1010–1013. [CrossRef] [PubMed]
104. Schmid-Schönbein, H.; Volger, E. Red-cell aggregation and red-cell deformability in diabetes. *Diabetes* **1976**, *25*, 897–902. [PubMed]
105. Fashing, P.; Kurzemann, S.; Wagner, B.; Wagner, O.; Waldhäusl, W.; Ehringer, H.; Koppensteiner, R. Only marginal influence of metabolic control on blood rheology in insulin-dependent diabetic patient without manifest angiopathy. *Clin. Hemorheol. Microcirc.* **1995**, *15*, 3–11. [CrossRef]
106. Stuart, J.; Juhan-Vague, I. Erythrocyte rheology in diabetes. *Biorheolology* **1986**, *23*, 216.
107. Schmid-Schönbein, H.; Teitel, P. In vitro assessment of "covertly" abnormal blood rheology: Critical appraisal of presently available microrheological methodology. A review focusing on diabetic retinopathy as a possible consequence of rheological occlusion. *Clin. Hemorheol. Microcirc.* **1987**, *7*, 203–238. [CrossRef]
108. Li, Q.; Yang, L.Z. Hemoglobin A1c level higher than 9.05% causes a significant impairment of erythrocyte deformability in diabetes mellitus. *Acta Endocrinol.* **2018**, *14*, 66–75. [CrossRef]
109. Blaslov, K.; Kruljac, I.; Mirošević, G.; Gaćina, P.; Kolonić, S.O.; Vrkljan, M. The prognostic value of red blood cell characteristics on diabetic retinopathy development and progression in type 2 diabetes mellitus. *Clin. Hemorheol. Microcirc.* **2019**, *71*, 475–481. [CrossRef]
110. Calderón-Salinas, J.V.; Muñoz-Reyes, E.G.; Guerrero-Romero, J.F.; Rodríguez-Morán, M.; Bracho-Riquelme, R.L.; Carrera-Gracia, M.A.; Quintanar-Escorza, M.A. Eryptosis and oxidative damage in type 2 diabetic mellitus patients with chronic kidney disease. *Mol. Cell. Biochem.* **2011**, *357*, 171–179. [CrossRef]
111. Restivo, I.; Attanzio, A.; Tesoriere, L.; Allegra, M. Suicidal Erythrocyte Death in Metabolic Syndrome. *Antioxidants* **2021**, *10*, 154. [CrossRef]
112. Khodabandehlou, T.; Zhao, H.; Vimeux, M.; Aouane, F.; Ledévéhat, C. Haemorheological consequences of hyperglycaemic spike in healthy volunteers and insulin-dependent diabetics. *Clin. Hemorheol. Microcirc.* **1998**, *19*, 105–114.
113. Babu, N.; Singh, M. Influence of hyperglycemia on aggregation, deformability and shape parameters of erythrocytes. *Clin. Hemorheol. Microcirc.* **2004**, *31*, 273–280.
114. Shin, S.; Ku, Y.H.; Suh, J.S.; Singh, M. Rheological characteristics of erythrocytes incubated in glucose media. *Clin. Hemorheol. Microcirc.* **2008**, *38*, 153–161. [PubMed]
115. Bigdelou, P.; Farnoud, A.M. Induction of eryptosis in red blood cells using a calcium ionophore. *J. Vis. Exp.* **2020**, *155*. [CrossRef]
116. Wong, P. The basis of echinocytosis of the erythrocyte by glucose depletion. *Cell Biochem. Funct.* **2011**, *29*, 708–711. [CrossRef]
117. Varlet-Marie, E.; Guiraudou, M.; Fédou, C.; Raynaud De Mauverger, E.; Durand, F.; Brun, J.F. Nutritional and metabolic determinants of blood rheology differ between trained and sedentary individuals. *Clin. Hemorheol. Microcirc.* **2013**, *55*, 39–54. [CrossRef]
118. Brun, J.F.; Fons, C.; Supparo, I.; Mallard, C.; Orsetti, A. Relationships between metabolic and hemorheologic modifications associated with overweight. *Clin. Hemorheol. Microcirc.* **1993**, *13*, 201–213. [CrossRef]

119. Forsyth, A.M.; Braunmuller, S.; Wan, J.; Franke, T.; Stone, H.A. The effects of membrane cholesterol and simvastatin on red blood cell deformability and ATP release. *Microvasc. Res.* **2012**, *83*, 347–351. [CrossRef]
120. Guézennec, C.Y.; Nadaud, J.F.; Satabin, P.; Leger, F.; Lafargue, P. Influence of polyunsaturated fatty acid diet on the hemorheological response to physical exercise in hypoxia. *Int. J. Sports Med.* **1989**, *10*, 286–291. [CrossRef]
121. Léger, C.L.; Guézennec, C.Y.; Kadri-Hassani, N.; Satabin, P. Les acides gras phospholipidiques membranaires au cours de l'effort physique de longue durée avec ou sans apport nutritionnel d'huiles de poissons. *Cah. Nutr. Diététique* **1992**, *27*, 82–89.
122. Bakker, N.; Schoorl, M.; Demirkiran, A.; Cense, H.A.; Houdijk, A.P. Erythrocyte deformability and aggregation in morbidly obese women undergoing laparoscopic gastric bypass surgery and effects of oral omega-3 fatty acid supplementation. *Clin. Hemorheol. Microcirc.* **2020**, *75*, 303–311. [CrossRef] [PubMed]
123. Bakker, N.; Schoorl, M.; Stoutjesdijk, E.; Houdijk, A.P. Erythrocyte deformability and aggregability in patients undergoing colon cancer surgery and effects of two infusions with omega-3 fatty acids. *Clin. Hemorheol. Microcirc.* **2020**, *74*, 287–297. [CrossRef] [PubMed]
124. Thoth, K.; Ernst, E.; Habon, T.; Horvath, I.; Juricskay, I.; Mozsic, G. Hemorheological and hemodynamical effects of fish oil (Ameu) in patients with ischemic heart disease and hyperlipoproteinemia. *Clin. Hemorheol. Microcirc.* **1995**, *15*, 867–875. [CrossRef]
125. Schütz, E.; Schuff-Werner, P.; Armstrong, V.W.; Senger, I.; Güttner, Y. Haemorheological changes during postprandial lipemia. In Proceedings of the VIIth European Conference on Clinical Haemorheology, Southampton, UK, 16–19 July 1991; Abstract book. p. 91.
126. Varlet-Marie, E.; Brun, J.F. Reciprocal relationships between blood lactate and hemorheology in athletes: Another hemorheologic paradox? *Clin. Hemorheol. Microcirc.* **2004**, *30*, 331–337.
127. Brun, J.F.; Fons, C.; Raynaud, E.; Fédou, C.; Orsetti, A. Influence of circulating lactate on blood rheology during exercise in professional football players. *Rev. Port Hemorheol.* **1991**, *5*, 219–229.
128. Connes, P.; Caillaud, C.; Bouix, D.; Kippelen, P.; Mercier, J.; Varray, A.; Préfaut, C.; Brun, J.F. Red cell rigidity paradoxically decreases during maximal exercise in endurance athletes unless they are prone to exercise-induced hypoxaemia. *J. Mal. Vasc.* **2000**, *25* (Suppl. B), 165.
129. Peyreigne, C.; Bouix, D.; Aïssa Benhaddad, A.; Raynaud, E.; Perez-Martin, A.; Mercier, J.; Brun, J.F. Hemorheologic effects of a short-term ketogenetic diet. *Clin. Hemorheol. Microcirc.* **1999**, *21*, 147–153. [PubMed]
130. McMahon, T.J. Red Blood Cell Deformability, Vasoactive Mediators, and Adhesion. *Front. Physiol.* **2019**, *10*, 1417. [CrossRef]
131. Simmonds, M.J.; Detterich, J.A.; Connes, P. Nitric oxide, vasodilation and the red blood cell. *Biorheology* **2014**, *51*, 121–134. [CrossRef]
132. Chen, K.; Popel, A.S. Nitric oxide production pathways in erythrocytes and plasma. *Biorheology* **2009**, *46*, 107–119. [CrossRef]
133. Ulker, P.; Sati, L.; Celik-Ozenci, C.; Meiselman, H.J.; Baskurt, O.K. Mechanical stimulation of nitric oxide synthesizing mechanisms in erythrocytes. *Biorheology* **2009**, *46*, 121–132. [CrossRef]
134. Baskurt, O.K.; Uyuklu, M.; Meiselman, H.J. Protection of erythrocytes from sub-hemolytic mechanical damage by nitric oxide mediated inhibition of potassium leakage. *Biorheology* **2004**, *41*, 79–89. [PubMed]
135. Bor-Kucukatay, M.; Wenby, R.B.; Meiselman, H.J.; Baskurt, O.K. Effects of nitric oxide on red blood cell deformability. *Am. J. Physiol. Heart Circ. Physiol.* **2003**, *284*, H1577–H1584. [CrossRef] [PubMed]
136. Korbut, R.; Gryglewski, R.J. Nitric oxide from polymorphonuclear leukocytes modulates red blood cell deformability in vitro. *Eur. J. Pharmacol.* **1993**, *234*, 17–22. [CrossRef]
137. Mesquita, R.; Picarra, B.; Saldanha, C.; Martins, E.; Silva, J. Nitric oxide effects on human erythrocytes structural and functional properties–an in vitro study. *Clin. Hemorheol. Microcirc.* **2002**, *27*, 137–147.
138. Bor-Kucukatay, M.; Meiselman, H.J.; Başkurt, O.K. Modulation of density-fractionated RBC deformability by nitric oxide. *Clin. Hemorheol. Microcirc.* **2005**, *33*, 363–367.
139. Rey, J.; Buffet, P.A.; Ciceron, L.; Milon, G.; Mercereau-Puijalon, O.; Safeukui, I. Reduced erythrocyte deformability associated with hypoargininemia during Plasmodium falciparum malaria. *Sci. Rep.* **2014**, *4*, 3767. [CrossRef] [PubMed]
140. Grau, M.; Mozar, A.; Charlot, K.; Lamarre, Y.; Weyel, L.; Suhr, F.; Collins, B.; Jumet, S.; Hardy-Dessources, M.D.; Romana, M.; et al. High red blood cell nitric oxide synthase activation is not associated with improved vascular function and red blood cell deformability in sickle cell anaemia. *Br. J. Haematol.* **2015**, *168*, 728–736. [CrossRef] [PubMed]
141. Mozar, A.; Connes, P.; Collins, B.; Hardy-Dessources, M.D.; Romana, M.; Lemonne, N.; Bloch, W.; Grau, M. Red blood cell nitric oxide synthase modulates red blood cell deformability in sickle cell anemia. *Clin. Hemorheol. Microcirc.* **2016**, *64*, 47–53. [CrossRef]
142. Bor-Küçükatay, M.; Yalçin, O.; Gökalp, O.; Kipmen-Korgun, D.; Yesilkaya, A.; Baykal, A.; Ispir, M.; Senturk, U.K.; Kaputlu, I.; Başkurt, O.K. Red blood cell rheological alterations in hypertension induced by chronic inhibition of nitric oxide synthesis in rats. *Clin. Hemorheol. Microcirc.* **2000**, *22*, 267–275.
143. Tai, Y.H.; Chu, Y.H.; Wu, H.L.; Lin, S.M.; Tsou, M.Y.; Huang, C.H.; Chang, H.H.; Lu, C.C. High-dose nitroglycerin administered during rewarming preserves erythrocyte deformability in cardiac surgery with cardiopulmonary bypass. *Microcirculation* **2020**, *27*, e12608. [CrossRef]
144. Shibuya, N.; Kimura, H. Production of hydrogen sulfide from d-cysteine and its therapeutic potential. *Front. Endocrinol.* **2013**, *4*, 87. [CrossRef] [PubMed]

145. Muravyov, A.V.; Tikhomirova, I.A.; Avdonin, P.V.; Bulaeva, S.V.; Malisheva, J.V. Comparative efficiency of three gasotransmitters (nitric oxide, hydrogen sulfide and carbon monoxide): Analysis on the model of red blood cell microrheological responses. *J. Cell. Biotechnol.* **2021**, *7*, 1–9. [CrossRef]
146. Wong, W.W.; Jack, R.M.; Smith, J.A.; Kennedy, C.A.; Fearon, D.T. Rapid purification of the human C3b/C4b receptor (CR1) by monoclonal antibody affinity chromatography. *J. Immunol. Methods* **1985**, *82*, 303–313. [CrossRef]
147. Sung, L.A.; Kabat, E.A.; Chien, S. Interaction of lectins with membrane receptors on erythrocyte surfaces. *J. Cell Biol.* **1985**, *101*, 646–651. [CrossRef]
148. Bryszewska, M.; Leyko, W. Effect of insulin on human erythrocyte membrane fluidity in diabetes mellitus. *Diabetologia* **1983**, *24*, 311–313. [CrossRef]
149. Dutta-Roy, A.K.; Ray, T.K.; Sinha, A.K. Control of erythrocyte membrane microviscosity by insulin. *Biochim. Biophys. Acta* **1985**, *816*, 187–190. [CrossRef]
150. Rahmani-Jourdheuil, D.; Mourayre, Y.; Vague, P.; Boyer, J.; Juhan-Vague, I. In vivo insulin effect on ATPase activities in erythrocyte membrane from insulin-dependent diabetics. *Diabetes* **1987**, *36*, 991–995. [CrossRef]
151. Linde, T.; Sandhagen, B.; Berne, C.; Lind, L.; Reneland, R.; Hanni, A.; Lithell, H. Erythrocyte deformability is related to fasting insulin and declines during euglycaemic hyperinsulinaemia in hypertensive patients. *J. Hum. Hypertens.* **1999**, *13*, 285–286. [CrossRef]
152. Aursnes, I.; Dahl-Jørgensen, K.; Hanssen, K.F. ATP-concentrations in erythrocytes influenced by insulin levels in plasma. *Clin. Hemorheol. Microcirc.* **1986**, *12*, 429–433. [CrossRef]
153. Forst, T.; De La Tour, D.D.; Kunt, T.; Pfutzner, A.; Goitom, K.; Pohlmann, T.; Schneider, S.; Johansson, B.L.; Wahren, J.; Lobig, M.; et al. Effects of proinsulin C-peptide on nitric oxide, microvascular blood flow and erythrocyte Na+,K+-ATPase activity in diabetes mellitus type I. *Clin. Sci.* **2000**, *98*, 283–290. [CrossRef]
154. Catanese, V.M.; Grigorescu, F.; King, G.L.; Kahn, C.R. The human erythrocyte insulin-like growth factor I receptor: Characterization and demonstration of ligand-stimulated autophosphorylation. *J. Clin. Endocrinol. Metab.* **1986**, *62*, 692–699. [CrossRef] [PubMed]
155. Monnier, J.F.; Benhaddad, A.A.; Micallef, J.P.; Mercier, J.; Brun, J.F. Relationships between blood viscosity and insulin-like growth factor I status in athletes. *Clin. Hemorheol. Microcirc.* **2000**, *22*, 277–286. [PubMed]
156. Valensi, P.; Gaudey, F.; Attali, J.R. La déformabilité érythrocytaire est réduite par le glucagon. *Diabetes Metab.* **1986**, *12*, 281.
157. Komatsu, R.; Tsushima, N.; Matsuyama, T. Effects of glucagon administration on microcirculation and blood rheology. *Clin. Hemorheol. Microcirc.* **1997**, *17*, 271–277.
158. Schnyder, L.; Walter, R.; Rohrer, A.; Contesse, J.; Reinhart, W.H. No influence of C-peptide, insulin, and glucagon on blood viscosity in vitro in healthy humans and patients with diabetes mellitus. *Clin. Hemorheol. Microcirc.* **2001**, *24*, 65–74. [PubMed]
159. Angel, R.C.; Botta, J.A.; Farias, R.N. High affinity L-triiodothyronine binding to right-side-out and inside-out vesicles from rat and human erythrocyte membrane. *J. Biol. Chem.* **1989**, *264*, 19143–19146. [CrossRef]
160. Larsson, H.; Valdemarsson, S.; Hedner, P.; Odeberg, H. Blood viscosity in hyperthyroidism: Significance of erythrocyte changes and hematocrit. *Clin. Hemorheol. Microcirc.* **1988**, *8*, 257–265. [CrossRef]
161. Zhang, Y.; Chua, S., Jr. Leptin Function and Regulation. *Compr. Physiol.* **2017**, *8*, 351–369. [CrossRef]
162. Tsuda, K.; Kimura, K.; Nishio, I. Leptin improves membrane fluidity of erythrocytes in humans via a nitric oxide-dependent mechanism: An electron paramagnetic resonance investigation. *Biochem. Biophys. Res. Commun.* **2002**, *297*, 672–681. [CrossRef]
163. Brun, J.F.; Perez-Martin, A.; Raynaud, E.; Mercier, J. Correlation between plasma viscosity, erythrocyte disaggregability and leptin. *Biorheology* **1999**, *36*, 156–157.
164. Brun, J.F.; Varlet-Marie, E.; Vachoud, L.; Marion, B.; Roques, C.; Raynaud de Mauverger, E.; Mercier, J. Is leptin a regulator of erythrocyte rheology? *Clin. Hemorheol. Microcirc.* **2021**, *in press*.
165. Faulkner, J.L.; Belin de Chantemèle, E.J. Leptin and Aldosterone. *Vitam. Horm.* **2019**, *109*, 265–284. [CrossRef]
166. Faulkner, J.L.; Bruder-Nascimento, T.; Belin de Chantemèle, E.J. The regulation of aldosterone secretion by leptin: Implications in obesity-related cardiovascular disease. *Curr. Opin. Nephrol. Hypertens.* **2018**, *27*, 63–69. [CrossRef] [PubMed]
167. Reinhart, W.H. Molecular biology and self-regulatory mechanisms of blood viscosity: A review. *Biorheology* **2001**, *38*, 203–212.
168. Brun, J.F. Hormones, metabolism and body composition as major determinants of blood rheology: Potential pathophysiological meaning. *Clin. Hemorheol. Microcirc.* **2002**, *26*, 63–79.
169. Muravyov, A.V.; Cheporov, S.V.; Kislov, N.V.; Bulaeva, S.V.; Maimistova, A.A. Comparative efficiency and hemorheological consequences of hemotransfusion and epoetin therapy in anemic cancer patients. *Clin. Hemorheol. Microcirc.* **2010**, *44*, 115–123. [CrossRef] [PubMed]
170. Vota, D.M.; Maltaneri, R.E.; Wenker, S.D.; Nesse, A.B.; Vittori, D.C. Differential erythropoietin action upon cells induced to eryptosis by different agents. *Cell Biochem. Biophys.* **2013**, *65*, 145–157. [CrossRef]
171. Zyznar, E.S.; Pietri, A.O.; Harris, V.; Unger, R.H. Evidence for the hormonal status of somatostatin in man. *Diabetes* **1981**, *30*, 883–886. [CrossRef]
172. Savi, L.; Cardillo, C.; Bombardieri, G. Somatostatin and peripheral blood flow in man. *Angiology* **1985**, *36*, 511–515. [CrossRef]
173. Fuse, I.; Ito, S.; Takagi, A.; Shibata, A. Different effects of three kinds of somatostatin (15–28, 1–14, 1–28) on rabbit's platelet aggregation. *Life Sci.* **1985**, *36*, 2047–2052. [CrossRef]

174. Brun, J.F.; Rauturier, M.; Ghanem, Y.; Orsetti, A. In vitro effects of somatostatin on red cell filterability measured by three methods. *J. Mal. Vasc.* **1991**, *16*, 49–52.
175. Berker, M.; Dikmenoglu, N.; Bozkurt, G.; Ergönül, Z.; Özgen, T. Hemorheology, melatonin and pinealectomy. What's the relationship? An experimental study. *Clin. Hemorheol. Microcirc.* **2004**, *30*, 47–52.
176. Vazan, R.; Plauterova, K.; Porubska, G.; Radosinska, J. Changes in erythrocyte deformability during day and possible role of melatonin. *Endocr. Regul.* **2018**, *52*, 17–20. [CrossRef]
177. Freyburger, G.; Larrue, J.; Boisseau, M.R. Effect of two arachidonic acid metabolites of white blood cells on red blood cell filterability studied by cell transit-time analysis. *Clin. Hemorheol. Microcirc.* **1987**, *7*, 536.
178. Mary, A.; Modat, G.; Gal, L.; Bonne, C. Leukotriene B4 decreases RBC deformability as assessed by increased filtration index. *Clin. Hemorheol. Microcirc.* **1989**, *9*, 209–217. [CrossRef]
179. Elgatit, A.M.; Rashid, M.A.; Belboul, A.M.; Ramirez, J.J.; Roberts, D.G.; Olsson, G.W. Effects of aprostadil (PGE1) on red and white cell deformability during cardiopulmonary bypass. In Proceedings of the VIIth European Conference on Clinical Haemorheology, Southampton, UK, 16–19 July 1991; Abstract book. p. 81.
180. Langer, R.; Rossmanith, K.; Henrich, H.A. Hemorheological actions of the prostaglandins D2, E1, E2, F1a, F2a and Iloprost. *Clin. Hemorheol. Microcirc.* **1995**, *15*, 829–839. [CrossRef]
181. Brun, J.F.; Bouchahda, C.; Aissa-Benhaddad, A.; Sagnes, C.; Granat, M.C.; Bor Kucukatay, M.; Baskurt, O.; Mercier, J. Hemorheological aspects of leuko-platelet activation in atheromatous diseases: Clinical applications. *J. Mal. Vasc.* **2000**, *25*, 349–355.
182. Gonçalves, I.; Saldanha, C.; Martins, E.; Silva, J. Beta-estradiol effect on erythrocyte aggregation—A controlled in vitro study. *Clin. Hemorheol. Microcirc.* **2001**, *25*, 127–134.
183. Grau, M.; Cremer, J.M.; Bloch, W. Comparisons of blood parameters, red blood cell deformability and circulating nitric oxide between males and females considering hormonal contraception: A longitudinal gender study. *Front. Physiol.* **2018**, *9*, 1835. [CrossRef]
184. Swanepoel, A.C.; Emmerson, O.; Pretorius, E. Effect of Progesterone and Synthetic Progestins on Whole Blood Clot Formation and Erythrocyte Structure. *Microsc. Microanal.* **2017**, *23*, 607–617. [CrossRef]
185. Swanepoel, A.C.; Pretorius, E. Erythrocyte-platelet interaction in uncomplicated pregnancy. *Microsc. Microanal.* **2014**, *20*, 1848–1860. [CrossRef]
186. Kon, K.; Maeda, N.; Shiga, T. Functional impairments of human red cells, induced by dehydroepiandrosterone sulfate. *Pflugers Arch.* **1982**, *394*, 279–286. [CrossRef]
187. Kartal, H.; Comu, F.M.; Kucuk, A.; Polat, Y.; Dursun, A.D.; Arslan, M. Effect of apelin-13 on erythrocyte deformability during ischaemia-reperfusion injury of heart in diabetic rats. *Bratisl Lek Listy* **2017**, *118*, 133–136. [CrossRef]
188. Sundquist, J.; Blas, S.D.; Hogan, J.E.; Davis, F.B.; Davis, P.J. The alpha 1-adrenergic receptor in human erythrocyte membranes mediates interaction in vitro of epinephrine and thyroid hormone at the membrane Ca(2+)-ATPase. *Cell. Signal.* **1992**, *4*, 795–799. [CrossRef]
189. Borgese, F.; Garcia-Romeu, F.; Motais, R. Control of cell volume and ion transport by beta-adrenergic catecholamines in erythrocytes of rainbow trout, Salmo gairdneri. *J. Physiol.* **1987**, *382*, 123–144. [CrossRef] [PubMed]
190. Pfafferott, C.; Zaninelli, R.; Bauersachs, R.; Volger, E. In vitro effect of norepinephrine and isoprenaline on erythrocyte filterability under pathologic conditions. *Clin. Hemorheol. Microcirc.* **1987**, *7*, 409.
191. Hilario, S.; Saldanha, C.; Martins-Silva, J. The effect of adrenaline upon human erythrocyte. Sex-related differences? *Biorheology* **1999**, *36*, 124.
192. Chiocchia, G.; Motais, R. Effect of catecholamines on deformability of red cells from trout: Relative roles of cyclic AMP and cell volume. *J. Physiol.* **1989**, *412*, 321–332. [CrossRef]
193. Hilario, S.; Saldanha, C.; Martins-Silva, J. An in vitro study of adrenaline effect on human erythrocyte properties in both genders. *Clin. Hemorheol. Microcirc.* **2003**, *28*, 89–98. [PubMed]
194. Oonishi, T.; Sakashita, K.; Uyesaka, N. Regulation of red blood cell filterability by Ca2+ influx and cAMP-mediated signaling pathways. *Am. J. Physiol.* **1997**, *273*, C1828–C1834. [CrossRef] [PubMed]
195. Tikhomiroiva, I.A.; Muravyov, A.V.; Kruglova, E.V. Hormonal control mechanisms of the rheological properties of erythrocytes under physical exercises and stress. *Biorheology* **2008**, *45*, 49.
196. Muravyov, A.V.; Tikhomirova, I.A.; Maimistova, A.A.; Bulaeva, S.V. Extra- and intracellular signaling pathways under red blood cell aggregation and deformability changes. *Clin. Hemorheol. Microcirc.* **2009**, *43*, 223–232. [CrossRef] [PubMed]
197. Berent, H.; Wocial, B.; Kuczyńska, K.; Kochmański, M.; Ignatowska-Switalska, H.; Januszewicz, A.; Lapiński, M.; Lewandowski, J.; Januszewicz, W. Evaluation of blood rheology indices in patients with pheochromocytoma. *Pol. Arch. Med. Wewn.* **1996**, *95*, 190–197. [PubMed]
198. Mokrushnikov, P.V.; Panin, L.E.; Zaitsev, B.N. The action of stress hormones on the structure and function of erythrocyte membrane. *Gen. Physiol. Biophys.* **2015**, *34*, 311–321. [CrossRef]
199. Windberger, U.; Bartholovitsch, A. Hemorheology in spontaneous animal endocrinopathies. *Clin. Hemorheol. Microcirc.* **2004**, *31*, 207–215. [PubMed]
200. Lowe, H.; Toyang, N.; Steele, B.; Bryant, J.; Ngwa, W. The endocannabinoid system: A potential target for the treatment of various diseases. *Int. J. Mol. Sci.* **2021**, *22*, 9472. [CrossRef]
201. Piazza, P.V.; Cota, D.; Marsicano, G. The CB1 receptor as the cornerstone of exostasis. *Neuron* **2017**, *93*, 1252–1274. [CrossRef]

202. Cavuoto, P.; Wittert, G.A. The role of the endocannabinoid system in the regulation of energy expenditure. *Best. Pract. Res. Clin. Endocrinol. Metab.* **2009**, *23*, 79–86. [CrossRef]
203. Sakashita, K.; Oonishi, T.; Ishioka, N.; Uyesaka, N. Endothelin-1 improves the impaired filterability of red blood cells through the activation of protein kinase C. *Jpn. J. Physiol.* **1999**, *49*, 113–120. [CrossRef] [PubMed]
204. Walter, R.; Mark, M.; Gaudenz, R.; Harris, L.G.; Reinhart, W.H. Influence of nitrovasodilators and endothelin-1 on rheology of human blood in vitro. *Br. J. Pharmacol.* **1999**, *128*, 744–750. [CrossRef]
205. Mark, M.; Walter, R.; Harris, L.G.; Reinhart, W.H. Influence of parathyroid hormone, calcitonin, 1,25$(OH)_2$ cholecalciferol, calcium, and the calcium ionophore A23187 on erythrocyte morphology and blood viscosity. *J. Lab. Clin. Med.* **2000**, *135*, 347–352. [CrossRef] [PubMed]
206. Bester, J.; Pretorius, E. Effects of IL-1β, IL-6 and IL-8 on erythrocytes, platelets and clot viscoelasticity. *Sci. Rep.* **2016**, *6*, 32188. [CrossRef]
207. Olumuyiwa-Akeredolu, O.O.; Pretorius, E. Platelet and red blood cell interactions and their role in rheumatoid arthritis. *Rheumatol. Int.* **2015**, *35*, 1955–1964. [CrossRef] [PubMed]
208. Page, M.J.; Bester, J.; Pretorius, E. Interleukin-12 and its procoagulant effect on erythrocytes, platelets and fibrin(nogen): The lesser-known side of inflammation. *Brit. J. Hematol.* **2018**, *180*, 110–118. [CrossRef]
209. Page, M.J.; Bester, J.; Pretorius, E. The inflammatory effects of TNF-alpha and complement component 3 on coagulation. *Sci. Rep.* **2018**, *8*, 1812. [CrossRef]
210. Dhabhar, F.S. The short-term stress response—Mother nature's mechanism for enhancing protection and performance under conditions of threat, challenge, and opportunity. *Front. Neuroendocrinol.* **2018**, *49*, 175–192. [CrossRef] [PubMed]
211. Finnerty, F.A.; Buchholz, J.H.; Guillaudeu, R.L. The blood volumes and plasma protein during levarterenol-induced hypertension. *J. Clin. Invest.* **1958**, *37*, 425–429. [CrossRef]
212. Ehrly, A.M.; Landgraf, H.; Hessler, J.; Saeger-Lorenz, K. Influence of videofilm-induced emotional stress on the flow properties of blood. *Angiology* **1988**, *39*, 341–344. [CrossRef]
213. Sakai, A.; Nakano, H.; Ohira, T.; Hosoya, M.; Yasumura, S.; Ohtsuru, A.; Satoh, H.; Kawasaki, Y.; Suzuki, H.; Takahashi, A.; et al. Fukushima Health Management Survey Group. Persistent prevalence of polycythemia among evacuees 4 years after the Great East Japan Earthquake: A follow-up study. *Prev. Med. Rep.* **2017**, *5*, 251–256. [CrossRef]
214. Falvo, M.J.; Chen, Y.; Klein, J.C.; Ndirangu, D.; Condon, M.R. Abnormal rheological properties of red blood cells as a potential marker of Gulf War Illness: A preliminary study. *Clin. Hemorheol. Microcirc.* **2018**, *68*, 361–370. [CrossRef]
215. Qian, W.; Klein-Adams, J.C.; Ndirangu, D.S.; Chen, Y.; Falvo, M.J.; Condon, M.R. Hemorheological responses to an acute bout of maximal exercise in Veterans with Gulf War Illness. *Life Sci.* **2021**, *280*, 119714. [CrossRef] [PubMed]
216. Brun, J.F.; Varlet-Marie, E.; Richou, M.; Mercier, J.; Raynaud de Mauverger, E. Blood rheology as a mirror of endocrine and metabolic homeostasis in health and disease. *Clin. Hemorheol. Microcirc.* **2018**, *69*, 239–265. [CrossRef] [PubMed]
217. Saha, A.K.; Schmidt, B.R.; Wilhelmy, J.; Nguyen, V.; Abugherir, A.; Do, J.K.; Nemat-Gorgani, M.; Davis, R.W.; Ramasubramanian, A.K. Red blood cell deformability is diminished in patients with Chronic Fatigue Syndrome. *Clin. Hemorheol. Microcirc.* **2019**, *71*, 113–116. [CrossRef] [PubMed]
218. Brenu, E.W.; Staines, D.R.; Baskurt, O.K.; Ashton, K.J.; Ramos, S.B.; Christy, R.M.; Marshall-Gradisnikal, S.M. Immune and hemorheological changes in chronic fatigue syndrome. *J. Transl. Med.* **2010**, *8*, 1. [CrossRef] [PubMed]
219. Kennedy, G.; Spence, V.A.; McLaren, M.; Hill, A.; Underwood, C.; Belch, J.J. Oxidative stress levels are raised in chronic fatigue syndrome and are associated with clinical symptoms. *Free Radic. Biol. Med.* **2005**, *39*, 584–589. [CrossRef]
220. Kennedy, G.; Khan, F.; Hill, A.; Underwood, C.; Belch, J.J. Biochemical and vascular aspects of pediatric chronic fatigue syndrome. *Arch. Pediatr. Adolesc. Med.* **2010**, *164*, 817–823. [CrossRef] [PubMed]
221. Kim, Y.H.; Choi, S.U.; Youn, J.M.; Cha, S.H.; Shin, H.J.; Ko, E.J.; Lim, C.H. Effects of remote ischemic preconditioning on the deformability and aggregation of red blood cells in a rat endotoxemia model. *Clin. Hemorheol. Microcirc.* **2021**. *Online ahead of print*. [CrossRef]
222. Ko, E.; Youn, J.M.; Park, H.S.; Song, M.; Koh, K.H.; Lim, C.H. Early red blood cell abnormalities as a clinical variable in sepsis diagnosis. *Clin. Hemorheol. Microcirc.* **2018**, *70*, 355–363. [CrossRef]
223. Piagnerelli, M.; Boudjeltia, K.Z.; Vanhaeverbeek, M.; Vincent, J.L. Red blood cell rheology in sepsis. *Intensive Care Med.* **2003**, *29*, 1052–1061. [CrossRef] [PubMed]
224. Donadello, K.; Piagnerelli, M.; Reggiori, G.; Gottin, L.; Scolletta, S.; Occhipinti, G.; Zouaoui Boudjeltia, K.; Vincent, J.L. Reduced red blood cell deformability over time is associated with a poor outcome in septic patients. *Microvasc. Res.* **2015**, *101*, 8–14. [CrossRef] [PubMed]
225. Totsimon, K.; Biro, K.; Szabo, Z.E.; Toth, K.; Kenyeres, P.; Marton, Z. The relationship between hemorheological parameters and mortality in critically ill patients with and without sepsis. *Clin. Hemorheol. Microcirc.* **2017**, *65*, 119–129. [CrossRef]
226. Bateman, R.M.; Jagger, J.E.; Sharpe, M.D.; Ellsworth, M.L.; Mehta, S.; Ellis, C.G. Erythrocyte deformability is a nitric oxide-mediated factor in decreased capillary density during sepsis. *Am. J. Physiol. Heart Circ. Physiol.* **2001**, *280*, H2848–H2856. [CrossRef]
227. Bateman, R.M.; Sharpe, M.D.; Singer, M.; Ellis, C.G. The Effect of Sepsis on the Erythrocyte. *Int. J. Mol. Sci.* **2017**, *18*, 1932. [CrossRef]

228. Poraicu, D.; Mogoseanu, A.; Tomescu, N.; Bota, C.; Menessy, I. Decrease of red blood cell filterability seen in intensive care I. The correlation of low erythrocyte filterability with mortality and its return to normal values in critically ill patients under parenteral nutrition. *Resuscitation* **1983**, *10*, 291–303. [CrossRef]
229. Silva-Herdade, A.S.; Andolina, G.; Faggio, C.; Calado, A.; Saldanha, C. Erythrocyte deformability - A partner of the inflammatory response. *Microvasc. Res.* **2016**, *107*, 34–38. [CrossRef]
230. Korbut, R.; Gryglewski, R.J. The effect of prostacyclin and nitric oxide on deformability of red blood cells in septic shock in rats. *J. Physiol. Pharmacol.* **1996**, *47*, 591–599.
231. Caimi, G.; Montana, M.; Canino, B.; Calandrino, V.; Lo Presti, R.; Hopps, E. Erythrocyte deformability, plasma lipid peroxidation and plasma protein oxidation in a group of OSAS subjects. *Clin. Hemorheol. Microcirc.* **2016**, *64*, 7–14. [CrossRef] [PubMed]
232. Canino, B.; Hopps, E.; Calandrino, V.; Montana, M.; Lo Presti, R.; Caimi, G. Nitric oxide metabolites and erythrocyte deformability in a group of subjects with obstructive sleep apnea syndrome. *Clin. Hemorheol. Microcirc.* **2015**, *59*, 45–52. [CrossRef] [PubMed]
233. Valentim-Coelho, C.; Vaz, F.; Antunes, M.; Neves, S.; Martins, I.L.; Osório, H.; Feliciano, A.; Pinto, P.; Bárbara, C.; Penque, D. Redox-Oligomeric State of Peroxiredoxin-2 and Glyceraldehyde-3-Phosphate Dehydrogenase in Obstructive Sleep Apnea Red Blood Cells under Positive Airway Pressure Therapy. *Antioxidants* **2020**, *9*, 1184. [CrossRef] [PubMed]
234. Buono, M.; Rostomily, K. Acute normobaric hypoxia does not increase blood or plasma viscosity. *Clin. Hemorheol. Microcirc.* **2021**, *78*, 461–464. [CrossRef] [PubMed]
235. Jung, F. COVID-19. *Clin. Hemorheol. Microcirc.* **2020**, *74*, 347–348. [CrossRef]
236. Jung, F.; Krüger-Genge, A.; Franke, R.P.; Hufert, F.; Küpper, J.-H. COVID-19 and the endothelium. *Clin. Hemorheol. Microcirc.* **2020**, *75*, 7–11. [CrossRef]
237. Farber, P.L. Can erythrocytes behavior in microcirculation help the understanding the physiopathology and improve prevention and treatment for covid-19? *Clin. Hemorheol. Microcirc.* **2021**, *78*, 41–47. [CrossRef] [PubMed]
238. Ackermann, M.; Verleden, S.E.; Kuehnel, M.; Haverich, A.; Welte, T.; Laenger, F.; Vanstapel, A.; Werlein, C.; Stark, H.; Tzankov, A.; et al. Pulmonary Vascular Endothelialitis, Thrombosis, and Angiogenesis in Covid-19. *N. Engl. J. Med.* **2020**, *383*, 120–128. [CrossRef] [PubMed]
239. Goshua, G.; Pine, A.B.; Meizlish, M.L.; Chang, C.H.; Zhang, H.; Bahel, P.; Baluha, A.; Bar, N.; Bona, R.D.; Burns, A.J.; et al. Endotheliopathy in COVID-19-associated coagulopathy: Evidence from a single-centre, cross-sectional study. *Lancet Haematol.* **2020**, *7*, e575–e582. [CrossRef]
240. Grobler, C.; Maphumulo, S.C.; Grobbelaar, L.M.; Bredenkamp, J.C.; Laubscher, G.J.; Lourens, P.J.; Steenkamp, J.; Kell, D.B.; Pretorius, E. Covid-19: The Rollercoaster of Fibrin(Ogen), D-Dimer, Von Willebrand Factor, P-Selectin and Their Interactions with Endothelial Cells, Platelets and Erythrocytes. *Int. J. Mol. Sci.* **2020**, *21*, 5168. [CrossRef]
241. Pretorius, E.; Venter, C.; Laubscher, G.J.; Lourens, P.J.; Steenkamp, J.; Kell, D.B. Prevalence of readily detected amyloid blood clots in 'unclotted' Type 2 Diabetes Mellitus and COVID-19 plasma: A preliminary report. *Cardiovasc. Diabetol.* **2020**, *19*, 193. [CrossRef] [PubMed]
242. Roberts, I.; Muelas, M.W.; Taylor, J.M.; Davison, A.S.; Xu, Y.; Grixti, J.M.; Gotts, N.; Sorokin, A.; Goodacre, R.; Kell, D.B. Untargeted metabolomics of COVID-19 patient serum reveals potential prognostic markers of both severity and outcome. *MedRxiv* **2020**. [CrossRef]
243. Venter, C.; Bezuidenhout, J.A.; Laubscher, G.J.; Lourens, P.J.; Steenkamp, J.; Kell, D.B.; Pretorius, E. Erythrocyte, Platelet, Serum Ferritin, and P-Selectin Pathophysiology Implicated in Severe Hypercoagulation and Vascular Complications in COVID-19. *Int. J. Mol. Sci.* **2020**, *21*, 8234. [CrossRef] [PubMed]
244. Bobrova, L.; Kozlovskaya, N.; Korotchaeva, Y.; Bobkova, I.; Kamyshova, E.; Moiseev, S. Microvascular COVID-19 lung vessels obstructive thromboinflammatory syndrome (MicroCLOTS): A new variant of thrombotic microangiopathy? *Crit. Care Resusc.* **2020**, *22*, 284. [PubMed]
245. Ciceri, F.; Beretta, L.; Scandroglio, A.M.; Colombo, S.; Landoni, G.; Ruggeri, A.; Peccatori, J.; D'angelo, A.; De Cobelli, F.; Rovere-Querini, P.; et al. Microvascular COVID-19 lung vessels obstructive thromboinflammatory syndrome (MicroCLOTS): An atypical acute respiratory distress syndrome working hypothesis. *Crit. Care Resusc.* **2020**, *22*, 95–97. [CrossRef] [PubMed]
246. Renzi, S.; Landoni, G.; Zangrillo, A.; Ciceri, F. MicroCLOTS pathophysiology in COVID 19. *Korean J. Intern. Med.* **2020**. [CrossRef] [PubMed]
247. Akhter, N.; Ahmad, S.; Alzahrani, F.A.; Dar, S.A.; Wahid, M.; Haque, S.; Bhatia, K.; Sr Almalki, S.; Alharbi, R.A.; Sindi, A.A. Impact of COVID-19 on the cerebrovascular system and the prevention of RBC lysis. *Eur. Rev. Med. Pharmacol. Sci.* **2020**, *24*, 10267–10278. [CrossRef]
248. Berzuini, A.; Bianco, C.; Paccapelo, C.; Bertolini, F.; Gregato, G.; Cattaneo, A.; Erba, E.; Bandera, A.; Gori, A.; Lamorte, G.; et al. Red cell-bound antibodies and transfusion requirements in hospitalized patients with COVID-19. *Blood* **2020**, *136*, 766–768. [CrossRef]
249. Lam, L.M.; Murphy, S.J.; Kuri-Cervantes, L.; Weisman, A.R.; Ittner, C.A.; Reilly, J.P.; Pampena, M.B.; Betts, M.R.; Wherry, E.J.; Song, W.C.; et al. Erythrocytes Reveal Complement Activation in Patients with COVID-19. *medRxiv* **2020**. [CrossRef]
250. Pretorius, E.; Vlok, M.; Venter, C.; Bezuidenhout, J.A.; Laubscher, G.J.; Steenkamp, J.; Kell, D.B. Persistent clotting protein pathology in Long COVID/Post-Acute Sequelae of COVID-19 (PASC) is accompanied by increased levels of antiplasmin. *Cardiovasc. Diabetol.* **2021**, *20*, 172. [CrossRef] [PubMed]

251. Grobbelaar, L.M.; Venter, C.; Vlok, M.; Ngoepe, M.; Laubscher, G.J.; Lourens, P.J.; Steenkamp, J.; Kell, D.B.; Pretorius, E. Sars-Cov-2 Spike Protein S1 Induces Fibrin(Ogen) Resistant To Fibrinolysis: Implications For Microclot Formation In Covid-19. *Biosci. Rep.* **2021**, *41*, BSR20210611. [CrossRef]
252. Başkurt, O.K.; Edremitlioğlu, M.; Temiz, A. In Vitro Effects of in Vivo Activated Leukocytes on RBC Filterability and Lipid Peroxidation. *Clin. Hemorheol. Microcirc.* **1994**, *14*, 591–596. [CrossRef]
253. Nash, G.B.; Watts, T.; Thornton, C.; Barigou, M. Red cell aggregation as a factor influencing margination and adhesion of leukocytes and platelets. *Clin. Hemorheol. Microcirc.* **2008**, *39*, 303–310. [CrossRef] [PubMed]
254. Tigno, X.T.; Selaru, I.K.; Angeloni, S.V.; Hansen, B.C. Is microvascular flow rate related to ghrelin, leptin and adiponectin levels? *Clin. Hemorheol. Microcirc.* **2003**, *29*, 409–416. [PubMed]
255. Piecuch, J.; Mertas, A.; Nowowiejska–Wiewiora, A.; Sławomir, G.; Zenon, C.; Wiewiora, M. The relationship between the rheological behavior of RBCs and angiogenesis in the morbidly obese. *Clin. Hemorheol. Microcirc.* **2019**, *71*, 95–102. [CrossRef] [PubMed]
256. Murohara, T.; Asahara, T. Nitric oxide and angiogenesis in cardiovascular disease. *Antioxid Redox Signal.* **2002**, *4*, 825–831. [CrossRef]
257. Kondo, T.; Kobayashi, K.; Murohara, T. Nitric oxide signaling during myocardial angiogenesis. *Mol. Cell. Biochem.* **2004**, *264*, 25–34. [CrossRef]
258. Gaudard, A.; Varlet-Marie, E.; Bressolle, F.; Mercier, J.; Brun, J.F. Nutrition as a determinant of blood rheology and fibrinogen in athletes. *Clin. Hemorheol. Microcirc.* **2004**, *30*, 1–8.
259. Brun, J.F.; Varlet-Marie, E.; Romain, A.J.; Guiraudou, M.; Raynaud De Mauverger, E. Exercise hemorheology: Moving from old simplistic paradigms to a more complex picture. *Clin. Hemorheol. Microcirc.* **2013**, *55*, 15–27. [CrossRef]
260. Brun, J.F.; Varlet-Marie, E.; Chevance, G.; Pollatz, M.; Fedou, C.; Raynaud de Mauverger, E. Versatility of 'hemorheologic fitness' according to exercise intensity: Emphasis on the "healthy primitive lifestyle". *Korea-Aust. Rheol. J.* **2014**, *26*, 249–253. [CrossRef]

metabolites

MDPI

Article

The Effect of Glucose and Poloxamer 188 on Red-Blood-Cell Aggregation

Alicja Szołna-Chodór and Bronisław Grzegorzewski *

Department of Biophysics, Collegium Medicum in Bydgoszcz, Nicolaus Copernicus University, ul. Jagiellońska 15, 85-067 Bydgoszcz, Poland; alicja.szolna@cm.umk.pl
* Correspondence: grzego@cm.umk.pl

Abstract: Glucose metabolism disorders contribute to the development of various diseases. Numerous studies show that these disorders not only change the normal values of biochemical parameters but also affect the mechanical properties of blood. To show the influence of glucose and poloxamer 188 (P188) on the mechanical properties of a red-blood-cell (RBC) suspension, we studied the aggregation of the cells. To show the mechanisms of the mechanical properties of blood, we studied the effects of glucose and poloxamer 188 (P188) on red-blood-cell aggregation. We used a model in which cells were suspended in a dextran 70 solution at a concentration of 2 g/dL with glucose and P188 at concentrations of 0–3 g/dL and 0–3 mg/mL, respectively. RBC aggregation was determined using an aggregometer, and measurements were performed every 4 min for 1 h. Such a procedure enabled the incubation of RBCs in solution. The aggregation index determined from the obtained syllectograms was used as a measure of aggregation. Both the presence of glucose and that of P188 increased the aggregation index with the incubation time until saturation was reached. The time needed for the saturation of the aggregation index increased with increasing glucose and P188 concentrations. As the concentrations of these components increased, the joint effect of glucose and P188 increased the weakening of RBC aggregation. The mechanisms of the observed changes in RBC aggregation in glucose and P188 solutions are discussed.

Keywords: RBC aggregation; glucose; poloxamer 188

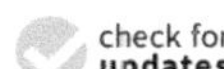

Citation: Szołna-Chodór, A.; Grzegorzewski, B. The Effect of Glucose and Poloxamer 188 on Red-Blood-Cell Aggregation. *Metabolites* **2021**, *11*, 886. https://doi.org/10.3390/metabo11120886

Academic Editor: Norbert Nemeth

Received: 25 November 2021
Accepted: 16 December 2021
Published: 18 December 2021

1. Introduction

The metabolic diseases plaguing today's civilization result from damage to normal metabolisms, especially in cases such as diabetes mellitus, in which the carbohydrate pathway is disrupted by inappropriate insulin secretion or abnormal responses by the body. With this disruption of the carbohydrate pathway, there is an elevated level of glucose in the blood, called hyperglycemia. The glucose consumption of most cells is regulated by insulin. Red blood cells (RBCs) do not need insulin; however, in the absence of this hormone, they are exposed to high glucose levels during hyperglycemia. Glucose has an effect on RBCs, changing their biochemical properties and the ability of these cells to aggregate. Changes in these processes depend on the period of time the cells are exposed to glucose and on the glucose concentration, but also on the composition of the solution in which the RBCs are suspended. Polymers are widely known among the substances that change the aggregation of RBCs. Thus, it is particularly important to understand the simultaneous effects of glucose and drug polymers on RBC aggregation.

Biochemical tests, which are the basis for the diagnosis of diabetes, are increasingly supplemented with analyses of the rheological properties of blood [1,2]. Many factors influence the rheology of whole blood, including RBC aggregation. One of the first documented studies on the changes in RBC aggregation caused by elevated glucose levels was conducted in 1956 [3]. The in vivo effect of increased RBC aggregation under hyperglycemia was confirmed in the following years using various research methods and under various experimental conditions [4–8]. It has been shown that increased aggregation in patients with elevated glucose levels in type 2 DM decreases after the application of glycemic control, and this effect is visible both for RBCs suspended in autologous plasma and cells suspended in dextran solutions [8]. In general, the incubation of RBCs in glucose solutions in vitro causes a decrease in the aggregation of the cells. In order to demonstrate the effect in vitro, cells of healthy volunteers incubated in high glucose concentrations were studied [9–12]. RBC aggregation, as measured by the Aggregate Shape Parameter (ASP) method after two hours of incubation [10] and by the Myrenne method after half an hour of incubation [11], showed a downward trend.

In the case of blood, RBCs are suspended in the plasma, which may contain medicinal substances. One of these substances may be the nonionic surfactant poloxamer 188 (P188), which has many uses as a pharmaceutical excipient [13–15]. Extensive research is ongoing into its potential use as a biological membrane sealant [16–18], while other research is focused on its hemorheological effects [19–22]. In vitro and in vivo studies of the influence of P188 on hemorheological parameters in patients with acute myocardial infarction showed that the aggregation of RBCs in autologous plasma decreased after the application of this surfactant [19]. Conversely, the aggregation of RBCs of healthy volunteers measured in dextran 70 solution after in vitro treatment with P188 decreased, and the effect was dependent on the concentration of P188 [21]. The in vitro activity of P188 tested in dextran 70 solution was confirmed by studies of cells from healthy donors and donors with sickle cell anemia [18]. Another study showed a reduction in RBC deformability by 10% only after 2 h of incubation in P188 [22]. The determination of the influence of P188 on the aggregation of RBCs from healthy donors in dextran 70 solution, used as a medium for cells, allows for the exclusion of the effects of interindividual differences in plasma composition.

Knowledge of the possible joint effect of P188 and glucose is important for both the development of dosage forms and the prediction of enhancing effects in therapeutic applications that are still under development. There are studies investigating the synergistic effect of P188 and tinidazole on the solubility and drug release profile [23] or on the treatment of certain diseases [24], but the effect on RBC aggregation is a new area of research. Knowledge about the possible additional enhancement of the anti-aggregation effect of P188 is important, especially in light of reports about the need to administer P188 as soon as possible in order to support endogenous cell repair systems [25]. In this work, the effect of P188 on RBC aggregation in the presence of glucose was investigated. We present the results of a new approach where RBC aggregation was tested during the incubation of cells in dextran solutions with glucose and with or without P188. The aggregation of RBCs is usually measured immediately after sample preparation. In this article, we use a working hypothesis that the aggregation of RBCs changes during the incubation of these cells in solutions containing glucose and P188. Thus, we propose the measurement of temporal changes in RBC aggregation. This working hypothesis may be useful as the basis for further research into the effects of various substances on erythrocyte aggregation.

2. Results

For each of the solutions used, the RBC aggregation index was measured based on the time. Figure 1 shows the data for the 16 solutions used in which RBCs were suspended. For every solution, six samples were prepared, and the results shown in Figure 1 represent the mean values of the aggregation index from the samples; the error bars represent the standard deviations.

The first result (in the top left corner) shows data for RBCs suspended in dextran only. The first row contains the results for RBCs suspended in solutions with increasing concentrations of glucose. The next rows contain results for solutions containing P188 with increasing concentrations of glucose. The first column contains results for RBCs suspended in solutions with increasing P188 concentrations. The following columns contain results for glucose solutions with increasing P188 concentrations. As can be observed, the first result is the only one for which the aggregation index did not change over time. In the other cases, the aggregation index increased and became saturated over time. This saturation reflects a certain equilibrium state achieved by the system. The time it took for the system to reach equilibrium varied with the concentration of glucose and P188. In addition, depending on the concentrations of glucose and P188, the starting and ending values of the aggregation index changed.

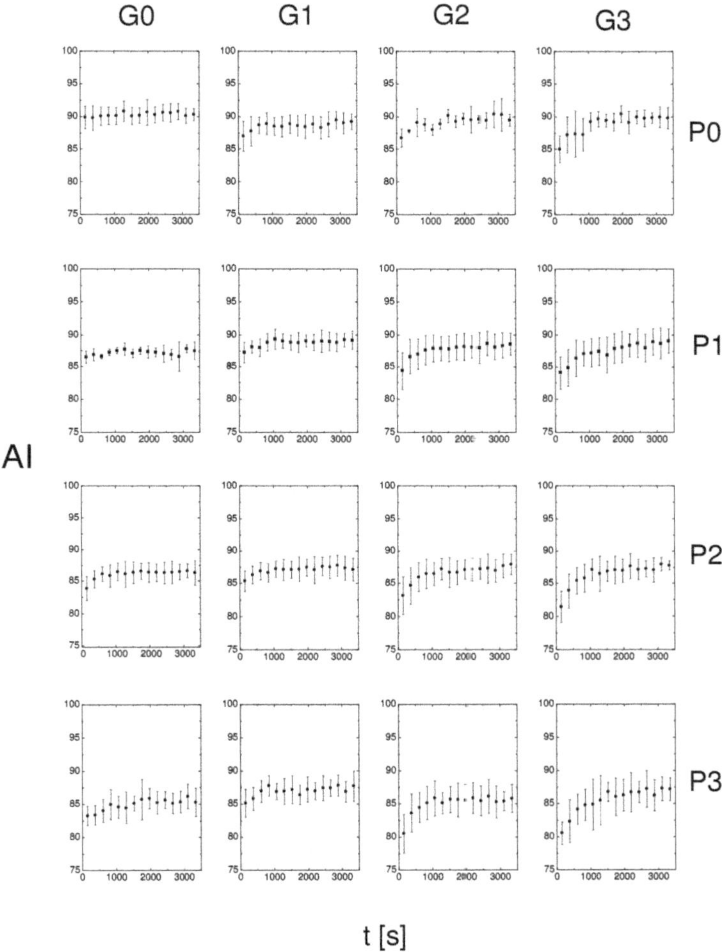

Figure 1. Aggregation index (AI) as a function of time for RBCs incubated in a dextran solution with P188 with concentrations ranging from 0 to 3 mg/mL (P1–P3) and glucose with concentrations ranging from 0 to 3 g/dL (G1–G3).

As can be seen from Figure 1, the changes in the aggregation parameters are particularly visible for the results distributed on the diagonal. Figure 2 shows these results together with the fit, allowing the time taken for the system to reach equilibrium to be determined. The inset in Figure 2 shows how the time needed to reach equilibrium, Teq, increased as the concentration of glucose and P188 increased. Note that this equilibrium time increased both for RBCs suspended only in glucose solutions (row 1 in Figure 1) and for RBCs suspended only in P188 solutions (column 1 in Figure 1).

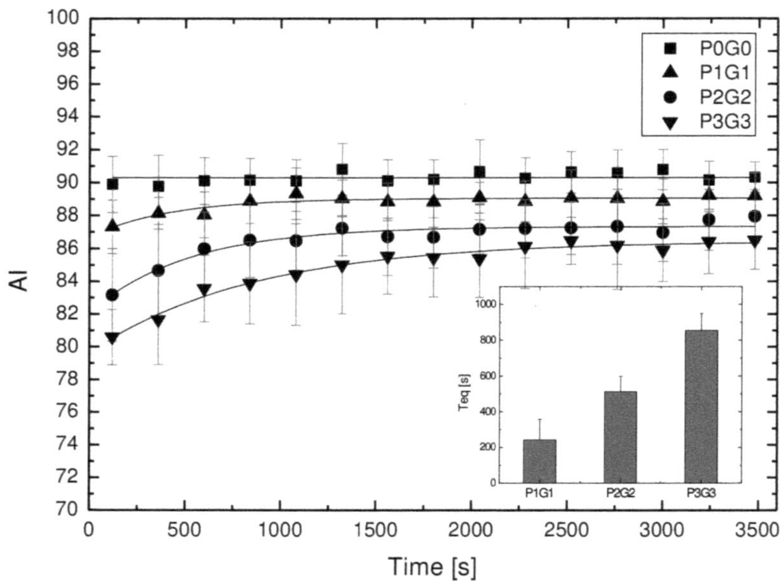

Figure 2. The aggregation index (AI) as a function of incubation time. P0G0 means RBCs suspended in dextran 70 only. P1G1 denotes addition of P188 at a concentration of 1 mg/mL and glucose at 1 g/dL. P2G2 and P3G3 represent the higher concentrations of P188 and glucose. The fits were made to obtain the Teq parameter. In the inset of this figure, the dependence of the time necessary to reach an equilibrium by the system Teq on the concentrations of P188 and glucose is shown.

The data presented in Figure 1 make it possible to determine the initial aggregation index corresponding to the first measurement and the final aggregation index corresponding to the last measurement in a given series. This initial aggregation index for the RBCs suspended in the test solutions is shown in Figure 3. The column groups contain the results for the specific glucose values and, for each glucose value, the values for the different P188 concentrations. In general, it can be seen that an increase in glucose concentration resulted in a decrease in the initial aggregation index. The addition of P188 exacerbated this effect, causing a further reduction in the initial aggregation index. Except for the results for a glucose concentration of 1 g/dL, in all the other cases, the addition of increasing concentrations of P188 resulted in a decrease in the initial aggregation index values.

The final aggregation index for the RBCs suspended in the test solutions is shown in Figure 4. As previously stated, the column groups contain the results for the specific glucose values and, for each glucose value, the values for the different P188 concentrations. It can be seen that the addition of P188 caused a reduction in the final aggregation index. As previously stated, except for the results for a glucose concentration of 1 g/dL, in all the other cases, the addition of increasing concentrations of P188 resulted in a decrease in the final aggregation index values.

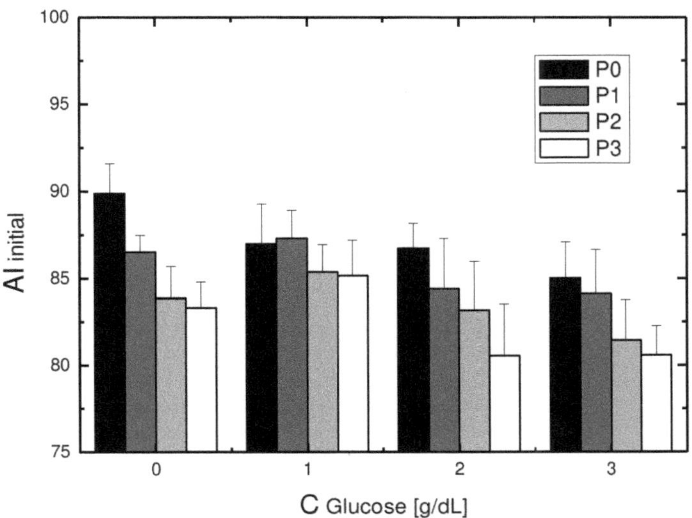

Figure 3. The dependence of the aggregation index (AI), obtained for the first measurement during incubation (2 min after the start of the experiment), on the concentration of glucose and P188. A decrease of the initial aggregation index with glucose concentration is observed. The presence of P 188 causes an additional decrease of the aggregation index.

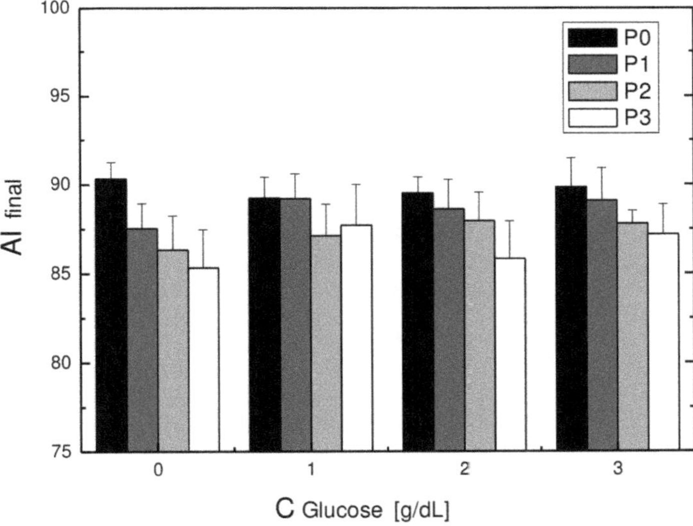

Figure 4. The dependence of the aggregation index (AI) obtained at 1 h of the incubation on the concentration of glucose and P188. The presence of P 188 causes a decrease of the aggregation index.

3. Discussion

Metabolic processes are time dependent, and their courses are modified by the presence of drugs and other substances in the body. Both of these elements are taken into account in this article. The subject of this study was the time dependence of RBC aggregation parameters in solutions containing glucose and the influence of P188 added to these solutions on the aggregation parameters. Thus, in addition to the temporal effect, the joint effect of glucose and P188 on RBC aggregation was investigated.

Studying the influence of glucose on the rheological properties of blood has a long history and results from the need to understand the mechanisms involved in diabetes [26]. Research shows that incubation of RBCs in glucose solutions causes oxidation of membrane lipids and glycation of proteins, which leads to a reduction of RBC aggregation [9,12]. Shin et al. suggest that these changes occur even after 1 h of incubation. We conduct research depending on time. Thus, oxidation and glycation affect changes in RBC aggregation; however, the results presented in this paper suggest that glucose transport to erythrocytes plays a key role in the temporal changes in the aggregation of these cells. Elevated blood glucose also affects red-blood-cell aggregation, but the in vivo and in vitro effects are not the same and are still not fully understood; thus, this issue still needs to be investigated [3–11]. The results presented in this study show that, for RBCs placed in glucose solutions, the aggregation index increases with the time of incubation in these solutions. After the growth period, the saturation of this parameter is observed. The time required for saturation increases as the glucose concentration in the solution increases. The aggregation index values measured immediately after placing the RBCs in solutions containing glucose decrease with the concentration of glucose. However, the aggregation index measured at the time of saturation does not show such a marked decrease as a function of the glucose concentration. This means that the time of the incubation of the RBCs in the glucose solution has a significant influence on the course of aggregation and may lead to differing results [9–12].

The phenomenon of transport occurs for RBCs placed in glucose solutions. Glucose enters the RBCs, and this transport does not require the presence of insulin [27]. As a result, the concentration of glucose in the solution decreases while the concentration of glucose inside the RBCs increases. Lowering the glucose concentration in the solution in which the RBCs are suspended results in a lower viscosity of the solution [28,29]. This happens until equilibrium is achieved—when the transportation back and forth is the same. The study of this transport showed that the values of the parameters of this transport significantly differ depending on whether the transport takes place inside the RBC or in the solution [12]. This immediately indicates that the issue of the incubation time and cell cleaning after incubation can cause major changes in aggregation parameters. Finally, a question arises about the mechanisms of the change in aggregation during incubation in glucose solutions. Research shows that the biochemical changes of RBCs incubated in glucose solutions occur slowly. Thus, it can be assumed that, for the results shown here, the biochemical changes had little effect on the changes in aggregation. This allows for a hypothesis that, in the cases considered here, mechanical changes in the cell membrane and changes in the viscosity of the substance in which the RBCs are dissolved determine the changes in the aggregation of these cells.

For RBCs placed in P188 solutions, as in the case of RBCs placed in solutions containing glucose, we observed an increase in the aggregation index with the incubation time. The saturation effect was also observed. In this case, both the initial and final aggregation index values decreased with increasing P188 concentrations in the solution. This RBC aggregation behavior corresponding to the initial aggregation index values studied here was observed using the microscopic aggregation index (MAI) and Myrenne method [21]. As in the previous case, a question arises about the mechanism of the aggregation process in the case of RBCs in solutions containing P188. The reduction in the aggregation index appears to be due to the presence of a low-molecular-weight polymer in the solution. According to the depletion theory of aggregation, such polymers reduce the aggregation capacity of RBCs [30]. The question remains open, however, of why the presence of polaxamer causes temporary changes in the RBC aggregation index similar to the temporal changes in the aggregation index of RBCs placed in glucose solutions.

Consider the problem of the synergistic effect of glucose and P188 on RBC aggregation. According to the definition, a synergistic effect takes place when the effect of processes interacting together is greater than the effects of the individual processes. The results presented in this study show that the presence of P188 enhances the process of suppressing glucose-dependent RBC aggregation. We showed this effect, expected by clinicians, only in vitro. The mechanism of lowering aggregation in this case results from the aggregation mechanisms for glucose solutions and the aggregation mechanism for P188 solutions. It is not certain, however, whether the mechanism of the synergistic effect is a simple assembly of partial mechanisms.

Each measurement technique has certain limitations. This is also the case for RBC aggregation measurements. When choosing the measurement technique, we were guided by the need to simultaneously measure the aggregation and incubation of cells. In addition, we were looking for a technique that would allow for relatively fast repeatable measurements. The choice fell on a photometric technique that meets the above-mentioned requirements. The approach proposed in this paper enables RBC aggregation measurements with a resolution of 4 min.

In conclusion, we performed a study of the effects of P188 in the presence of glucose on RBC aggregation. The aggregation was studied as a function of time. The study showed a synergistic effect of glucose and P188 on RBC aggregation and the temporary changes in this aggregation. To the best of our knowledge, the synergistic effect of P188 and glucose, as well as temporal changes, on aggregation are presented for the first time. The research method applied and the results obtained appear to be promising for further research in this area.

4. Materials and Methods

4.1. Materials

Human venous blood was obtained from 96 healthy adult volunteers. The blood was collected in sterile tubes containing the K3-EDTA anticoagulant and was maintained at 4 °C until processing, which was performed as soon as possible. The RBCs were separated from whole blood via centrifugation at 3000× *g* rpm for 5 min at 4 °C; next, the buffy coat and plasma were discarded. The fractionated RBCs were washed three times with phosphate-buffered saline (PBS), pH 7.4, under the same centrifugation conditions as before. Dextran with a molecular mass of 70 kDa (Dextran from *Leuconostoc* spp., Sigma-Aldrich, MO, USA), glucose (D-(+)-Glucose, Sigma-Aldrich, MO, USA) and P188 (Poloxamer 188 Solid, Alfa Aesar, TX, USA) were used to make the solutions in PBS. The sixteen solutions for RBCs were prepared, in which dextran was always at the same concentration, while glucose and P188 were combined at different concentrations. The final concentrations were as follows: dextran: 2 g/dL; glucose: 0, 1, 2 or 3 g/dL; and P188: 0, 1, 2 or 3 mg/mL. The RBCs were suspended in these solutions immediately before measurement. The hematocrit of these suspensions was adjusted to 40%. The measurements were performed at room temperature (22 ± 1 °C). All the experiments were performed according to the guidelines of the Bioethics Commission of Collegium Medicum Nicolaus Copernicus University.

4.2. Method

RBC aggregation was determined using an aggregometer, as shown in Figure 5. The investigated suspension was placed in a transparent cylinder with an internal diameter of 32 mm. A rotor with an outer diameter of 30.9 mm was placed in the cylinder. The layer of the suspension was illuminated by laser diode light at a wavelength of 840 nm. The backscattered light was detected by a photodiode, and the backscattered light's intensity was recorded as a function of time.

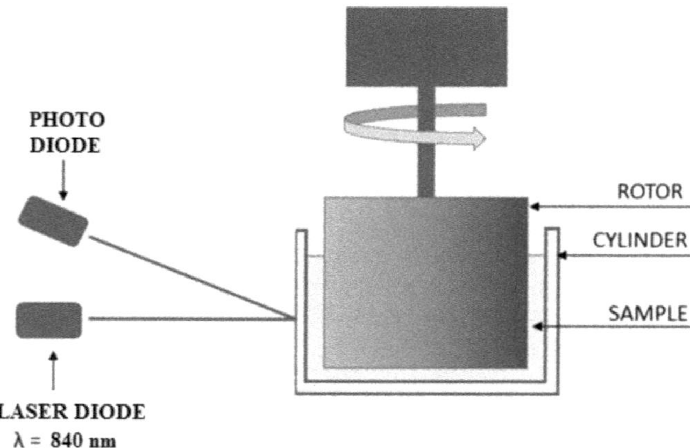

Figure 5. Experimental set up.

The measurement was carried out for one hour. During this time, the rotor was operated in cycles of 2 min of rotation and 2 min of rest. Figure 6 shows the light intensity of the backscattered light for the first cycle for RBCs suspended in dextran. During the first part of the cycle, the intensity remained constant. Due to the rotor operation, the RBC suspension was subjected to a shear stress at a shear rate of 165 s^{-1}. During the first part of the cycle, red blood cells were completely disaggregated and deformed through elongation. Stopping the rotor first caused the light intensity to increase rapidly and then to slowly decrease. The rapid increase in light intensity reflects the return of the RBCs to their original shape. The decrease in light intensity was the result of RBC aggregation. The light intensity recorded in this last part of the cycle was recorded as a syllectogram. The aggregation index $AI = 100 \times A/(A + B)$ was taken as a measurement of RBC aggregation, where surfaces A and B are shown in Figure 6.

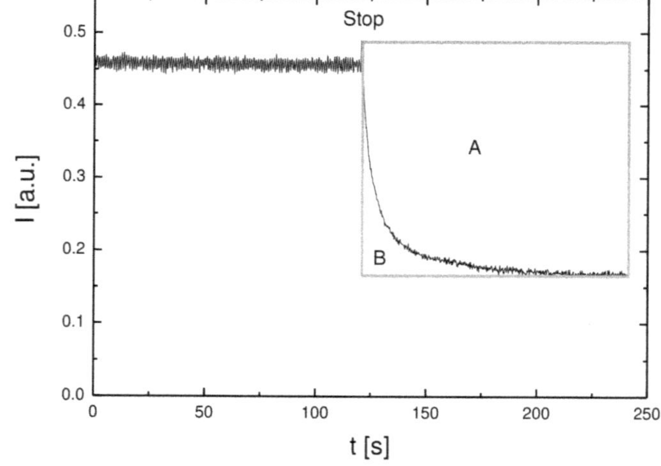

Figure 6. Syllectogram of RBC aggregation in dextran.

The classical measurement of RBC aggregation was carried out on the basis of the first rotor cycle. In this paper, we present a new approach to measuring RBC aggregation by tracking this process during 15 rotor cycles. The procedure for measuring the time dependence of aggregation is shown in Figure 7. The first column in this figure shows the light intensity recorded over 1 h, sequentially, for RBCs suspended in dextran, for RBCs suspended in glucose solution, and for RBCs suspended in P188 solution. As can be observed, the upper light intensity envelope was constant for RBCs suspended in dextran and for RBCs suspended in P188 solution, while it changed with time for RBCs suspended in glucose solution. For this reason, the light intensity in all cases was divided by the intensity value corresponding to the upper envelope. In this way, the data presented in the second column were obtained. Based on these data, the aggregation index was determined for each syllectogram. This resulted in time-dependent aggregation index values. These values are presented in the third column. The presented method makes it possible to determine the aggregation index with a resolution of 4 min. An additional advantage of this method is that, during the measurement, RBCs were incubated in the solution in which they were suspended, and the effect of this incubation on aggregation could be recorded.

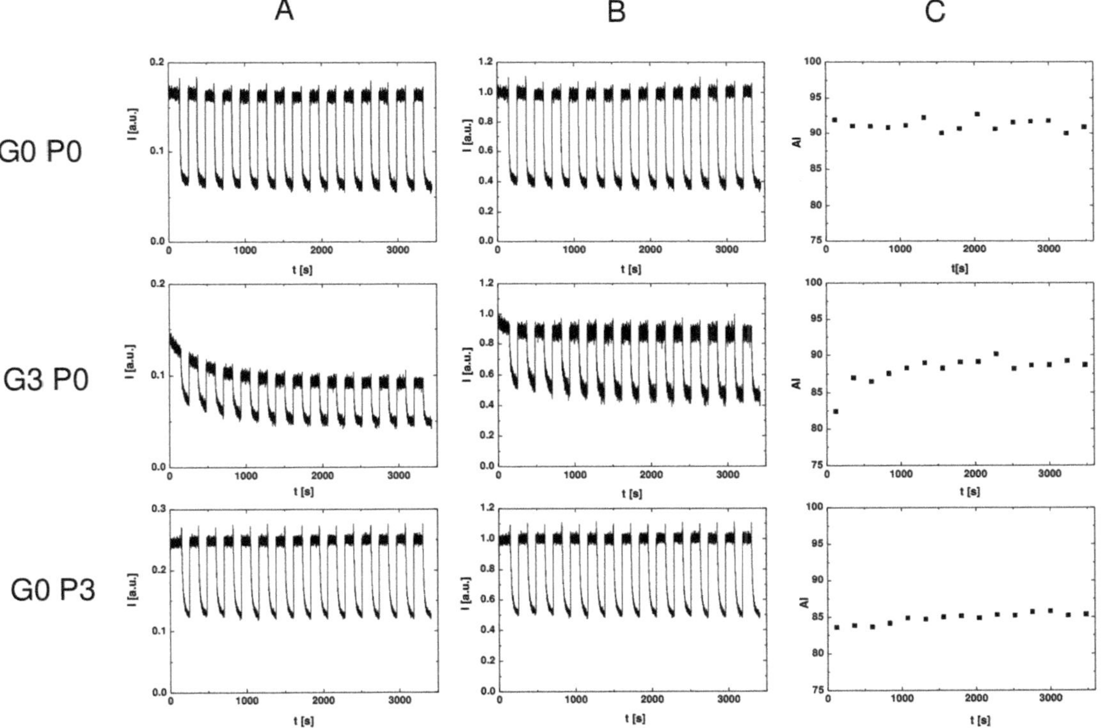

Figure 7. Scheme of data analysis. Column (**A**) shows the intensity obtained from the measurement. Column (**B**) shows the normalized intensity, and column (**C**) contains the values of the aggregation index over time. The first row represents RBCs suspended in dextran, the second row represents RBCs suspended in glucose solution, and the last row represents RBCs suspended in P188.

Author Contributions: Conceptualization, B.G. and A.S.-C.; methodology, B.G.; software, B.G.; validation, B.G. and A.S.-C.; formal analysis, B.G.; investigation, A.S.-C.; resources, B.G. and A.S.-C.; data curation, B.G. and A.S.-C.; writing—original draft preparation, A.S.-C.; writing—review and editing, B.G.; visualization, B.G. and A.S.-C.; supervision, B.G. All authors have read and agreed to the published version of the manuscript.

Funding: This research received no external funding.

Institutional Review Board Statement: This study was conducted according to the guidelines of the Declaration of Helsinki and was approved by the Bioethics Commission of Collegium Medicum Nicolaus Copernicus University (protocol code: KB 91/2003; date of approval: 25 October 2016).

Informed Consent Statement: Informed consent was obtained from all the subjects involved in the study.

Data Availability Statement: The data presented in this study are available on request from the corresponding author according to local policies.

Conflicts of Interest: The authors declare no conflict of interest. The funders had no role in the design of the study; in the collection, analyses, or interpretation of data; in the writing of the manuscript; or in the decision to publish the results.

References

1. Antonova, N.; Tsiberkin, K.; Podtaev, S.; Paskova, V.; Velcheva, I.; Chaushev, N. Comparative study between microvascular tone regulation and rheological properties of blood in patients with type 2 diabetes mellitus. *Clin. Hemorheol. Microcirc.* **2016**, *64*, 837–844. [CrossRef] [PubMed]
2. Le Dévéhat, C.; Vimeux, M.; Khodabandehlou, T. Blood rheology in patients with diabetes mellitus. *Clin. Hemorheol. Microcirc.* **2004**, *30*, 297–300. [PubMed]
3. Ditzel, J. Angioscopic Changes in the Smaller Blood Vessels in Diabetes Mellitus and their Relationship to Aging. *Circulation* **1956**, *14*, 386–397. [CrossRef] [PubMed]
4. Schmid-Schönbein, H.; Volger, E. Red-cell aggregation and red-cell deformability in diabetes. *Diabetes* **1976**, *25*, 897–902.
5. Van Haeringen, N.J.; Oosterhuis, J.A.; Terpstra, J.; Glasius, E. Erythrocyte aggregation in relation to diabetic retinopathy. *Diabetologia* **1973**, *9*, 20–24. [CrossRef]
6. Demiroğlu, H.; Gürlek, A.; Barişta, I. Enhanced erythrocyte aggregation in type 2 diabetes with late complications. *Exp. Clin. Endocrinol. Diabetes* **1999**, *107*, 35–39. [CrossRef] [PubMed]
7. Babu, N.; Singh, M. Influence of hyperglycemia on aggregation, deformability and shape parameters of erythrocytes. *Clin. Hemorheol. Microcirc.* **2004**, *31*, 273–280.
8. Chong-Martinez, B.; Buchanan, T.A.; Wenby, R.B.; Meiselman, H.J. Decreased red blood cell aggregation subsequent to improved glycaemic control in Type 2 diabetes mellitus. *Diabet. Med.* **2003**, *20*, 301–306. [CrossRef]
9. Resmi, H.; Akhunlar, H.; Temiz Artmann, A.; Güner, G. In vitro effects of high glucose concentrations on membrane protein oxidation, G-actin and deformability of human erythrocytes. *Cell. Biochem. Funct.* **2005**, *23*, 163–168. [CrossRef] [PubMed]
10. Riquelme, B.; Foresto, P.; D'Arrigo, M.; Valverde, J.; Rasia, R. A dynamic and stationary rheological study of erythrocytes incubated in a glucose medium. *J. Biochem. Biophys. Methods* **2005**, *62*, 131–141. [CrossRef] [PubMed]
11. Shin, S.; Ku, Y.-H.; Suh, J.-S.; Singh, M. Rheological characteristics of erythrocytes incubated in glucose media. *Clin. Hemorheol. Microcirc.* **2008**, *38*, 153–161. [PubMed]
12. Viskupicova, J.; Blaskovic, D.; Galiniak, S.; Soszynski, M.; Bartosz, G.; Horakova, L.; Sadowska-Bartosz, I. Effect of high glucose concentrations on human erythrocytes in vitro. *Redox Biol.* **2015**, *5*, 381–387. [CrossRef]
13. Shubhra, Q.T.H.; Tóth, J.; Gyenis, J.; Feczkó, T. Poloxamers for Surface Modification of Hydrophobic Drug Carriers and Their Effects on Drug Delivery. *Polym. Rev.* **2014**, *54*, 112–138. [CrossRef]
14. Moghimi, S.; Hunter, A. Poloxamers and poloxamines in nanoparticle engineering and experimental medicine. *Trends Biotechnol.* **2000**, *18*, 412–420. [CrossRef]
15. Patel, H.R.; Patel, R.P.; Patel, M.M. Poloxamers: A pharmaceutical excipients with therapeutic behaviors. *Int. J. Pharmtech Res.* **2009**, *1*, 299–303.
16. Collins, J.M.; Despa, F.; Lee, R.C. Structural and functional recovery of electropermeabilized skeletal muscle in-vivo after treatment with surfactant poloxamer 188. *Biochim. Biophys. Acta* **2007**, *1768*, 1238–1246. [CrossRef]
17. Maskarinec, S.; Hannig, J.; Lee, R.C.; Lee, K.Y.C. Direct Observation of Poloxamer 188 Insertion into Lipid Monolayers. *Biophys. J.* **2002**, *82*, 1453–1459. [CrossRef]
18. Sandor, B.; Marin, M.; Lapoumeroulie, C.; Rabaï, M.; Lefevre, S.D.; Lemonne, N.; El Nemer, W.; Mozar, A.; Francais, O.; Le Pioufle, B.; et al. Effects of Poloxamer 188 on red blood cell membrane properties in sickle cell anaemia. *Br. J. Haematol.* **2016**, *173*, 145–149. [CrossRef]
19. Toth, K.; Bogar, L.; Juricskay, I.; Keltai, M.; Yusuf, S.; Haywood, L.J.; Meiselman, H.J. The effect of RheothRx Injection on the hemorheological parameters in patients with acute myocardial infarction. *Clin. Hemorheol. Microcirc.* **1997**, *17*, 117–125.
20. Lechmann, T.; Reinhart, W.H. The non-ionic surfactant Poloxamer 188 (RheothRx) increases plasma and whole blood viscosity. *Clin. Hemorheol. Microcirc.* **1998**, *18*, 31–36. [PubMed]
21. Toth, K.; Wenby, R.B.; Meiselman, H.J. Inhibition of polymer-induced red blood cell aggregation by poloxamer 188. *Biorheology* **2000**, *37*, 301–312. [PubMed]
22. Guzniczak, E.; Jimenez, M.; Irwin, M.; Otto, O.; Willoughby, N.; Bridle, H. Impact of poloxamer 188 (Pluronic F-68) additive on cell mechanical properties, quantification by real-time deformability cytometry. *Biomicrofluidics* **2018**, *12*, 044118. [CrossRef]

23. Shaikh, A.; Yeole, P.G.; Iyer, D. Formulation and Evaluation of Stable Solid Dispersion of Anti-Protozoal Drug. *Res. J. Pharm. Biol. Chem. Sci.* **2013**, *4*, 1495–1509.
24. Jain, D.; Gangshettiwar, A. Combination of lycopene, quercetin and poloxamer 188 alleviates anxiety and depression in 3-nitropropionic acid-induced Huntington's disease in rats. *J. Intercult. Ethnopharmacol.* **2014**, *3*, 186–191. [CrossRef]
25. Houang, E.M.; Bartos, J.; Hackel, B.J.; Lodge, T.P.; Yannopoulos, D.; Bates, F.S.; Metzger, J.M. Cardiac Muscle Membrane Stabilization in Myocardial Reperfusion Injury. *JACC Basic Transl. Sci.* **2019**, *4*, 275–287. [CrossRef] [PubMed]
26. Zhang, Z.; Meiselman, H.J.; Neu, B. Red blood cell adhesion can be reduced by non-reactive macromolecules. *Colloids Surf. B Biointerfaces* **2019**, *174*, 168–173. [CrossRef] [PubMed]
27. Ebeling, P.; Koistinen, H.A.; Koivisto, V.A. Insulin-independent glucose transport regulates insulin sensitivity. *FEBS Lett.* **1998**, *436*, 301–303. [CrossRef]
28. Brahm, J. Kinetics of glucose transport in human erythrocytes. *J. Physiol.* **1983**, *339*, 339–354. [CrossRef]
29. Cloherty, E.K.; Heard, K.S.; Carruthers, A. Human Erythrocyte Sugar Transport is Incompatible with Available Carrier Models. *Biochemistry* **1996**, *35*, 10411–10421. [CrossRef]
30. Baskurt, O.; Neu, B.; Meiselman, H.J. *Red Blood Cell Aggregation*; CRC Press: Boca Raton, FL, USA, 2012; Chapter 2.2.1; p. 12.

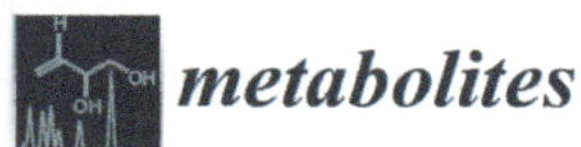

MDPI

Article

Hemorheological Parameters in Diabetic Patients: Role of Glucose Lowering Therapies

Katalin Biro [1], Gergely Feher [2], Judit Vekasi [3], Peter Kenyeres [1], Kalman Toth [1] and Katalin Koltai [1,*]

[1] 1st Department of Medicine, Medical School, University of Pecs, Szigeti u 12, 7624 Pecs, Hungary; biro.katalin@pte.hu (K.B.); kenyeres.peter@pte.hu (P.K.); toth.kalman@pte.hu (K.T.)

[2] Centre for Occupational Medicine, Medical School, University of Pecs, Szigeti u 12, 7624 Pecs, Hungary; feher.gergely@pte.hu

[3] Department of Ophthalmology, Medical School, University of Pecs, Szigeti u 12, 7624 Pecs, Hungary; vekasi.judit@pte.hu

* Correspondence: koltai.katalin@pte.hu

Abstract: Diabetes mellitus influences several important hemorheological parameters including blood viscosity, erythrocyte aggregation and deformability. In the present study, 159 type-2 diabetic patients and 25 healthy controls were involved. Patient's age, body weight, body mass index (BMI), smoking habits, physical activity, history of cardiovascular diseases, current antidiabetic therapy and concomitant medication were recorded. Patients were grouped according to their antidiabetic treatment with insulin, or with one or more of the following antidiabetic drugs: metformin, sulfonylureas, acarbose, or no antidiabetic therapy. Hemorheological measurements (hematocrit, erythrocyte aggregation, plasma fibrinogen, whole blood and plasma viscosity), von Willebrand factor activity, and platelet aggregation measurements were performed. Platelet aggregation was investigated with the method of Born. Plasma viscosity and red blood cell aggregation were significatly higher in diabetes. No significant difference was found in hemorheological parameters between different antidiabetic regimens. Whole blood and plasma viscosity and red blood cell aggregation correlated with glucose levels but not with HbA1C levels. In conclusion, plasma and whole blood viscosity, as well as red blood cell aggregation appear to be associated with concurrent hyperglycemia, but not with the quality of glycemic control or the applied antidiabetic treatment. Platelet aggregation induced by ADP or epinephrine does not seem to be associated with diabetes even at subthreshold doses.

Keywords: hemorheology; viscosity; red blood cell aggregation; diabetes; insulin; metformin; sulfonylureas

Citation: Biro, K.; Feher, G.; Vekasi, J.; Kenyeres, P.; Toth, K.; Koltai, K. Hemorheological Parameters in Diabetic Patients: Role of Glucose Lowering Therapies. *Metabolites* **2021**, *11*, 806. https://doi.org/10.3390/metabo11120806

Academic Editor: Norbert Nemeth

Received: 7 October 2021
Accepted: 24 November 2021
Published: 27 November 2021

1. Introduction

Diabetes mellitus resulting in micro-and macrovascular complications is one of the major risk factors for cardiovascular disease. Hemorheological alterations have been associated with diabetes mellitus and diabetes related conditions as hyperglycemia, hyperinsulinemia and insulin resistance [1–3]. Increased red blood cell aggregation in diabetic patients has been described by several studies. Poor glycemic control was found to be an important determinant of excessive erythrocyte aggregation. Our previous results showed significant elevation in red blood cell aggregability in patients with abnormal glucose tolerance in oral glucose tolerance tests during hyperglycemia [4]. Significantly higher whole blood viscosity has been found in patients with diabetes by several researchers [5]. Elevated blood viscosity plays an important role in diabetic microangiopathy by adversely affecting microcirculation [6].

Both type-1 and type-2 diabetes are characterized by diabetic thrombocytopathy. Hyperglycemia, insulin resistance or deficiency, cellular abnormalities as increased generation of thrombin and thromboxane A2, accelerated platelet turnover and associated metabolic conditions as obesity, dyslipidemia and inflammation have been identified as mechanisms of platelet dysfunction [7]. Platelet abnormalities in diabetes include reduced membrane fluidity, altered platelet shape, secretion and aggregation, increased formation of platelet-derived microparticles, increased expression of surface receptors and adhesion molecules as well as increased platelet-dependent activation of coagulation [8].

Compared to the large amount of data concerning the hemorheological consequences of diabetes, studies on the effects of antidiabetic drugs on these parameters are much less abundant.

Sulfonylureas are insulin secretagogues that stimulate insulin release from pancreatic beta cells. Early filterability studies did not show different effect on blood filterability from various kinds of sulfonylureas and insulin [9]. An inhibitory effect of certain sulfonylureas on platelet aggregation has been described by a number of authors. Gliclazide, glibenclamide, glyburide and to a lesser extent glimepiride were suggested to have platelet aggregation inhibitory properties [8,10]. On the other hand, a study by Larkins et al. reported no significant effect of gliclazide on platelet function in insulin-treated and non-insulin-treated diabetic patients [11]. A novel investigation of two population-based cohorts found decreased mean platelet volume (MPV), a cell trait partially associated with markers of platelet activity in sulfonylurea treated patients [12].

Metformin is the most prescribed drug for type 2 diabetes mellitus (T2DM) treatment. Studies investigating hemorheological properties in metformin treatment are scarce and limited in sample size. Metformin has been found to improve endothelial function [13]. Schiapaccassa et al. reported no effect of metformin treatment on blood viscosity [14]. A number of studies found that metformin affected platelet activation [15] and aggregation [16]. Metformin has been reported to decrease mean platelet volume (MPV) [12,17]. However, it was suggested that these effects were caused by improved glycemic control rather than any specific effect of metformin.

Acarbose, an α- glucosidase inhibitor is not widely used nowadays due to its relatively modest impact and the potential for significant gastrointestinal adverse effects, diarrhea and flatulence. Very limited research data is available on possible hemorheological or microcirculatory effects of acarbose therapy. In a study by Shimbakuro et al. postprandial endothelial dysfunction defined by peak forearm blood flow response and total reactive hyperemic flow was improved by a prior use of acarbose [18].

The first investigations on the hemorheological effects of insulin were performed in the 70ies and the 80ies, however, most of these studies had a limited sample size and used varying methodologies. Results concerning the effect of insulin on hemorheological parameters remained contradictory [9]. Insulin therapy reduced erythrocyte aggregation in gestational diabetes [19]. In a study by Jennings et al. the effect of intensified dietary measures and subsequent insulin therapy upon haemorheological measures was studied in Type 2 diabetic patients. Increased levels of the platelet release proteins beta-thromboglobulin and platelet factor 4 but no hemorheological changes were found [20].

2. Results

Diabetic patients had a significantly higher BMI than controls. They were 8 years older on average, and exercised less (Table 1). They had higher glucose and lower HDL cholesterol levels than controls. Red blood cell aggregation and von Willebrand factor activity were significantly higher in the diabetic group. Plasma viscosity was significantly higher in diabetic patients. The difference in whole blood viscosity and in fibrinogen levels between diabetic patients and healthy controls was not statistically significant. Glucose levels, but not HbA1c levels, were correlated with plasma viscosity ($p < 0.01$), whole blood viscosity ($p < 0.05$), red blood cell aggregation (M: $p < 0.05$, M1: $p < 0.001$), and von Willebrand factor activity ($p < 0.05$). In the subgroup of diabetic patients who were not treated with antiplatelet agents, while there was a trend towards higher platelet aggregability compared to controls, the difference was significant only with the lowest applied doses of collagen inducer (Table 2).

The examined hemorheological parameters of insulin-treated patients did not differ significantly from those of patients treated with oral antidiabetic drugs. Insulin-treated patients had higher blood glucose, fructosamine, HbA1c, and CRP levels compared to patients treated with oral antidiabetic drugs (Table 3).

Platelet aggregation was higher in insulin-treated patients, compared to patients treated with oral antidiabetic drugs who were not receiving any antiplatelet therapy, in cases when low doses of collagen inducer were applied (Table 4).

We found no significant differences in hemorheological parameters (Figure 1), von Willebrand factor activity, or lipid and glycemic parameters between patients treated with different antidiabetic therapy (Table 5).

Table 1. Hemorheological and selected other parameters in diabetic patients and non-diabetic controls. NS: not significant.

Variables	Diabetic Patients (n = 159)	Healthy Controls (n = 25)	p
Body weight (kg)	83 (75–95)	72 (63–92)	$p < 0.05$
BMI (kg/m^2)	29.4 (26.1–32.9)	24.9 (21.8–29.3)	$p < 0.001$
Physical activity/week	1 (1–2)	3 (2–3.5)	$p < 0.001$
Glucose (mmol/L)	8.0 (6.7–11.3)	4.7 (4.4–4.9)	$p < 0.001$
Triglyceride (mmol/L)	1.90 (1.30–2.82)	0.94 (0.70–1.29)	$p < 0.001$
Total cholesterol (mmol/L)	4.99 (4.31–5.71)	4.91 (4.37–5.45)	NS
HDL cholesterol (mmol/L)	1.12 (0.96–1.37)	1.37 (1.17–1.69)	$p < 0.001$
Uric acid (µmol/L)	288 (234–342)	279 (242–341)	NS
CRP (mg/L)	3.55 (1.6–6.1)	1.95 (1.0–3.43)	$p < 0.001$
Hematocrit (%)	40.8 (38.7–43.3)	41.5 (38.9–44.8)	NS
Platelet count (G/L)	240 (189–280)	260 (206–325)	NS
Whole blood viscosity (mPAS)	4.74 (4.22–5.21)	4.5 (4.13–4.94)	NS
Plasma viscosity (mPAS)	1.33 (1.28–1.40)	1.28 (1.24–1.3)	$p < 0.001$
Erythrocyte aggregation index M	11.35 (9.48–13.3)	8.05 (7.15–10.73)	$p < 0.01$
Erythrocyte aggregation index M1	24.13 (21.86–26.21)	19.95 (16.95–21.73)	$p < 0.001$
Fibrinogen (g/L)	3.44 (3.03–3.94)	3.27 (2.93–3.65)	NS
vWf activity	138 (98–182)	102 (75–117)	$p < 0.01$

Table 2. Platelet aggregation induced with different doses of ADP, collagen and epinephrine in diabetic patients who were not under antiplatelet therapy and in non-diabetic controls. NS: not significant.

Inducer Concentrations	Diabetic Patients (n = 45)	Control Group (n = 25)	p (Mann-Whitney)
ADP 10 µM	79 (69–84)	79 (69–89)	NS
ADP 5 µM	78 (64–82)	76 (62–86)	NS
ADP 2.5 µM	68 (58–76)	67 (82–80)	NS
ADP 1 µM	57 (16–77)	37 (14–72)	NS
ADP 0.5 µM	12 (4–62)	11 (2–64)	NS
Collagen 2 µg/mL	75 (67–81)	78 (69–89)	NS
Collagen 1 µg/mL	68 (60–74)	70 (35–77)	NS
Collagen 0.5 µg/mL	66 (44–75)	21 (2–76)	NS
Collagen 0.2 µg/mL	58 (6–70)	2 (0–53)	$p < 0.01$
Epinephrine 10 µM	81 (68–89)	85 (64–92)	NS
Epinephrine 5 µM	71 (55–77)	71 (43–79)	NS
Epinephrine 2.5 µM	68 (39–83)	68 (13–73)	NS
Epinephrine 1 µM	65 (29–77)	60 (5–70)	NS
Spont. aggr	4 (1–19)	2 (0–4)	NS

Table 3. Hemorheological and other parameters in diabetic patients who were either on insulin, or on oral antidiabetic treatment. NS: not significant.

Variables	Insulin Therapy (n = 33)	Oral Antidiabetic Therapy (n = 124)	p (Mann-Whitney)
Age (years)	59 (50.5–67)	60.5 (54–68)	NS
Sex	61% male	52% male	NS (Chi Square Test)
Body weight (kg)	85 (78–94)	82 (73–95)	NS
BMI (kg/m^2)	30.4 (27.2–33.1)	29.3 (25.9–32.9)	NS
Physical activity/week	1.5 (1–2.25)	1 (1–2)	NS
Glucose (mmol/L)	11.1 (7.7–13.4)	7.7 (6.5–9.6)	$p < 0.001$
HbA1c (%)	7.89 (6.65–8.74)	6.56 (5.91–7.57)	$p = 0.001$
Triglyceride (mmol/L)	1.76 (1.19–2.30)	1.90 (1.41–2.86)	NS
Total cholesterol (mmol/L)	5.11 (4.61–5.89)	4.98 (4.26–5.68)	NS
HDL cholesterol (mmol/L)	1.14 (1.03–1.54)	1.10 (0.92–1.30)	NS
Uric acid (µmol/L)	297 (240–338)	286 (234–350)	NS
CRP (mg/L)	5.0 (3.0–8.1)	3.0 (1.3–6.0)	$p < 0.01$
Hematocrit (%)	41.5 (40.0–44.2)	40.8 (38.6–43.1)	NS
Platelet (G/L)	244 (192–283)	235 (189–279)	NS
Whole blood viscosity (mPAS)	4.8 (4.36–5.39)	4.72 (4.29–5.21)	NS
Plasma viscosity (mPAS)	1.35 (1.28–1.44)	1.33 (1.28–1.39)	NS
Erythrocyte aggregation index M	10.7 (9.5–12.7)	11.5 (9.5–13.6)	NS
Erythrocyte aggregation index M1	24.7 (20.0–25.6)	24.0 (22.0–26.7)	NS
Fibrinogen (g/L)	3.58 (3.16–4.25)	3.40 (2.96–3.83)	NS
vWf activity	138 (119–200)	138 (87–178)	NS

Table 4. Platelet aggregation in diabetic patients who were not under antiplatelet therapy and were treated either with insulin or with oral antidiabetic medicines. NS: not significant.

Inducer Concentrations	Insulin Therapy (n = 15)	Oral Antidiabetic Therapy (n = 31)	p (Mann-Whitney)
ADP 10 µM	79 (70–85)	78 (68–84)	NS
ADP 5 µM	80 (72–84)	76 (63–82)	NS
ADP 2.5 µM	72 (59–76)	66 (51–76)	NS
ADP 1 µM	67 (40–77)	36 (14–70)	NS
ADP 0.5 µM	10 (6–59)	13 (4–65)	NS
Collagen 2 µg/mL	71 (66–79)	76 (68–83)	NS
Collagen 1 µg/mL	75 (66–82)	66 (52–72)	$p < 0.05$
Collagen 0.5 µg/mL	72 (59–76)	66 (31–71)	NS
Collagen 0.2 µg/mL	65 (45–78)	14 (3–65)	$p < 0.05$
Epinephrine 10 µM	76 (68–88)	84 (77–90)	NS
Epinephrine 5 µM	74 (69–78)	70 (32–76)	NS
Epinephrine 2.5 µM	83 (68–84)	66 (13–76)	NS
Epinephrine 1 µM	70 (65–80)	65 (9–76)	NS
Spont. aggr	6 (1–42)	4 (1–17)	NS

Table 5. Selected hemorheological, laboratory, and clinical parameters of patients treated with different oral antidiabetic regimens.

Variables	Metformin n = 16	Sulfonylureas n = 41	Acarbose n = 15	Combined Oral Antidiabetic Therapy n = 42	No Antidiabetic Therapy n = 12
Whole blood viscosity (mPAS)	4.75 (4.44–5.26)	4.80 (4.29–5.28)	4.56 (4.15–5.11)	4.67 (4.27–5.08)	4.48 (4.10 –5.58)
Plasma viscosity (mPAS)	1.35 (1.28–1.40)	1.34 (1.29–1.44)	1.31 (1.27–1.35)	1.34 (1.27–1.42)	1.26 (1.24–1.38)
Aggregation index M	11.8 (9.3–14.1)	10.4 (8.8–12.9)	11.6 (9.8–14.9)	11.5 (9.0–13.6)	12.9 (10.1–14.9)
Aggregation index M1	25.5 (22.9–27.7)	23.8 (22.6–26.2)	25.8 (19.2–27.3)	23.9 (20.8–25.5)	25.0 (23.0–29.7)
Plasma fibrinogen (g/L)	3.51 (2.96–3.74)	3.48 (3.05–4.06)	3.42 (3.01–3.61)	3.40 (2.89–3.85)	3.53 (3.06–3.78)
Triglyceride (mmol/L)	2.15 (1.37–3.08)	2.09 (1.24–3.01)	1.84 (1.46–2.57)	1.90 (1.43–2.87)	2.04 (1.51–3.94)
Cholesterol (mmol/L)	4.72 (4.19–5.26)	5.19 (4.42–5.72)	4.78 (4.16–5.74)	4.80 (3.96–5.62)	4.48 (4.04–6.03)
HDL cholesterol (mmol/L)	1.18 (1.06–1.31)	1.05 (0.88–1.44)	1.07 (0.98–1.45)	1.07 (0.88–1–24)	1.15 (1.02–1.52)
vWf activity	115 (35–171)	132 (118–185)	141 (78–240)	129 (88–177)	118 (36–156)
Glucose (mmol/L)	8.04 (5.6–9.3)	7.7 (6.8–10.0)	6.5 (5.9–8.2)	7.9 (6.7–11.8)	6.8 (6.4–8.3)
HbA1c (%)	6.7 (5.08–7.3)	6.5 (6.1–7.2)	6.03 (5.6–8.1)	6.6 (6.1–8.0)	7.9 (6.6–8.7)

Table 5. *Cont.*

Variables	Metformin $n = 16$	Sulfonylureas $n = 41$	Acarbose $n = 15$	Combined Oral Antidiabetic Therapy $n = 42$	No Antidiabetic Therapy $n = 12$
Hematocrit (%)	40.8 (39.7–44.0)	41.7 (38.6–43.1)	40.3 (38.2–45.6)	40.2 (38.0–41.7)	42.5 (38.3–45.9)
Body weight (kg)	75 (70–92)	80 (75–94)	77 (64–92)	87 (77–105)	75 (63–94)
BMI (kg/m^2)	27.1 (25.2–31.7)	29.4 (26.0–32.5)	27.6 (25.1–30.1)	30.5 (27.5–35.8)	27.7 (23.9–32.8)
Previous diseases (%)					
Hypertension	87	92	83	90	100
Myocardial infarction	18	22	21	12	0
Angina pectoris	43	35	35	46	27
Stroke	56	32	14	37	54
Transient ischemic attack	31	12	28	23	18
Peripheral artery disease	6	7	0	7	0
Carotid stenosis	0	2	0	0	0
Venous thromboembolism	0	2	0	0	18
Average of years since diagnosis of diabetes	11	7	2	10	5

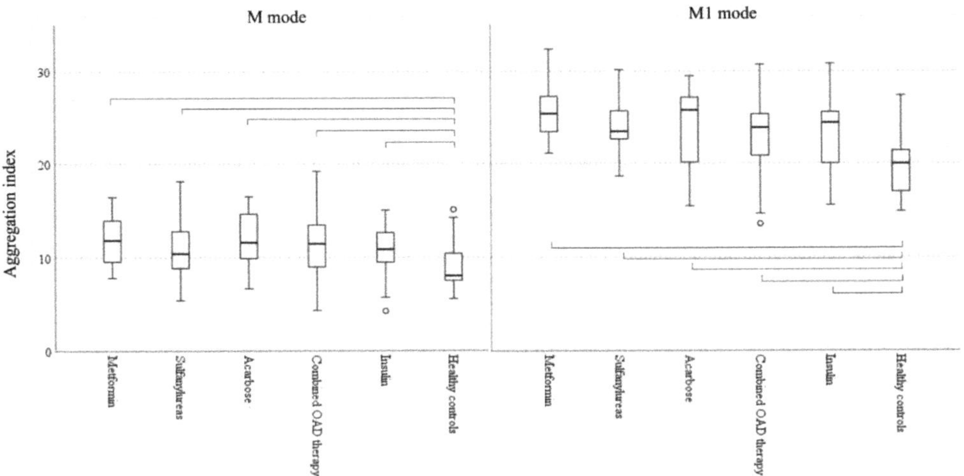

Figure 1. Red blood cell aggregation in patients with different antidiabetic regimens and in healthy controls (Boxplots: median, IQR and 5–95 percentiles; circle: outliers; parentheses show significant difference among groups according to Kruskal-Wallis test, $p < 0.05$). M mode: aggregation at stasis, M1 mode: aggregation at low shear.

3. Discussion

Similar to results commonly found in earlier studies, we found higher plasma viscosity and red blood cell aggregation in diabetic patients compared to healthy controls [5,21]. Whole blood viscosity is mainly determined by hematocrit, plasma viscosity, erythrocyte aggregation and red blood cell deformability [5], which was reflected in correlations between these parameters in the present study. On the other hand, we did not find a clear difference in whole blood viscosity between diabetic patients and healthy controls, unlike several pioneer researchers of hemorheology in earlier decades [5]. This might be associated with basic changes in the therapy of diabetes and concomitant vascular diseases when compared to the 70ies, such as the use of statins and the much wider use of antiplatelet agents. Fibrinogen is an acute phase reactant and independent cardiovascular risk factor. Although fibrinogen has been found to be elevated in type-2 diabetic patients in several studies [22–26], in our present study we did not find a significant difference in fibrinogen levels between diabetic patients and controls, despite the presence of several other aggravating factors in the diabetic group, such as higher age, a higher percentage of vascular diseases, and higher body weight and BMI. Concomitant lipid-lowering therapies may influence fibrinogen levels in diabetic patients, as both statin and fibrate monotherapies have been found to decrease fibrinogen levels [27].

De Silva et al. found that patients treated with sulfonylureas had higher fibrinogen concentration than patients who were treated with insulin, biguanides, or sulfonylurea plus biguanides [28]. Contrary to these results, no significant difference in fibrinogen levels was found between different antidiabetic regimens in the present study.

High platelet reactivity in diabetes has been described in several studies. Platelets in diabetic patients were reported to undergo rapid consumption due to hyperreactivity even to subthreshold stimuli [29]. While chronic hyperglycemia has been associated with in vivo platelet activation and platelet hyperreactivity, tight metabolic control has been found to reduce urinary thromboxane metabolites. Increased levels of von Willebrand factor in the circulatory system seem to correlate with an increase in platelet activation in diabetes [30]. In our study, we applied such subthreshold doses of inducers, however, a significant difference in platelet aggregation between diabetic and healthy people was seen only with the lowest applied dose of collagen. In a recent study performed with light aggregometry, partially similar to our results, platelet aggregation induced by ADP, collagen, or epinephrine did not appear to be consistently associated with diabetes [12].

4. Materials and Methods

4.1. Patients and Methods

A total of 159 type-2 diabetic patients (83 females, 76 males, mean age: 60 $\pm$ 10 years) and 25 healthy controls (15 females, 10 males, mean age: 52 $\pm$ 12 years) were involved in this study. Patient's age, body weight, BMI, smoking habits, physical activity, case history of cardiovascular diseases, family history of diabetes and cardiovascular diseases, current antidiabetic therapy, and concomitant medication were recorded. Physical activity/week was assessed by self-report. Hours spent engaging in exercise at least equivalent in intensity to vigorous walking were recorded. Patients were either treated with insulin or with one or more of the following antidiabetic drugs: metformin, sulfonylureas, acarbose. A total of 12 diabetic patients were involved, who were not treated with any antidiabetic drugs. Patients who were treated both with insulin and with oral antidiabetic drugs were excluded from the study.

4.2. Hemorheological Measurements

Hematocrit, erythrocyte aggregation, and von Willebrand factor measurements were performed at room temperature (22 $\pm$ 1 °C). Whole blood viscosity, plasma viscosity, and platelet aggregation measurements were performed at 37 °C within two hours after venipuncture. Plasma fibrinogen levels were measured by the Clauss method.

4.2.1. Plasma and Whole Blood Viscosity

Venous blood was collected into lithium-heparin coated Vacutainer tubes. Plasma and whole blood viscosity were measured in a Hevimet 40 capillary viscosimeter (Hemorex Ltd., Budapest, Hungary). Samples were centrifugated at 1500$\times$ g for ten minutes to obtain plasma. A 1.0 mL sample was injected into the capillary tube of the viscosimeter. The flow of the fluid was detected optoelectronically. Shear rate and shear stress were calculated from the flow curve. Viscosity was calculated as a function of these parameters, according to Casson's principle. Whole blood viscosity values, calculated at 90 s^{-1} shear rate, were used for the presentation of results.

4.2.2. RBC Aggregation

Venous blood samples were collected into lithium-heparin coated Vacutainer tubes. RBC aggregation was assessed in a Myrenne aggregometer (MA-1 Aggregometer, Myrenne GmbH, Roetgen, Germany), according to the light transmission method of Schmid-Schonbein et al. To assess M and M1 mode aggregation, 30 μL of blood was first sheared at 600 s^{-1} in order to disperse pre-existing aggregates. M mode was detected by decreasing the shear rate rapidly to zero. In the case of M1 mode, low shear rate was achieved. The aggregation indices, M and M1, were calculated from the surface area below the light-intensity curve, in a 10 s measurement period.

4.2.3. Hematocrit

Venous blood was collected into lithium-heparin coated Vacutainer tubes. Samples were centrifugated in hematocrit capillaries at 12,000 rpm for five minutes, in a microhematocrit centrifuge (Hemofuge, Heraeus Instr., Hanau, Germany).

4.3. Measurement of von Willebrand Factor

Von Willebrand factor (vWf) activity was assessed by quantitative direct-enzyme immunoassay (Shield Diagnostics Ltd., Dundee, UK). The wells of microtiter strips were coated with a preparation of a purified monoclonal antibody that binds to a functional epitope of the vWf antigen. Plasma was added, and vWf bound to the plates. After further incubation and washing, a third layer was introduced that was comprised of a horseradish-peroxidase-labelled mouse anti-human monoclonal anti-vWf conjugate to vWf. After a washing step, a substrate solution was added to trace the specifically bound antibody. A stop solution terminated the reaction. The amount of conjugate bound was measured in absorbance units. A dose-response curve was prepared from a calibrator set, according to the 4th International Standard. Activity of vWf was estimated by interpolation.

4.4. Measurement of Platelet Aggregability

Blood samples were collected into sodium-citrate coated Vacutainer tubes between 8 a.m. and 10 a.m., after an overnight fast. Platelet-rich plasma (PRP) was obtained by centrifugation at 150× g for 10 min. PRP was carefully removed. Platelet-poor plasma (PPP) was prepared by further centrifugation of the remaining samples at 2500× g for 10 min. Platelet aggregation was measured according to Born's method in a 4-channel optical aggregometer (Carat TX-4, Carat Diagnostics Ltd., Budapest, Hungary), at 37 °C, within two hours after vein puncture. Platelet aggregation was evaluated considering the maximal percentage of platelet aggregation in response to 0.5, 1, 2.5, 5, and 10 μM ADP; 0.2, 0.5, 1, and 2 μg/mL collagen; and 1, 2.5, 5, and 10 μM epinephrine inducers. Spontaneous aggregation was also assessed. In the diabetic group, platelet aggregability was investigated only in a subgroup of those patients who were not under antiplatelet therapy (55 patients, 24 males, 31 females). None of the control patients were on antiplatelet therapy.

4.5. Statistical Analysis

The Kolmogorov–Smirnov test was used to investigate normal distribution of parameters. As the results showed mostly non-normal distribution, nonparametric tests were used. Variables were presented as median (interquartile range) or mean ± SD. The Mann–Whitney U test was used to detect differences between diabetic patients and controls, as well as patients taking oral antidiabetic drugs or insulin. The Kruskall-Wallis test was used to test hemorheological and blood-chemistry parameters between different regimens of oral antidiabetic treatments. A p value < 0.05 was considered statistically significant. Relationships between dichotomic variables were tested with the Chi-square test. Spearman's correlation was used between hemorheological parameters, von Willebrand factor activity, and glucose and HbA1c levels.

Author Contributions: Conceptualization, K.K.; methodology, K.T.; validation, K.T.; formal analysis, K.K. and P.K.; investigation, K.K., G.F. and J.V.; resources, K.T.; data curation, K.K.; writing—original draft, K.K.; writing—review and editing, K.B., visualization, K.K. and P.K., supervision, K.T.; funding acquisition, K.T. All authors have read and agreed to the published version of the manuscript.

Funding: This research received no external funding.

Institutional Review Board Statement: The study was conducted according to the guidelines of the Declaration of Helsinki and approved by the Ethics Committee of the University of Pecs.

Informed Consent Statement: Informed consent was obtained from all subjects involved in the study.

Data Availability Statement: Data are available on request from the corresponding author according to local policies.

Acknowledgments: We are thankful to Kornelia Tapasztone Fazekas for her technical support.

Conflicts of Interest: The authors declare no conflict of interest.

References

1. Irace, C.; Carallo, C.; Scavelli, F.; De Franceschi, M.S.; Esposito, T.; Gnasso, A. Blood viscosity in subjects with normoglycemia and prediabetes. *Diabetes Care* **2014**, *37*, 488–492. [CrossRef]
2. Khodabandehlou, T.; Zhao, H.; Vimeux, M.; Aouane, F.; Le Devehat, C. Haemorheological consequences of hyperglycaemic spike in healthy volunteers and insulin-dependent diabetics. *Clin. Hemorheol. Microcirc.* **1998**, *19*, 105–114. [PubMed]
3. Brun, J.F.; Aloulou, I.; Varlet-Marie, E. Hemorheological aspects of the metabolic syndrome: Markers of insulin resistance, obesity or hyperinsulinemia? *Clin. Hemorheol. Microcirc.* **2004**, *30*, 203–209. [PubMed]
4. Koltai, K.; Feher, G.; Kesmarky, G.; Keszthelyi, Z.; Czopf, L.; Toth, K. The effect of blood glucose levels on hemorheological parameters, platelet activation and aggregation in oral glucose tolerance tests. *Clin. Hemorheol. Microcirc.* **2006**, *35*, 517–525. [CrossRef]
5. Cho, Y.I.; Mooney, M.P.; Cho, D.J. Hemorheological disorders in diabetes mellitus. *J. Diabetes Sci. Technol.* **2008**, *2*, 1130–1138. [CrossRef]
6. Vekasi, J.; Marton, Z.S.; Kesmarky, G.; Cser, A.; Russai, R.; Horvath, B. Hemorheological alterations in patients with diabetic retinopathy. *Clin. Hemorheol. Microcirc.* **2001**, *24*, 59–64. [PubMed]
7. Patti, G.; Cavallari, I.; Andreotti, F.; Calabrò, P.; Cirillo, P.; Denas, G.; Galli, M.; Golia, E.; Maddaloni, E.; Marcucci, R.; et al. Working Group on Thrombosis of the Italian Society of Cardiology. Prevention of atherothrombotic events in patients with diabetes mellitus: From antithrombotic therapies to new-generation glucose-lowering drugs. *Nat. Rev. Cardiol.* **2019**, *16*, 113–130. [CrossRef]
8. Forst, T.; Weber, M.M.; Löbig, M.; Lehmann, U.; Müller, J.; Hohberg, C.; Friedrich, C.; Fuchs, W.; Pfützner, A. Pioglitazone in addition to metformin improves erythrocyte deformability in patients with Type 2 diabetes mellitus. *Clin. Sci. (Lond.)* **2010**, *119*, 345–351. [CrossRef] [PubMed]
9. Ehrly, A.M. Specific Part. In *Therapeutic Hemorheology*, 1st ed.; Springer: Berlin/Heidelberg, Germany, 1991; pp. 207–211. [CrossRef]
10. Konya, H.; Hasegawa, Y.; Hamaguchi, T.; Satani, K.; Umehara, A.; Katsuno, T.; Ishikawa, T.; Miuchi, M.; Kohri, K.; Suehiro, A.; et al. Effects of gliclazide on platelet aggregation and the plasminogen activator inhibitor type 1 level in patients with type 2 diabetes mellitus. *Metabolism* **2010**, *59*, 1294–1299. [CrossRef] [PubMed]
11. Larkins, R.G.; Jerums, G.; Taft, J.L.; Godfrey, H.; Smith, I.L.; Martin, T.J. Lack of effect of gliclazide on platelet aggregation in insulin-treated and non-insulin-treated diabetes: A two-year controlled study. *Diabetes Res. Clin. Pract.* **1988**, *4*, 81–87. [CrossRef]
12. Rodriguez, B.A.T.; Johnson, A.D. Platelet Measurements and Type 2 Diabetes: Investigations in Two Population-Based Cohorts. *Front. Cardiovasc. Med.* **2020**, *7*, 118. [CrossRef]
13. Vitale, C.; Mercuro, G.; Cornoldi, A.; Fini, M.; Volterrani, M.; Rosano, G.M. Metformin improves endothelial function in patients with metabolic syndrome. *J. Intern. Med.* **2005**, *258*, 250–256. [CrossRef]
14. Schiapaccassa, A.; Maranhão, P.A.; de Souza, M.D.G.C.; Panazzolo, D.G.; Neto, J.F.N.; Bouskela, E.; Kraemer-Aguiar, L.G. 30-days effects of vildagliptin on vascular function, plasma viscosity, inflammation, oxidative stress, and intestinal peptides on drug-naïve women with diabetes and obesity: A randomized head-to-head metformin-controlled study. *Diabetol. Metab. Syndr.* **2019**, *23*, 70. [CrossRef]
15. Formoso, G.; De Filippis, E.A.; Michetti, N.; Di Fulvio, P.; Pandolfi, A.; Bucciarelli, T.; Ciabattoni, G.; Nicolucci, A.; Davì, G.; Consoli, A. Decreased in vivo oxidative stress and decreased platelet activation following metformin treatment in newly diagnosed type 2 diabetic subjects. *Diabetes Metab. Res. Rev.* **2008**, *24*, 231–237. [CrossRef] [PubMed]
16. Gin, H.; Freyburger, G.; Boisseau, M.; Aubertin, J. Study of the effect of metformin on platelet aggregation in insulin-dependent diabetics. *Diabetes Res. Clin. Pract.* **1989**, *6*, 61–67. [CrossRef]
17. Dolasık, I.; Sener, S.Y.; Celebı, K.; Aydın, Z.M.; Korkmaz, U.; Canturk, Z. The effect of metformin on mean platelet volume in dıabetıc patients. *Platelets* **2013**, *24*, 118–121. [CrossRef] [PubMed]
18. Shimabukuro, M.; Higa, N.; Chinen, I.; Yamakawa, K.; Takasu, N. Effects of a single administration of acarbose on postprandial glucose excursion and endothelial dysfunction in type 2 diabetic patients: A randomized crossover study. *J. Clin. Endocrinol. Metab.* **2006**, *91*, 837–842. [CrossRef] [PubMed]
19. Tranquilli, A.L.; De Tommaso, G.; Boemi, M.; Arduini, D.; Garzetti, G.G.; Fumelli, P.; Romanini, C. Glycemic control by insulin reduces erythrocyte aggregation in gestational diabetes. *J. Matern.-Fetal Investig.* **1995**, *5*, 110–112.
20. Jennings, A.M.; Ford, I.; Murdoch, S.; Greaves, M.; Preston, F.E.; Ward, J.D. The effects of diet and insulin therapy on coagulation factor VII, blood viscosity, and platelet release proteins in diabetic patients with secondary sulphonylurea failure. *Diabet. Med.* **1991**, *8*, 346–353. [CrossRef]
21. Fuller, J.H.; Keen, H.; Jarrett, R.J.; Omer, T.; Meade, T.W.; Chakrabarti, R.R.; North, W.R.; Stirling, Y. Haemostatic variables associated with diabetes and its complications. *Br. Med. J.* **1979**, *2*, 964–966. [CrossRef] [PubMed]

22. Kannel, W.; D'Agostino, R.B.; Wilson, P.W.; Belanger, A.J.; Gagnon, D.R. Diabetes, fibrinogen and risk of cardiovascular disease: The Framingham experience. *Am. Heart J.* **1990**, *120*, 672–676. [CrossRef]
23. Ganda, O.; Arkin, C.F. Hyperfibrinogenemia: An important risk factor for vascular complications in diabetes. *Diabetes Care* **1992**, *15*, 1245–1250. [CrossRef]
24. De Feo, P.; Volpi, E.; Lucidi, P.; Cruciani, G.; Reboldi, G.; Siepi, D.; Mannarino, E.; Santeusanio, F.; Brunetti, P.; Bolli, G.B. Physiological increments in plasma insulin concentrations have selective and different effects on synthesis of hepatic proteins in normal humans. *Diabetes* **1993**, *42*, 995–1002. [CrossRef] [PubMed]
25. O'Riordain, M.G.; Ross, J.A.; Fearon, K.C.; Maingay, J.; Farouk, M.; Garden, O.J.; Carter, D.C. Insulin and counterregulatory hormones influence acute-phase protein production in human hepatocytes. *Am. J. Physiol.* **1995**, *269*, E323–E330. [CrossRef]
26. Barazzoni, R.; Kiwanuka, E.; Zanetti, M.; Cristini, M.; Vettore, M.; Tessari, P. Insulin acutely increases fibrinogen production in individuals with type 2 diabetes but not in individuals without diabetes. *Diabetes* **2003**, *52*, 1851–1856. [CrossRef]
27. Sahebkar, A.; Serban, M.C.; Mikhailidis, D.P.; Toth, P.P.; Muntner, P.; Ursoniu, S.; Mosterou, S.; Glasser, S.; Martin, S.S.; Jones, S.R.; et al. Lipid and Blood Pressure Meta-analysis Collaboration (LBPMC) Group. Head-to-head comparison of statins versus fibrates in reducing plasma fibrinogen concentrations: A systematic review and meta-analysis. *Pharmacol. Res.* **2016**, *103*, 236–252. [CrossRef]
28. De Silva, S.R.; Shawe, J.E.; Patel, H.; Cudworth, A.G. Plasma fibrinogen in diabetes mellitus. *Diabete Metab.* **1979**, *5*, 201–206. [PubMed]
29. Santilli, F.; Simeone, P.; Liani, R. The role of platelets in diabetes mellitus. In *Platelets*, 4th ed.; Michelson, A.D., Ed.; Elsevier: Cambridge, MA, USA, 2019; pp. 469–503.
30. Santilli, F.; Simeone, P.; Liani, R.; Davì, G. Platelets and diabetes mellitus. *Prostaglandins Other Lipid Mediat.* **2015**, *120*, 28–39. [CrossRef]

Article

Changes of Hematological and Hemorheological Parameters in Rabbits with Hypercholesterolemia

Bence Tanczos [1,2], Viktoria Somogyi [1], Mariann Bombicz [3], Bela Juhasz [3], Norbert Nemeth [1] and Adam Deak [1,*]

1 Department of Operative Techniques and Surgical Research, Faculty of Medicine, University of Debrecen, Moricz Zsigmond u. 22, H-4032 Debrecen, Hungary; tanczos.bence@med.unideb.hu (B.T.); sogor.viktoria@med.unideb.hu (V.S.); nemeth@med.unideb.hu (N.N.)

2 Doctoral School of Clinical Medicine, University of Debrecen, Nagyerdei krt. 98, 4032 Debrecen, Hungary

3 Department of Pharmacology and Pharmacotherapy, Faculty of Medicine, University of Debrecen, Nagyerdei krt. 98, 4032 Debrecen, Hungary; bombicz.mariann@pharm.unideb.hu (M.B.); juhasz.bela@med.unideb.hu (B.J.)

* Correspondence: deak.adam@med.unideb.hu; Tel./Fax: +36-52-416-915

Abstract: Hypercholesterolemia plays an important role in the development of atherosclerosis, leading to endothelial dysfunction, ischemic events, and increased mortality. Numerous studies suggest the pivotal role of rheological factors in the pathology of atherosclerosis. To get a more detailed hematological and hemorheological profile in hypercholesterolemia, we carried out an experiment on rabbits. Animals were divided into two groups: the control group (Control) was kept on normal rabbit chow, the high-cholesterol diet group (HC) was fed with special increased cholesterol-containing food. Hematological parameters (Sysmex K-4500 automate), whole blood and plasma viscosity (Hevimet-40 capillary viscometer), red blood cell (RBC) aggregation (Myrenne MA-1 aggregometer), deformability and mechanical stability (LoRRca MaxSis Osmoscan ektacytometer) were tested. The white blood cell and platelet count, mean corpuscular volume, and mean corpuscular hemoglobin were significantly higher in the HC group, while the RBC count, hemoglobin, and hematocrit values were lower than the Control data. Viscosity values corrected to 40% hematocrit were higher in the HC group. The RBC aggregation significantly increased in the HC vs. the Control. The HC group showed significantly worse results both in RBCs' deformability and membrane stability. In conclusion, the atherogenic diet worsens the hematological and macro- and micro-rheological parameters, affecting blood flow properties and microcirculation.

Keywords: hypercholesterolemia; rabbit model; hemorheology; atherosclerosis

Citation: Tanczos, B.; Somogyi, V.; Bombicz, M.; Juhasz, B.; Nemeth, N.; Deak, A. Changes of Hematological and Hemorheological Parameters in Rabbits with Hypercholesterolemia. *Metabolites* **2021**, *11*, 249. https://doi.org/10.3390/metabo11040249

Academic Editor: Amedeo Lonardo

Received: 20 March 2021
Accepted: 16 April 2021
Published: 17 April 2021

1. Introduction

Atherosclerosis is a generalized disease of the arterial wall, characterized by thickening of the intimal layer and accumulation of fat, partly caused by hyperlipidemia (high concentration of lipids and/or lipoproteins) and lipid oxidation (as low-density lipoprotein [LDL] oxidation) [1]. An increasing number of international multi-center studies (Edinburgh Artery Study, Puerto Rico Study, Caerphilly and Speedwell Study, Northwick Park Heart Study) have shown that the development of atherosclerosis and thrombotic predisposition is associated with changes in hemorheological factors [2–5]. There are several mechanisms by which hemorheological factors can promote atherogenesis. These include the hypercoagulability, which predisposes to thrombosis, the decreased blood flow due to rheological changes, and the increased concentration of fibrinogen and its metabolites [6,7].

The rabbit is a widely used animal model for the study of human metabolic diseases [8]. The lipid metabolism of rabbits makes these animals particularly suitable for the study of the pathophysiology of lipoprotein metabolism, atherosclerosis, and metabolic syndromes [9–12]. Furthermore, this species' cardiac physiology (actin-myosin structure, ion channel characteristics) [13,14] is comparable to humans, defining the rabbit as an ideal model to study heart diseases as cardiac arrhythmia, myocardial infarction, heart failure, ischemic heart disease. A considerable number of articles have presented data about atherosclerotic rabbits [10,11], but the number of papers presenting detailed hematological and hemorheological data of healthy [15,16] and atherosclerotic [17] rabbits is relatively few.

The aim of our research was to evaluate how the atherogenic diet affects hematological, macro-rheological (whole blood cell and plasma viscosity) and micro-rheological (erythrocytes' aggregation, deformability, and mechanical membrane stability) parameters in a rabbit model of cardiovascular disease.

2. Results

2.1. Bodyweight Changes

The bodyweight of the animals before starting the follow-up period with diet was 2898 ± 111 g in the Control and 2923 ± 133 g in the HC group. At the time of the blood sampling the weight of Control group animals was 3087 ± 56 g (p = 0.004 vs. base), and 4131 ± 61 g in the HC group (p = 0.002 vs. base, $p < 0.001$ vs. Control).

2.2. Hematological Parameters

Table 1 summarizes the hematological results. The white blood cell count, the mean corpuscular volume and platelet count were significantly higher in the HC group than in the Control group. Mean corpuscular hemoglobin did not differ significantly. The red blood cell count, the hemoglobin value and the mean corpuscular hemoglobin concentration significantly decreased in HC group versus to the Control group.

Table 1. Hematology parameters in the Control and the atherogenic groups (HC). Means ± SEM.

Hematological Parameter	Control (n = 6)	HC (n = 6)	p Value vs. Control
White blood cell count [$\times 10^9$/L]	8.375 ± 0.270	23.59 ± 4.762	0.0042
Red blood cell count [$\times 10^{12}$/L]	6.863 ± 0.125	3.758 ± 0.245	<0.0001
Hemoglobin [g/dL]	15.25 ± 0.272	8.392 ± 0.558	<0.0001
Hematocrit [%]	46.37 ± 0.929	29.38 ± 1.729	<0.0001
Mean corpuscular volume [fL]	67.58 ± 0.472	78.88 ± 2.426	0.0001
Mean corpuscular hemoglobin [pg]	22.22 ± 0.258	22.50 ± 0.757	n.s.
Mean corpuscular hemoglobin concentration [g/L]	32.90 ± 0.270	28.82 ± 1.270	0.0047
Platelet count [$\times 10^9$/L]	254.8 ± 27.54	481.5 ± 38.73	<0.0001

2.3. Hemorheological Parameters

2.3.1. Whole Blood and Plasma Viscosity

The whole blood viscosity values corrected to 40% hematocrit were significantly increased vs. the Control group's data (p = 0.0051). In the plasma viscosity no considerable changes were detected (Figure 1).

2.3.2. Red Blood Cell Aggregation

In the HC group all the aggregation index values increased. The changes were significant at stasis (M 5 s: $p < 0.001$, M 10 s: $p < 0.001$) and at low shear rate (3 s^{-1}, M1 10 s: p = 0.0251) (Figure 2).

Figure 3 shows the changes in erythrocyte aggregation values determined by the LoRRca device. The difference in aggregation index (AI [%]) values was highly significant (p = 0.0003) between Control vs. HC groups. The amplitude values (Amp [au]) were lower in the HC group than in the Control group ($p < 0.0001$). The $t_{1/2}$ [s], which describes the kinetics of RBC aggregation, in the HC group values presented a non-significant decrease compared to the Control group.

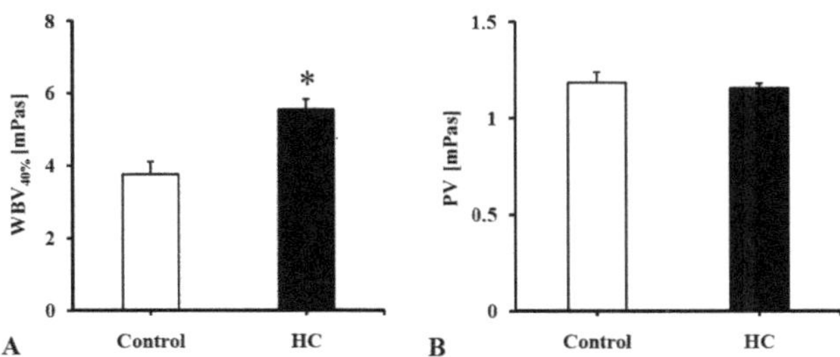

Figure 1. (**A**): whole blood viscosity corrected to 40% hematocrit (WBV [mPas]) and (**B**): plasma viscosity (PV [mPas]) values in the Control and the atherogenic groups (HC). Means ± SEM; * $p < 0.05$ vs. Control.

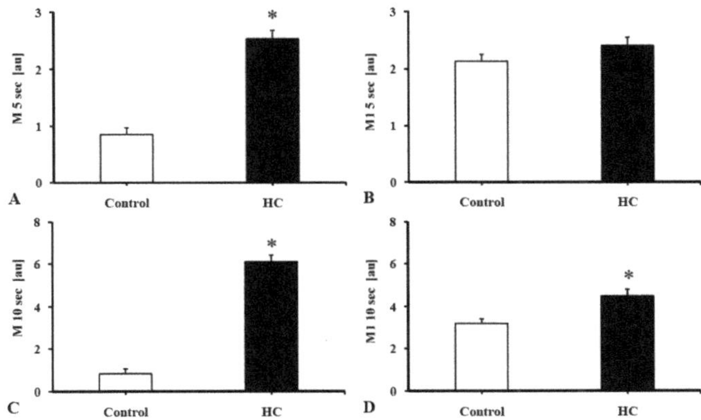

Figure 2. The red blood cell aggregation indices (**A**): M 5 s, (**B**): M1 5 s, (**C**): M 10 s, (**D**): M1 10 s [au]) measured by the Myrenne MA-1 aggregometer in the Control and the atherogenic group (HC). In M mode (shear rate = 0 s^{-1}) and M1 mode (shear rate = 3 s^{-1}) the index values are expressed at the 5th or at the 10th second of the aggregation. Means ± SEM; * $p < 0.05$ vs. Control.

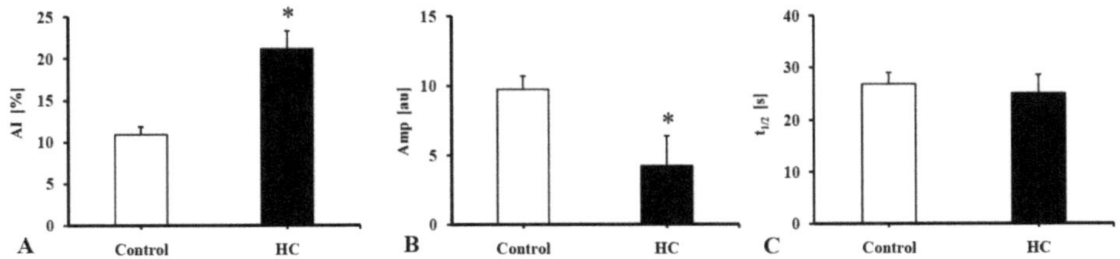

Figure 3. The red blood cell aggregation parameters measured by LoRRca rotational ectacytometer in the Control and the atherogenic group (HC). (**A**): aggregation index (AI [%]), (**B**): amplitude (Amp [au]) of the aggregation syllectogram (maximal-minimal intensity), (**C**): $t_{1/2}$ [s] representing the aggregation time at half Amp. Means ± SEM; * $p < 0.05$ vs. Control.

2.3.3. Red Blood Cell Deformability

The deformability of red blood cells was impaired in the HC group. The elongation index (EI)—shear stress (SS) curves showed remarkable differences, as the EI values of the HC groups were significantly lower compared to the Control group (Figure 4).

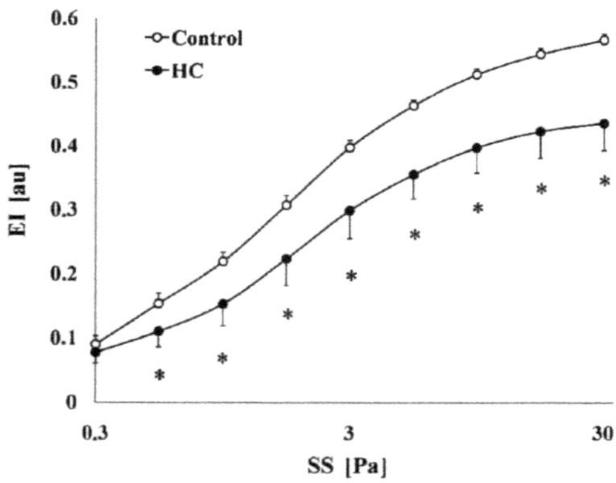

Figure 4. The red blood cell deformability describing elongation index (EI [au]) in the function of shear stress (SS [Pa]) of the Control and the atherogenic groups (HC); Means ± SEM; * $p < 0.05$ vs. Control.

The calculated parameters from the individual EI-SS curves also expressed the differences. The EI_{max} data were higher in the Control group and the differences were significant versus the HC group. EI values at 3 Pa were significantly lower in the HC group. $SS_{1/2}$ values were significantly increased in the HC group, while the $EI_{max}/SS_{1/2}$ ratio values of the HC group were lower than in the Control animals (Table 2).

Table 2. The red blood cell deformability measurements in the Control and the atherogenic groups (HC). EI at 3 Pa: elongation index at shear stress of 3 Pa, EI_{max}: the maximal elongation index, $SS_{1/2}$: shear stress belonging to the half of EI_{max}. Mean ± SEM.

Parameter	Control	HC	*p* Value vs. Control
EI at 3Pa	0.591 ± 0.009	0.448 ± 0.051	<0.0001
EI_{max}	0.397± 0.011	0.298 ± 0.043	<0.0001
$SS_{1/2}$ [Pa]	1.544 ± 0.172	2.125 ± 0.633	0.0272
$EI_{max}/SS_{1/2}$ [Pa^{-1}]	0.387 ± 0.044	0.228 ± 0.069	0.0007

2.3.4. Red Blood Cell Membrane (Mechanical) Stability

The EI-SS curves obtained before and after applied mechanical stress (100 Pa for 300 s) on the samples presented remarkable changes. Firstly, as the conventional deformability tests showed (see above), the HC group presented significantly lower EI values versus the Control with lower EI_{max} and higher $SS_{1/2}$ values. Secondly, after the mechanical stress, the decrease in EI values was more expressed in the HC group. The decrease in EI and EI_{max} values, as well as the increase in $SS_{1/2}$ values, were significant, compared to the Control group (Figure 5A,B, Table 3).

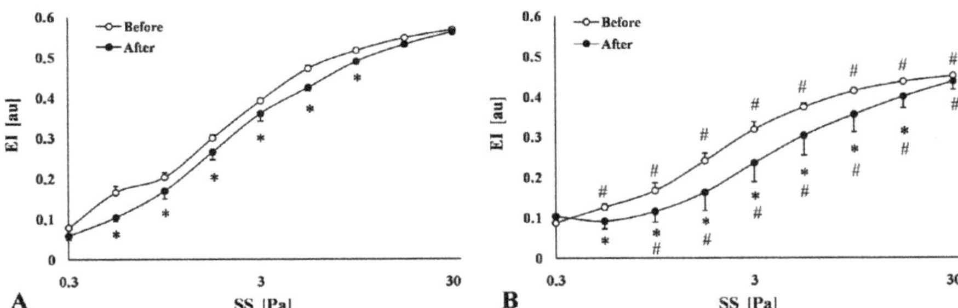

Figure 5. Elongation index (EI [au])—shear stress (SS [Pa]) curves in the mechanical stability test: before and after applying mechanical stress (100 Pa for 300 s) on the samples of the Control (**A**) and the atherogenic (HC) group (**B**). Means ± SEM; * $p < 0.05$ vs. before mechanical stress; # $p < 0.05$ vs. Control.

Table 3. Comparative parameters of red blood cell membrane stability test before and after applying the mechanical stress (100 Pa for 300 s). EI at 3 Pa: elongation index at shear stress of 3 Pa, EI_{max}: maximal elongation index, $SS_{1/2}$: shear stress belonging to the half of EI_{max}. Mean ± SEM.

Parameter	Test	Control	HC	*p* Value vs. Control, or vs. before (* Control, # HC)
EI at 3Pa	before (B)	0.392 ± 0.002	0.318 ± 0.015	0.002
	after (A)	0.360 ± 0.008 *	0.235 ± 0.019 #	0.0015; * 0.002; # 0.006
	ratio (A/B)	0.919 ± 0.019	0.734 ± 0.038	<0.0001
EI_{max}	before (B)	0.592 ± 0.003	0.461 ± 0.017	<0.0001
	after (A)	0.595 ± 0.004	0.458 ± 0.011	<0.0001
	ratio (A/B)	1.005 ± 0.007	0.999 ± 0.033	ns
$SS_{\frac{1}{2}}$ [Pa]	before (B)	1.690 ± 0.081	1.977 ± 0.227	ns
	after (A)	1.960 ± 0.149	4.540 ± 0.875 #	0.009; # 0.018
	ratio (A/B)	1.169 ± 0.089	2.228 ± 0.280	0.001
$EI_{max}/SS_{1/2}$ [Pa^{-1}]	before (B)	0.355 ± 0.021	0.249 ± 0.028	0.012
	after (A)	0.311 ± 0.019	0.124 ± 0.025 #	0.0318; # 0.002
	ratio (A/B)	0.883 ± 0.062	0.484 ± 0.062	<0.0001

3. Discussion

The hypercholesterolemic rabbit model is a preferred model to study human atherosclerosis and lipoprotein metabolism [18,19]. It is well-known that rabbits are sensitive to dietary cholesterol and rapidly develop severe hypercholesterolemia which drives aortic atherosclerosis. The hepatic LDL receptors in both humans and rabbits are down-regulated according to the level of cholesterol uptake in the liver. Very-low-density lipoprotein (VLDL) receptors are highly expressed in macrophages, this is also a similarity between rabbits and humans [11]. The larger arteries compared to the small rodent models allow the clinical evaluation: using MRI and ultrasound (echocardiography), morphological (plaque composition and structure) and functional changes (systolic, diastolic dysfunction) can be detected [18,19]. The atherosclerotic rabbit model is considered suitable to investigate numerous human diseases, however, it presents several limitations: e.g., the laboratory rabbits do not develop spontaneous atherosclerosis on a standard diet, because they have low cholesterol levels. The severe pathological changes, such as worsened ejection fraction and general deterioration of cardiac functions, with increased atherosclerotic plaque formation, infarct size, and increased mortality, develop only in rabbits receiving high cholesterol diets for a long period of time [20]. When feeding the rabbits with high-cholesterol-containing food, aortic lesions can develop, first in the aortic arch and then in the thoracic aorta. The abdominal aortic lesions, characteristic for humans, appear only when the whole aortic lesions are severe. Coronary atherosclerosis is also observed in cholesterol-fed rabbits (with predilection in the left arterial trunks). Another disadvantage is that the advanced lesion as fibrosis, hemorrhage, ulceration, or aortic aneurysms are not seen; the rabbits' plaque is characterized by foam cells with a fatty streak and they are rich in macrophages [11,18]. The so-called advanced lesions can develop following prolonged cholesterol feeding, but due to low hepatic lipase activity, this leads to increased hepatotoxicity [18].

In our study the used rabbit model was characterized by significant morphological, functional, and serological alterations. The area of the left atrium was enlarged; the weight of the left ventricle and relative wall thickness was increased. During the histological analysis, a foamy atherosclerotic plaque was observed on aortic sections, while in myocardial tissue interstitial fibrosis was determined. Symptoms of diastolic dysfunction were detected too. The serum lipid parameters, the atherogenic index, and ApoB/ApoA ratio were increased significantly in rabbits fed with additional 1% cholesterol and 1% saturated fat [19]. However, the limitations of the study include the low case number and the inter-species differences as mentioned above.

The atherogenic diet has affected numerous hematological parameters in our study. The red blood cell and the hemoglobin count were considerably decreased. This alteration was reported in experimental rabbits with high total cholesterol and increased LDL levels [21]. The low RBC and hemoglobin can appear even after 6 weeks of the experiment in rabbits, together with MCV, MCH, and MCHC changes [22,23]. In our investigation, in the HC group, the significantly increased MCV, unchanged MCH, and significantly decreased MCHC count show the signs of macrocytic hypochromic regenerative anemia. Anemia causes hypoxia due to decreased hemoglobin level, and there are several hemodynamic and non-hemodynamic compensatory mechanisms. The clinical and hemodynamic changes as a result of acute anemia are reversible, but chronic anemia drives progressive cardiac enlargement and left ventricular hypertrophy [24–26]. This cardiac alteration was detected in our rabbits, too [19]. The elevated MCV can be associated with the severity of atherosclerotic alterations and deficiency of vitamins related to atherosclerotic diseases as well [27].

Hypercholesterolemia stimulates platelet biogenesis through megakaryopoiesis, and leukocytosis by myelopoiesis, and increases platelet activation, by promoting platelet production and by direct impact on platelets [28–31]. The increased cholesterol level enhances the hyperaggregability of thrombocytes, too. Activated platelets can form aggregates with neutrophils and monocytes, and the subsequent crosstalk between platelets and leukocytes also plays an important role in the production of inflammatory cytokine, in the biosynthesis of leukotrienes and reactive oxygen species (ROS) [28]. The ROS can induce the production of inflammatory mediators such as C-reactive protein (CRP), which can activate the pro-thrombotic factors and platelets [32,33]. In our experimental animals, we detected significantly increased white blood cell and platelet count and CRP level [19]. These markers have shown the inflammatory character of atherosclerosis.

The high cholesterol level has direct effects on blood flow; this includes the growth of atherosclerotic plaques in the arterial system, reducing the lumen of coronary arteries, causing endothelial inflammation, and impaired endothelium-dependent vasorelaxation [34]. All together, these lead to an impairment of myocardial circulation and tissue perfusion [35]. Indirect effects of hypercholesterolemia involve blood rheology: a high level of cholesterol may increase whole blood viscosity by promoting the elevation of white blood cell and platelet count [29,36].

Remarkable changes in red blood cell aggregation using light-transmittance and syllectometry methods were observed. With light-transmittance, we detected that the HC group has significantly increased aggregation index values in each measuring mode (M and M1). Using the syllectometry method, in the HC group an increased aggregation index was accompanied by decreased aggregation amplitude, with unchanged time values. A similar tendency in aggregation index and syllectogram amplitude was reported in a clinical trial, performed with the same measuring method, on obese diabetic patients with hypercholesterolemia [37,38]. This study revealed that the total cholesterol level is correlated positively with the RBC aggregation index and negatively with aggregation half-time.

Red blood cell aggregation under low shear conditions is the main cause of increased blood viscosity [39–42]. The blood flow resistance during aggregation may decrease due to a reduced hematocrit but can increase at the same time by redistribution of red blood cells. Diminished blood flow can be caused by adhesion reactions between the blood cells and the endothelium of capillaries, such as the adhesion of white blood cells during inflammatory processes. Narrowing of the vessels due to atherosclerotic plaque can also adversely affect and increase cellular adhesion and flow resistance [43–45]. The increased aggregation can stimulate the axial migration of red blood cells which promotes plasma skimming. In this environment, a lower tissue hematocrit may cause decreased local viscosity at the marginal zone of blood vessels and could reduce the frictional resistance with the endothelium [46–49]. The axial migration promotes the phenomenon called margination described in white blood cells and platelets. The rigid RBC increases platelet marginalization which increases the tendency for thrombosis. Munn and Dupin [47] showed that the rouleaux formation of aggregating RBC is a more productive way to push the WBC to the vascular wall compared to a loosely associated group of cells. The WBC margination depends on the flow properties, axial migration of RBCs and RBC aggregates, local hematocrit as well as on blood cell deformability [48,49].

The mechanical stability of erythrocytes is essential to complete their function and to survive in the blood circulation. Several physiological and pathophysiological changes can affect the determining parameters of red blood cell deformability [45,50,51]. The stability of erythrocytes is directly related to LDL-cholesterol levels [52]. An excessive increase in the quantity of cholesterol in the erythrocyte membrane increases the rigidity and decreases the membrane deformability. This rigidity is correlated to the increase in relative cholesterol/phospholipid ratio [53]. The changes in cholesterol/phospholipid ratio can also affect RBCs' phosphatidylserine by reducing the exposure on the external surface of the cell in patients with hypercholesterolemia and spur cell anemia (in vitro study [54]. In high-fat diet fed mice, the levels of membrane cholesterol and phosphatidylserine externalization were increased, promoting erythrocytes-macrophage inflammatory interactions, and promoting macrophage phagocytosis in vitro [55]. Impairment of red blood cell deformability in hypercholesterolemia has been shown in clinical cases as well [56]. The phosphatidylserine serves as a trigger for macrophage recognition for senescent cells and plays an important role in erythrocyte's membrane stability [54,57].

Erythrocytes with excessively rigid membrane are less stable and more susceptible to lysis by mechanical forces especially when passing through narrow vessels (capillaries, spleen) [58,59]. In comparison to normal erythrocytes, the rigid ones are unable to deform well under shear stress, so the higher viscosity caused by increased shear force can also be attributed to impaired erythrocyte deformability [60]. Our result supports these findings: impaired deformability of red blood cells was accompanied by dramatically decreased membrane stability in the HC group. The examined sensitive parameters well expressed the differences in deformability (EI, EI_{max}, $SS_{1/2}$, and their ratio) and mechanical stability (before and after shear stress) deterioration of red blood cells [51]. The degradation of erythrocytes' deformability in the HC group was confirmed by the parameterization of EI-SS curves and was manifested in a decrease of maximal RBC elongation index and higher $SS_{1/2}$, and decreased $SS_{1/2}/EI_{max}$ values. The worsening of membrane stability was represented by EI-SS curves compared before and after mechanical stress application. We must remark that the HC group elongation index and the maximum of elongation have deteriorated significantly even before the application of the shear stress and this tendency exacerbates after the applied mechanical stress. This worsening was well presented by impaired $SS_{1/2}/EI_{max}$ ratio, too.

4. Materials and Methods

4.1. Experimental Animals

The animal experiments were approved by the University of Debrecen Committee of Animal Welfare and by the National Food Chain Safety Office (registration Nr.25/2013 UDCAW) in accordance with the national (Act XXVIII of 1998 on the protection and sparing of animals) and EU (Directive 2010/63/EU) regulations.

Male Californian-New Zealand hybrid (CAL/NZW) rabbits (n = 12), age 20 weeks and 2700–3000 g bodyweight, were involved in this study. The animals were kept in a conventional experimental animal facility, under a 12 h-12 h light-dark cycle. The rabbits (Jurasko Ltd., Debrecen, Hungary) received during the first two weeks of the adaptation (acclimatization) period commercial laboratory rabbit chow. After the acclimatization period the animals were randomly divided into Control (n = 6) and high-cholesterol diet (HC), atherogenic group (n = 6). During the next 16 weeks, the animals of the Control group were fed with standard rabbit chow, while a special "atherogenic" chow (additional 1% cholesterol and 1% saturated fat, formulated in the Department of Pharmaceutical Technology, Faculty of Pharmacy, University of Debrecen) were given in the HC group [19,61]. At the end of the follow-up period, blood samples were taken (Figure 6).

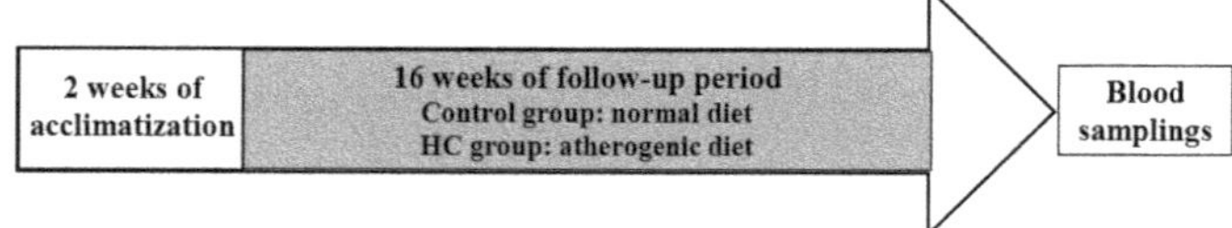

Figure 6. The timeline of the study.

4.2. Collection of Blood Samples

The blood samples were obtained from the marginal ear vein, with Vacutainer™-system, into a 3 mL BD Vacutainer® tube containing 1.8 mg/mL K3-EDTA as anticoagulant (Becton, Dickinson and Company, Franklin Lake, NJ, USA) 2 mL blood samples per animal, and kept on 20 °C for further lab analysis. All laboratory measurements were completed within 2 h.

4.3. Laboratory Methods

4.3.1. Hematological Parameters

A Sysmex K-4500 automate (TOA Medical Electronics Co., Ltd., Kobe, Japan) was used to determine the hematological parameters: red blood cell count (RBC [$10^{12}/\mu L$]), white blood cell count (WBC [$10^9/\mu L$]), hemoglobin concentration (Hgb [g/dL]), platelet count (Plt [$10^9/\mu L$]). The hematocrit (Hct [%]), mean corpuscular volume (MCV [fL]), mean corpuscular hemoglobin (MCH [pg]), and mean corpuscular hemoglobin concentration (MCHC [g/L] were calculated from the measured data.

4.3.2. Hemorheological Parameters

The changes in whole blood and plasma viscosity were measured by Hevimet-40 capillary viscometer (Hemorex Ltd., Budapest, Hungary) at 90 s^{-1} shear rates [16]. To calculate the whole blood viscosity the hematocrit count was normalized to 40% [62].

The erythrocytes' aggregation was measured by light-transmittance method using Myrenne MA-1 erythrocyte aggregometer (Myrenne GmbH, Germany). After disaggregation (600 s^{-1}) of 20 μL blood sample, at 5 or 10 s, the aggregation index values M mode (at a shear rate of 0 s^{-1}) and M1 mode (at a shear rate of 3 s^{-1}) were calculated [41,63]. The RBC aggregation was also tested with a LoRRca MaxSis Osmoscan ektacytometer (Mechatronics BV, The Netherlands). The device was operated with laser backscattering method. In the Couette-system, the blood sample is disaggregated by rotation, after this, the rotor stops promptly, and the changes in the intensity of the light reflected from the blood sample are measured [41,63]. The analyzed parameters were amplitude (Amp [au]), aggregation index (AI [%]) and half-amplitude time ($t_{1/2}$ [s]). The test requires 1 mL of blood.

Using LoRRca MaxSis Osmoscan ektacytometer, for the deformability and membrane stability of erythrocytes, for each measuring, 10 μL of blood was diluted in 2 mL of polyvinyl-pyrrolidone (PVP)/phosphate-buffered saline (PBS) solution (viscosity: 36.1 mPas, osmolarity: 300, mOsm/kg, pH: 7.3). The elongation index (EI) values of red blood cells were tested in the function of shear stress (SS [Pa], range: 0.3–30 Pa) [41,63]. For the comparison of the EI-SS curves EI values at 3 Pa, maximal elongation index (EI_{max}) and the shear stress belonging to the half of it ($SS_{1/2}$, [Pa]) and their ratio ($EI_{max}/SS_{1/2}$) were used. These values were calculated using Lineweaver-Burk equation [64]. The cell membrane (mechanical) stability test was performed by comparing two deformability measurements before and after mechanical stress (100 Pa, for 300 s) [51,65].

4.4. Statistical Analysis

All data are presented as the average of the data on the group (mean) +/− standard error of the mean (SEM). The D'Agostino–Pearson normality test was used to determine Gaussian distribution, and statistical analysis was then performed using unpaired Student's *t*-test or Mann–Whitney test (when normality test was not passed) between the groups, and two-way ANOVA was performed in case of the red blood cell deformability and membrane stability results. Analyses were carried out using GraphPad Prism software for Windows, version 8.0 (GraphPad Software Inc., La Jolla, CA, USA). Probability values (p) less than 0.05 were considered as statistically significantly different.

5. Conclusions

Macro- and micro-rheological parameters play an important role in determining tissue perfusion and shear stress-related endothelial functions and are influenced by numerous factors, involving metabolic changes too. Our study demonstrates that hypercholesterolemia can cause severe changes in hematological, macro-, and microrheological factors. The 16-week "atherogenic" diet altered not only the red blood cells' number and hemoglobin content but also decreased the deformability and membrane stability of the erythrocytes. The aggregation indices of erythrocytes were characterized by a significant deterioration in the high cholesterol group, and this was proven by two different measuring techniques too. Our results may provide additional information to better understand the processes taking place in the vascular system during atherosclerosis and might contribute to optimizing the therapy.

Author Contributions: Conceptualization, A.D. and N.N.; sample preparation and laboratory investigation, B.T., V.S., M.B.; data analysis, B.T.; writing—original draft preparation, B.T., A.D.; writing—review and editing, A.D., B.J. and N.N.; visualization, B.T.; supervision, A.D.; funding acquisition, N.N. and B.J. All authors have read and agreed to the published version of the manuscript.

Funding: The work was supported by The National Research, Development and Innovation Fund, Hungary NKFIH-1150-6/2019 and TKP-2019, ED_18-1-2019-0028 and co-financed by the European Union and the European Regional Development Fund GINOP-2.3.4-15-2016-00002 project, and by the Bridging Fund of the Faculty of Medicine, University of Debrecen.

Institutional Review Board Statement: The animal experiments were approved by the University of Debrecen Committee of Animal Welfare and by the National Food Chain Safety Office (registration Nr.25/2013/UDCAW) in accordance with the national (Act XXVIII of 1998 on the protection and sparing of animals) and EU (Directive 2010/63/EU) regulations.

Informed Consent Statement: Not applicable.

Data Availability Statement: The data presented in this study are available on request from the corresponding author.

Acknowledgments: Authors are grateful to the staff of the Department of Operative Techniques and Surgical Research and the Department of Pharmacology and Pharmacotherapy, Faculty of Medicine, University of Debrecen.

Conflicts of Interest: The authors declare no conflict of interest.

References

1. Rafieian-Kopaei, M.; Setorki, M.; Doudi, M.; Baradaran, A.; Nasri, H. Atherosclerosis: Process, indicators, risk factors and new hopes. *Int. J. Prev. Med.* **2014**, *5*, 927–946.
2. Baker, I.A.; Pickering, J.; Elwood, P.C.; Bayer, A.; Ebrahim, S. Fibrinogen, viscosity and white blood cell count predict myocardial, but not cerebral infarction: Evidence from the Caerphilly and Speedwell cohort. *Thromb. Haemost.* **2002**, *87*, 421–425.
3. Monsanto, H.A.; Renta-Muñoz, A.; Dones, W.; Comulada, A.; Cidre, C.; Orengo, J.C. The Puerto Rico Cardiovascular Risk-Estimation Study (PRCaRES): An exploratory assessment of new patients in physicians' offices. *P. R. Health Sci. J.* **2014**, *33*, 58–64. [CrossRef]
4. Pizzi, C.; De Stavola, B.L.; Meade, T.W. Long-term association of routine blood count (Coulter) variables on fatal coronary heart disease: 30-year results from the first prospective Northwick Park Heart Study (NPHS-I). *Int. J. Epidemiol.* **2009**, *39*, 256–265. [CrossRef] [PubMed]
5. Tzoulaki, I.; Murray, G.D.; Lee, A.J.; Rumley, A.; Lowe, G.D.; Fowkes, F.G. Relative value of inflammatory, hemostatic, and rheological factors for incident myocardial infarction and stroke: The Edinburgh Artery Study. *Circulation* **2007**, *115*, 2119–2127. [CrossRef]
6. Baskurt, O.K.; Meiselman, H.J. Blood rheology and hemodynamics. *Semin. Thromb. Hemost.* **2003**, *29*, 435–450. [PubMed]
7. Koscielny, J.; Jung, E.M.; Mrowietz, C.; Kiesewetter, H.; Latza, R. Blood fluidity, fibrinogen, and cardiovascular risk factors of occlusive arterial disease: Results of the Aachen study. *Clin. Hemorheol. Microcirc.* **2004**, *31*, 185–195. [PubMed]
8. Kolodgie, F.D.; Katocs, A.S., Jr.; Largis, E.E.; Wrenn, S.M.; Cornhill, J.F.; Herderick, E.E.; Lee, S.J.; Virmani, R. Hypercholesterolemia in the rabbit induced by feeding graded amounts of low-level cholesterol. *Arterioscler. Thromb. Vasc. Biol.* **1996**, *16*, 1454–1464. [CrossRef]
9. Yan, R.T.; Fernandes, V.; Yan, A.T.; Cushman, M.; Redheuil, A.; Tracy, R.; Vogel-Claussen, J.; Bahrami, H.; Nasir, K.; Bluemke, D.A.; et al. Fibrinogen and left ventricular myocardial systolic function: The Multi-Ethnic Study of Atherosclerosis (MESA). *Am. Heart J.* **2010**, *160*, 479–486. [CrossRef]
10. Dornas, W.C.; Oliveira, T.T.; Augusto, L.E.; Nagem, T.J. Experimental atherosclerosis in rabbits. *Arq. Bras. Cardiol.* **2010**, *95*, 272–278. [CrossRef] [PubMed]
11. Fan, J.; Kitajima, S.; Watanabe, T.; Xu, J.; Zhang, J.; Liu, E.; Chen, Y.E. Rabbit models for the study of human atherosclerosis: From pathophysiological mechanisms to translational medicine. *Pharmacol. Ther.* **2014**, *146*, 104–119. [CrossRef]
12. Lozano, W.M.; Arias-Mutis, O.J.; Calvo, C.J.; Chorro, F.J.; Zarzoso, M. Diet-induced rabbit models for the study of metabolic syndrome. *Animals* **2019**, *9*, 463. [CrossRef]
13. Milani-Nejad, N.; Janssen, P.M. Small and large animal models in cardiac contraction research: Advantages and disadvantages. *Pharmacol. Ther.* **2014**, *141*, 235–249. [CrossRef] [PubMed]
14. Conceição, G.; Heinonen, I.; Lourenço, A.P.; Duncker, D.J.; Falcão-Pires, I. Animal models of heart failure with preserved ejection fraction. *Neth. Heart J.* **2016**, *24*, 275–286. [CrossRef] [PubMed]
15. Windberger, U.; Bartholovitsch, A.; Plasenzotti, R.; Korak, K.J.; Heinze, G. Whole blood viscosity, plasma viscosity and erythrocyte aggregation in nine mammalian species: Reference values and comparison of data. *Exp. Physiol.* **2003**, *88*, 431–440. [CrossRef]
16. Nemeth, N.; Alexy, T.; Furka, A.; Baskurt, O.K.; Meiselman, H.J.; Furka, I.; Miko, I. Inter-species differences in hematocrit to blood viscosity ratio. *Biorheology* **2009**, *46*, 155–165. [CrossRef] [PubMed]
17. Abdelhalim, M.A.K.; Al-Ayed, S.M.; Moussa, S.A.A.; Al-Mohy, Y.H. The changes in serum and whole blood rheological properties of rabbits during the progression of atherosclerosis. *Pak. J. Pharm. Sci.* **2016**, *29*, 1053–1057.
18. Lee, Y.T.; Laxton, V.; Lin, H.Y.; Chan, Y.W.F.; Fitzgerald-Smith, S.; To, T.L.O.; Yan, B.P.; Liu, T.; Tse, G. Animal models of atherosclerosis. *Biomed. Rep.* **2017**, *6*, 259–266. [CrossRef] [PubMed]
19. Priksz, D.; Bombicz, M.; Varga, B.; Kurucz, A.; Gesztelyi, R.; Balla, J.; Toth, A.; Papp, Z.; Szilvassy, Z.; Juhasz, B. Upregulation of myocardial and vascular phosphodiesterase 9a in a model of atherosclerotic cardiovascular disease. *Int. J. Mol. Sci.* **2018**, *19*, 2882. [CrossRef]
20. Kertesz, A.; Bombicz, M.; Priksz, D.; Balla, J.; Balla, G.; Gesztelyi, R.; Varga, B.; Haines, D.D.; Tosaki, A.; Juhasz, B. Adverse impact of diet-induced hypercholesterolemia on cardiovascular tissue homeostasis in a rabbit model: Time-dependent changes in cardiac parameters. *Int. J. Mol. Sci.* **2013**, *14*, 19086–19108. [CrossRef]
21. Abdelhalim, M.A.K.; Moussa, S.A.A. Biochemical changes of hemoglobin and osmotic fragility of red blood cells in high fat diet rabbits. *Pak. J. Biol. Sci.* **2010**, *13*, 73–77. [CrossRef]
22. Króliczewska, B.; Miśta, D.; Zawadzki, W.; Wypchło, A.; Króliczewski, J. Effects of a skullcap root supplement on haematology, serum parameters and antioxidant enzymes in rabbits on a high-cholesterol diet. *J. Anim. Physiol. Anim. Nutr.* **2011**, *95*, 114–124. [CrossRef] [PubMed]
23. Karbiner, M.S.; Sierra, L.; Minahk, C.; Fonio, M.C.; Bruno, M.P.; Jerez, S. The role of oxidative stress in alterations of hematological parameters and inflammatory markers induced by early hypercholesterolemia. *Life Sci.* **2013**, *93*, 503–508. [CrossRef]

24. Mozos, I. Mechanisms linking red blood cell disorders and cardiovascular diseases. *Biomed. Res. Int.* **2015**, *2015*, 682054. [CrossRef]
25. Metivier, F.; Marchais, S.J.; Guerin, A.P.; Pannier, B.; London, G.M. Pathophysiology of anaemia: Focus on the heart and blood vessels. *Nephrol. Dial. Transplant.* **2000**, *15* (Suppl. 3), 14–18. [CrossRef] [PubMed]
26. Felker, G.M.; Adams, K.F., Jr.; Gattis, W.A.; O'Connor, C.M. Anemia as a risk factor and therapeutic target in heart failure. *J. Am. Coll. Cardiol.* **2004**, *44*, 959–966. [CrossRef] [PubMed]
27. Mueller, T.; Luft, C.; Haidinger, D.; Poelz, W.; Haltmayer, M. Erythrocyte mean corpuscular volume associated with the anatomical distribution in peripheral arterial disease. *VASA* **2002**, *31*, 81–85. [CrossRef]
28. Wang, N.; Tall, A.R. Cholesterol in platelet biogenesis and activation. *Blood* **2016**, *127*, 1949–1953. [CrossRef]
29. Tomaiuolo, G. Biomechanical properties of red blood cells in health and disease towards microfluidics. *Biomicrofluidics* **2014**, *8*, 051501. [CrossRef] [PubMed]
30. Huo, Y.; Ley, K.F. Role of platelets in the development of atherosclerosis. *Trends Cardiovasc. Med.* **2004**, *14*, 18–22. [CrossRef]
31. Barale, C.; Cavalot, F.; Frascaroli, C.; Bonomo, K.; Morotti, A.; Guerrasio, A.; Russo, I. Association between high on-aspirin platelet reactivity and reduced superoxide dismutase activity in patients affected by type 2 diabetes mellitus or primary hypercholesterolemia. *Int. J. Mol. Sci.* **2020**, *21*, 4983. [CrossRef]
32. Zhang, Z.; Yang, Y.; Hill, M.A.; Wu, J. Does C-reactive protein contribute to atherothrombosis via oxidant-mediated release of pro-thrombotic factors and activation of platelets? *Front. Physiol.* **2012**, *3*, 433. [CrossRef] [PubMed]
33. Braune, S.; Küpper, J.H.; Jung, F. Effect of prostanoids on human platelet function: An overview. *Int. J. Mol. Sci.* **2020**, *21*, 9020. [CrossRef]
34. Krüger-Genge, A.; Blocki, A.; Franke, R.P.; Jung, F. Vascular endothelial cell biology: An update. *Int. J. Mol. Sci.* **2019**, *20*, E4411. [CrossRef]
35. Buchwalow, I.B.; Cacanyiova, S.; Neumann, J.; Samoilova, V.E.; Boecker, W.; Kristek, F. The role of arterial smooth muscle in vasorelaxation. *Biochem. Biophys. Res. Commun.* **2008**, *377*, 504–507. [CrossRef] [PubMed]
36. Ho, C.H. White blood cell and platelet counts could affect whole blood viscosity. *J. Chin. Med. Assoc.* **2004**, *67*, 394–397. [PubMed]
37. Wiewiora, M.; Sosada, K.; Wylezol, M.; Slowinska, L.; Zurawinski, W. Red blood cell aggregation and deformability among patients qualified for bariatric surgery. *Obes. Surg.* **2007**, *17*, 365–371. [CrossRef]
38. Krüger-Genge, A.; Sternitzky, R.; Pindur, G.; Rampling, M.; Franke, R.P.; Jung, F. Erythrocyte aggregation in relation to plasma proteins and lipids. *J. Cell. Biotechnol.* **2019**, *5*, 65–70. [CrossRef]
39. Rampling, M.W. Hyperviscosity as a complication in a variety of disorders. *Semin. Thromb. Hemost.* **2003**, *29*, 459–465. [PubMed]
40. Baskurt, O.K.; Yalcin, O.; Ozdem, S.; Armstrong, J.K.; Meiselman, H.J. Modulation of endothelial nitric oxide synthase expression by red blood cell aggregation. *Am. J. Physiol. Heart Circ. Physiol.* **2004**, *286*, H222–H229. [CrossRef]
41. Baskurt, O.K.; Boynard, M.; Cokelet, G.C.; Connes, P.; Cooke, B.M.; Forconi, S.; Liao, F.; Hardeman, M.R.; Jung, F.; Meiselman, H.J.; et al. International Expert Panel for Standardization of Hemorheological Methods. New guidelines for hemorheological laboratory techniques. *Clin. Hemorheol. Microcirc.* **2009**, *42*, 75–97. [CrossRef] [PubMed]
42. Baskurt, O.K.; Neu, B.; Meiselman, H.J. Determinants of red blood cell aggregation. In *Red Blood Cell Aggregation*; Baskurt, O.K., Neu, B., Meiselman, H.J., Eds.; CRC Press: Boca Raton, FL, USA, 2012; pp. 9–29.
43. Geddes, J.B.; Carr, R.T.; Wu, F.; Lao, Y.; Maher, M. Blood flow in microvascular networks: A study in nonlinear biology. *Chaos* **2010**, *20*, 045123. [CrossRef]
44. Lipowsky, H.H. Microvascular rheology and hemodynamics. *Microcirculation* **2005**, *12*, 5–15. [CrossRef] [PubMed]
45. Baskurt, O.K. Mechanisms of blood rheology alterations. In *Handbook of Hemorheology and Hemodynamics*; Baskurt, O.K., Hardeman, M.R., Rampling, M.W., Meiselman, H.J., Eds.; IOS Press: Amsterdam, The Netherlands, 2007; pp. 170–190.
46. Muravyov, A.V.; Tikhomirova, I.A.; Maimistova, A.A.; Bulaeva, S.V.; Mikhailov, P.V.; Kislov, N.V. Red blood cell aggregation changes are depended on its initial value: Effect of long-term drug treatment and short-term cell incubation with drug. *Clin. Hemorheol. Microcirc.* **2011**, *48*, 231–240. [CrossRef]
47. Munn, L.L.; Dupin, M.M. Blood cell interactions and segregation in flow. *Ann. Biomed. Eng.* **2008**, *36*, 534–544. [CrossRef]
48. Nash, G.B.; Watts, T.; Thornton, C.; Barigou, M. Red cell aggregation as a factor influencing margination and adhesion of leukocytes and platelets. *Clin. Hemorheol. Microcirc.* **2008**, *39*, 303–310. [CrossRef] [PubMed]
49. Fedosov, D.; Gompper, G. White blood cell margination in microcirculation. *Soft Matter* **2014**, *10*, 2961–2970. [CrossRef]
50. Pretini, V.; Koenen, M.H.; Kaestner, L.; Fens, M.H.A.M.; Schiffelers, R.M.; Bartels, M.; Van Wijk, R. Red Blood Cells: Chasing Interactions. *Front. Physiol.* **2019**, *10*, 945. [CrossRef]
51. Nemeth, N.; Sogor, V.; Kiss, F.; Ulker, P. Interspecies diversity of erythrocyte mechanical stability at various combinations in magnitude and duration of shear stress, and osmolality. *Clin. Hemorheol. Microcirc.* **2016**, *63*, 381–398. [CrossRef] [PubMed]
52. Holm, T.M.; Braun, A.; Trigatti, B.L.; Brugnara, C.; Sakamoto, M.; Krieger, M.; Andrews, N.C. Failure of red blood cell maturation in mice with defects in the high-density lipoprotein receptor SR-BI. *Blood* **2002**, *99*, 1817–1824. [CrossRef]
53. Meurs, I.; Hoekstra, M.; van Wanrooij, E.J.; Hildebrand, R.B.; Kuiper, J.; Kuipers, F.; Hardeman, M.R.; Van Berkel, T.J.; Van Eck, M. HDL cholesterol levels are an important factor for determining the lifespan of erythrocytes. *Exp. Hematol.* **2005**, *33*, 1309–1319. [CrossRef] [PubMed]

54. van Zwieten, R.; Bochem, A.E.; Hilarius, P.M.; van Bruggen, R.; Bergkamp, F.; Hovingh, G.K.; Verhoeven, A.J. The cholesterol content of the erythrocyte membrane is an important determinant of phosphatidylserine exposure. *Biochim. Biophys. Acta* **2012**, *1821*, 1493–1500. [CrossRef] [PubMed]
55. Unruh, D.; Srinivasan, R.; Benson, T.; Haigh, S.; Coyle, D.; Batra, N.; Keil, R.; Sturm, R.; Blanco, V.; Palascak, M.; et al. Red blood cell dysfunction induced by high-fat diet: Potential implications for obesity-related atherosclerosis. *Circulation* **2015**, *132*, 1898–1908. [CrossRef]
56. Koter, M.; Franiak, I.; Strychalska, K.; Broncel, M.; Chojnowska-Jezierska, J. Damage to the structure of erythrocyte plasma membranes in patients with type-2 hypercholesterolemia. *Int. J. Biochem. Cell Biol.* **2004**, *36*, 205–215. [CrossRef]
57. Manno, S.; Takakuwa, Y.; Mohandas, N. Identification of a functional role for lipid asymmetry in biological membranes: Phosphatidylserine-skeletal protein interactions modulate membrane stability. *Proc. Natl. Acad. Sci. USA* **2002**, *99*, 1943–1948. [CrossRef]
58. Lee, C.Y.; Kim, K.C.; Park, H.W.; Song, J.H.; Lee, C.H. Rheological properties of erythrocytes from male hypercholesterolemia. *Microvasc. Res.* **2004**, *67*, 133–138. [CrossRef] [PubMed]
59. da Silva Garrote-Filho, M.; Bernardino-Neto, M.; Penha-Silva, N. Influence of erythrocyte membrane stability in atherosclerosis. *Curr. Atheroscler. Rep.* **2017**, *19*, 17. [CrossRef]
60. Vayá, A.; Riveram, L.; de la Espriella, R.; Sanchez, F.; Suescun, M.; Hernandez, J.L.; Fácila, L. Red blood cell distribution width and erythrocyte deformability in patients with acute myocardial infarction. *Clin. Hemorheol. Microcirc.* **2015**, *59*, 107–114. [CrossRef] [PubMed]
61. Maj, D.; Bieniek, J.; Łapa, P.; Sternstein, I. The effect of crossing New Zealand White with Californian rabbits on growth and slaughter traits. *Arch. Anim. Breed.* **2009**, *52*, 205–211. [CrossRef]
62. Matrai, A.; Whittington, R.B.; Ernst, E. A simple method of estimating whole blood viscosity at standardized hematocrit. *Clin. Hemorheol.* **1987**, *7*, 261–265. [CrossRef]
63. Hardeman, M.; Goedhart, P.; Shin, S. Methods in hemorheology. In *Handbook of Hemorheology and Hemodynamics*; Baskurt, O.K., Hardeman, M.R., Rampling, M.W., Meiselman, H.J., Eds.; IOS Press: Amsterdam, The Netherlands, 2007; pp. 242–266.
64. Baskurt, O.K.; Hardeman, M.R.; Uyuklu, M.; Ulker, P.; Cengiz, M.; Nemeth, N.; Shin, S.; Alexy, T.; Meiselman, H.J. Parameterization of red blood cell elongation index–shear stress curves obtained by ektacytometry. *Scand. J. Clin. Lab. Investig.* **2009**, *69*, 777–788. [CrossRef] [PubMed]
65. Sogor, V.; Tanczos, B.; Deak, A. Data interpretation of erythrocyte membrane mechanical stability test using the laser-assisted optical rotational cell analyzer. *Ser. Biomech.* **2016**, *30*, 27–34.

 metabolites

MDPI

Article

Alterations of Selected Hemorheological and Metabolic Parameters Induced by Physical Activity in Untrained Men and Sportsmen

Sandor Szanto [1,†], Tobias Mody [1,2,†], Zsuzsanna Gyurcsik [1], Laszlo Balint Babjak [3], Viktoria Somogyi [3], Barbara Barath [3], Adam Varga [3], Adam Attila Matrai [3] and Norbert Nemeth [3,*]

1 Department of Sports Medicine, Faculty of Medicine, University of Debrecen, Nagyerdei Park 12, H-4032 Debrecen, Hungary; szanto.sandor@med.unideb.hu (S.S.); modytobias@gmail.com (T.M.); gyurcsik.zsuzsanna@med.unideb.hu (Z.G.)
2 Doctoral School of Clinical Medicine, University of Debrecen, Nagyerdei krt. 98, H-4032 Debrecen, Hungary
3 Department of Operative Techniques and Surgical Research, Faculty of Medicine, University of Debrecen, Moricz Zsigmond u. 22, H-4002 Debrecen, Hungary; babjakl43@gmail.com (L.B.B.); sogor.viktoria@med.unideb.hu (V.S.); barath.barbara@med.unideb.hu (B.B.); varga.adam@med.unideb.hu (A.V.); matrai.adam@med.unideb.hu (A.A.M.)
* Correspondence: nemeth@med.unideb.hu; Tel./Fax: +36-52-416-915
† Sandor Szanto and Tobias Mody contributed equally to the work.

check for updates

Citation: Szanto, S.; Mody, T.; Gyurcsik, Z.; Babjak, L.B.; Somogyi, V.; Barath, B.; Varga, A.; Matrai, A.A.; Nemeth, N. Alterations of Selected Hemorheological and Metabolic Parameters Induced by Physical Activity in Untrained Men and Sportsmen. *Metabolites* **2021**, *11*, 870. https://doi.org/10.3390/metabo11120870

Academic Editor: German Perdomo

Received: 23 October 2021
Accepted: 13 December 2021
Published: 14 December 2021

Abstract: Optimal tissue oxygen supply is essential for proper athletic performance and endurance. It also depends on perfusion, so on hemorheological properties and microcirculation. Regular exercise is beneficial to the rheological status, depending on its type, intensity, and duration. We aimed to investigate macro and microrheological changes due to short, high-intensity exercise in professional athletes (soccer and ice hockey players) and untrained individuals. The exercise was performed on a treadmill ergometer during a spiroergometry examination. Blood samples were taken before and after exercise to analyze lactate concentration, hematological parameters, blood and plasma viscosity, and red blood cell (RBC) deformability and aggregation. Leukocyte, RBC and platelet counts, and blood viscosity increased with exercise, by the largest magnitude in the untrained group. RBC deformability slightly impaired after exercise, but showed better values in ice hockey versus soccer players. RBC aggregation increased with exercise, dominantly in ice hockey players. Lactate increased mostly in soccer players, and the respiratory exchange rate was the lowest in ice hockey players. Overall, short, high-intensity exercise altered macro and microrheological parameters, mostly in the untrained group. Significant differences were found between the two sports. The data can be useful in training status monitoring, selection, and in revealing the causes of physical loading symptoms.

Keywords: hemorheology; metabolites; physical activity; training; sport

1. Introduction

Hemorheological parameters play a pivotal role in tissue perfusion. Blood is a non-Newtonian fluid as its viscosity depends on the shear rate. Lowering the shear rate is associated with increasing viscosity values due to the red blood cell aggregation [1–3]. Whole blood viscosity is mainly determined by the plasma viscosity (a Newtonian fluid), number of blood cells, dominantly red blood cells (RBCs), and microrheological parameters of the RBCs, such as deformability and aggregation [1,3,4]. RBC deformability is influenced by the cells' volume, surface-to-volume ratio, morphology, intracellular viscosity, as well as by cell membrane properties [5,6]. RBC aggregation is determined by plasmatic factors (fibrinogen, other plasma proteins, and macromolecules) and cellular features (deformability, morphology, and the composition of the glycocalyx layer) as well as by the shearing forces [1,4–7]. It is well known that oxidative stress (free radical reactions), inflammatory processes, mechanical stress, osmolarity changes, oxygenation levels, temperature, nitric oxide, and metabolic and pH alterations may deteriorate these microrheological parameters [5,6,8–12]. Impaired RBC deformability and enhanced RBC aggregation increase viscosity, so decreasing blood fluidity, and result in disturbed tissue perfusion, so increasing vascular resistance, and a deterioration in the microcirculation [3,5,13].

Optimal tissue oxygen supply is essential for proper athletic performance and endurance, an important factor of which is tissue perfusion, and thus, hemorheological parameters and microcirculation characteristics. It is known that exercise, usual physical activity, has a beneficial effect on the rheological status, but it also depends on its type, intensity, and regularity [14–17].

Romain et al. in their meta-analysis study confirmed that regular exercise decreases hematocrit and RBC aggregation; however, there are still many controversial data and more studies are needed to further analyze these effects [18]. In general, it can be concluded that an unhealthy, sedentary lifestyle with obesity is associated with impaired RBC deformability, enhanced RBC aggregation, and increased hematocrit and plasma viscosity [16,17]. Physical activity is beneficial, but the effect strongly depends on the type of exercise, its intensity, regularity, and duration or volume [16,19–22]. Irregular, inappropriate, or heavy/intensive physical exercise may increase blood and plasma viscosity, and hematocrit, with bidirectional changes in the microrheological features. Regular, well-balanced exercise may lead to a kind of "hemorheological fitness" characterized with lower blood and plasma viscosity, lower hematocrit, improved RBC deformability, and decreased aggregation [22,23].

The question may rise whether untrained people, who perform physical activity in their leisure time, and professional athletes, sportsmen, may have different hemorheological conditions. What happens with these parameters if short, high-intensity exercises have to be performed? To investigate this question, a standardized condition of physical exercise load can be used, such as the spiroergometry test.

Spiroergometry is a diagnostic procedure to continuously measure respiration and gas metabolism during ergometer exercise [24,25]. The aim of a standard cardiopulmonary exercise testing (CPET) protocol is for the individual to be exposed to a load using an ergometer and incrementally increase workload for about 8 to 12 min until they can go no further. These are often referred to as an incremental ramp protocol to a volitional maximum. It enables judgment of function and performance capacity of the cardiopulmonary system and metabolism. This also makes it possible to determine the maximum exercise capacity, maximal oxygen uptake capacity as well as aerobic threshold and anaerobic/lactate threshold values [24,25].

The aim of our research was to evaluate the macro and microrheological parameters of the blood as well as lactate and respiratory parameters in professional athletes and in untrained men before and after a standardized physical exercise in spiroergometry. We also wished to investigate whether these parameters differ in various professional sports, with different training and competition loads, such as soccer and ice hockey.

2. Results

2.1. Hematological Parameters

Table 1 summarizes the changes in the general quantitative and qualitative hematological parameters. Short-term, high-intensity exercise by spiroergometry resulted in increased white blood cells (untrained Control: $p < 0.001$, Soccer players: $p < 0.001$, Ice hockey players: $p < 0.001$ vs. before; Control vs. Soccer players: $p < 0.001$; Soccer players vs. Ice hockey players: $p = 0.01$) and platelet counts (Control: $p < 0.001$, Soccer players: $p < 0.001$, Ice hockey players: $p < 0.001$ vs. before) with hemoconcentration (Hct, Control: $p < 0.001$, Soccer players: $p < 0.001$, Ice hockey players: $p < 0.001$ vs. before). The increase in the white blood cell count was the largest in its magnitude in the ice hockey players, while the red blood cell count and hematocrit increased slightly more in the untrained group.

Table 1. Changes of hematological parameters in untrained control group and groups of professional soccer players and ice hockey players before and after the standardized physical exercise load by spiroergometry.

Variable	UnTrained Control Group			Professional Soccer Players			Professional Ice Hockey Players		
	Before	After	Before/After Ratio	Before	After	Before/After Ratio	Before	After	Before/After Ratio
WBC	6.53	11.86	1.81	5.22	9.28	1.78	5.48	11.01	2.05
[$\times 10^9$/L]	± 0.99	± 2.34 *	± 0.22	± 1.21 #	± 2.39 *,#	± 0.37	± 0.84 #	± 2.08 *,+	± 0.52
Lymph [%]	37.36	46.24	1.28	32.39	43.47	1.31	34.12	35.01	1.04
	± 10.77	± 8.16 *	± 0.21	± 5.52	± 8.34 *	± 0.21	± 7.99 +	± 11.77 #,+	± 0.27 #,+
Gr+Mo [%]	10.54	10.21	0.97	9.08	8.43	1.06	13.79	11.44	0.82
	± 2.23	± 2.78	± 0.16	± 2.54 #	± 2.43 #	± 0.47	± 3.85 #,+	± 4.45 +	± 0.18 #,+
RBC	5.06	5.45	1.08	5.24	5.53	1.06	5.14	5.41	1.05
[$\times 10^{12}$/L]	± 0.25	± 0.25 *	± 0.03	± 0.19 #	± 0.22 *	± 0.03 #	± 0.33	± 0.41 *	± 0.03 #
Hct [%]	45.14	49.45	1.10	45.52	49.02	1.08	45.71	49.27	1.08
	± 2.11	± 2.29 *	± 0.03	± 1.56	± 1.42 *	± 0.03 #	± 1.74	± 2.78 *	± 0.03
Hgb [g/L]	15.46	16.77	1.08	15.68	16.53	1.06	15.65	17.86	1.14
	± 0.84	± 0.97 *	± 0.03	± 0.73	± 1.07 *	± 0.06 #	± 0.68	± 6.21 *	± 0.41 #
MCV [fL]	89.24	90.74	1.02	86.95	88.62	1.02	89.09	91.21	1.02
	± 3.05	± 2.87 *	± 0.01	± 3.07 #	± 3.06 *,#	± 0.01	± 3.25 +	± 3.90 *,+	± 0.01 #,+
MCH [pg]	30.57	30.76	1.01	29.95	29.91	1.00	30.51	31.98	1.05
	± 1.34	± 1.49	± 0.01	± 1.47	± 2.33	± 0.06	± 1.12	± 1.15	± 0.21
MCHC	34.24	33.90	0.99	34.44	33.75	0.98	34.24	33.71	0.98
[g/L]	± 0.72	± 0.85	± 0.02	± 0.89	± 2.19	± 0.06	± 0.54	± 0.64 *,+	± 0.02
Plt	231.45	301.59	1.31	215.46	287.25	1.31	224.75	288.67	1.29
[$\times 10^9$/L]	± 49.71	± 66.13 *	± 0.08	± 22.61	± 41.43 *	± 0.14	± 38.89	± 45.53 *	± 0.10
MPV	10.77	11.09	1.03	10.35	10.46	1.03	10.40	10.88	1.05
[fL]	± 0.77	± 0.91	± 0.03	± 1.23	± 1.19	± 0.03	± 0.92	± 1.00	± 0.04

Means ± S.D.; * $p < 0.05$ vs. before, # $p < 0.05$ vs. untrained group, and + $p < 0.05$ vs. professional soccer players.

2.2. Blood and Plasma Viscosity

Both blood and plasma viscosity increased after the high-intensity exercise. In the case of blood viscosity, the change was significant in all groups (untrained Control: $p < 0.001$, Soccer players: $p = 0.038$, and Ice hockey players: $p = 0.001$). However, the highest elevation was seen in the untrained group. The difference between the untrained groups and the groups of professional sportsmen was more obvious when the blood viscosity values were corrected for 40% hematocrit (Soccer players: $p = 0.03$, Ice hockey players: $p = 0.034$ vs. Control). Hematocrit/viscosity values significantly dropped in the untrained group ($p = 0.002$ vs. before, $p = 0.004$ vs. Soccer players, and $p = 0.018$ vs. ice hockey players) (Figure 1).

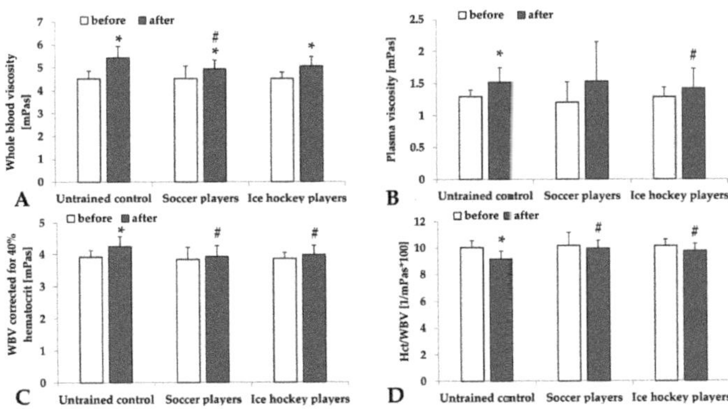

Figure 1. Changes of whole blood viscosity (WBV [mPas]) (**A**), plasma viscosity (PV [mPas]) (**B**), WBV corrected for 40% hematocrit (Hct) (**C**), and Hct/WBV ratio (**D**) in untrained control group and groups of professional soccer players and ice hockey players before and after the standardized physical exercise load by spiroergometry. Means ± S.D.: * $p < 0.05$ vs. before, # $p < 0.05$ vs. untrained.

2.3. Red Blood Cell Deformability

The elongation index at 3 Pa did not change with exercise; however, the values were the highest in ice hockey players ($p < 0.001$ vs. the untrained Control or Soccer players). The maximal elongation index showed a decrease with exercise, more dominantly in the untrained men ($p = 0.045$) and in a smaller manner in the sportsmen groups. The highest shear stress values belonging to the half-maximal elongation index ($SS_{1/2}$ [Pa]) were found in the Soccer player group ($p < 0.001$ vs. untrained Control or Soccer players), associated with the lowest ratio of these two parameters (Figure 2). The ratio of the values tested before and after the exercise are shown in Table 2 with the viscosity data.

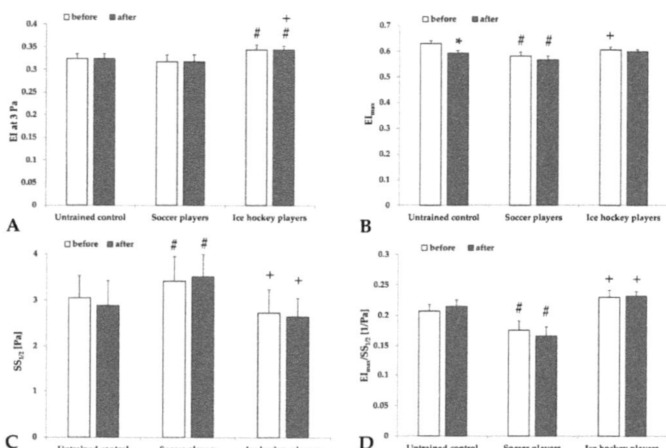

Figure 2. Changes of red blood cell deformability parameters: the elongation index at 3 Pa shear stress (EI at 3 Pa) (**A**), the maximal elongation index (EI_{max}) (**B**), the shear stress at half-EI_{max} ($SS_{1/2}$ [Pa]) (**C**), and $EI_{max}/SS_{1/2}$ ratio (**D**) in untrained control group and groups of professional soccer players and ice hockey players before and after the standardized physical exercise load by spiroergometry. Means ± S.D.; * $p < 0.05$ vs. before, # $p < 0.05$ vs. untrained group, and + $p < 0.05$ vs. professional soccer players.

Table 2. The ratio of blood and plasma viscosity and red blood cell deformability parameters tested before and after the standardized physical exercise load by spiroergometry in untrained control group and groups of professional soccer players and ice hockey players.

Before/After Ratio of	Untrained Control Group	Professional Soccer Players	Professional Ice Hockey Players
WBV [mPas]	1.20 ± 0.08	1.10 ± 0.10 #	1.12 ± 0.05 #
PV [mPas]	1.18 ± 0.19	1.41 ± 0.77	1.09 ± 0.11
$Hct_{40\%}$ [%]	1.08 ± 0.06	1.03 ± 0.13	1.03 ± 0.06 #
Hct/WBV [$mPas^{-1}$]	0.91 ± 0.046	0.99 ± 0.11 #	0.96 ± 0.04 #
EI at 3 Pa	1.02 ± 0.11	1.02 ± 0.05	1.02 ± 0.04
EI_{max}	1.13 ± 0.55	0.98 ± 0.04 #	0.1 ± 0.04
$SS_{1/2}$ [Pa]	1.02 ± 0.37	0.96 ± 0.14	0.96 ± 0.17
$EI_{max}/SS_{1/2}$ [Pa^{-1}]	1.34 ± 1.32	1.04 ± 0.11	1.06 ± 0.15

Means ± S.D.; # $p < 0.05$ vs. untrained control group.

Investigating the osmotic gradient deformability (osmoscan) parameters, we found that the maximal elongation index values were the highest in sportsmen (Soccer players: $p = 0.011$, Ice hockey players: $p < 0.001$ vs. Control). With exercise, their values decreased, while the values of ice hockey players increased. The minimal elongation index values were higher in soccer players ($p = 0.04$ vs. Control, $p = 0.015$ vs. Ice hockey players). The parameter derived from the area under the elongation index–osmolarity curves were higher in sportsmen (Soccer players: $p = 0.023$, Ice hockey players: $p = 0.008$ vs. Control), and with high-intensity exercise, these values decreased in soccer players and increased in ice hockey players (Table 3).

Table 3. Changes of osmotic gradient deformability (osmoscan) parameters of the red blood cells in untrained control group and groups of professional soccer players and ice hockey players before and after the standardized physical exercise load by spiroergometry.

Variable	Untrained Control Group			Professional Soccer Players			Professional Ice Hockey Players		
	Before	After	Before/ After Ratio	Before	After	Before/ After Ratio	Before	After	Before/ After Ratio
EI min	0.117 ± 0.007	0.12 ± 0.01	1.029 ± 0.069	0.13 ± 0.013 #	0.128 ± 0.009 #	0.992 ± 0.079	0.122 ± 0.01	0.119 ± 0.008 +	0.979 ± 0.074
EI_{max}	0.548 ± 0.011	0.547 ± 0.011	0.999 ± 0.034	0.569 ± 0.015#	0.558 ± 0.007 *,#	0.978 ± 0.022	0.554 ± 0.011 +	0.566 ± 0.008 *,#,+	1.022 ± 0.027 +
EI hyper	0.2742 ± 0.005	0.2738 ± 0.005	0.999 ± 0.033	0.284 ± 0.007 #	0.279 ± 0.004 *#	0.978 ± 0.021	0.277 ± 0.005 +	0.283 ± 0.004 *,#,+	1.021 ± 0.027 +
O min [mOsm/L]	139.09 ± 3.96	145.45 ± 4.69 *	1.04 ± 0.02	137.5 ± 4.381	141.08 ± 4.72 #	1.021 ± 0.03 #	138.66 ± 4.81	140.75 ± 4.22 #	1.015 ± 0.016 #
O (EI_{max}) [mOsm/L]	281.54 ± 16.38	278.64 ± 29.33	0.992 ± 0.115	288.64 ± 9.45	291.33 ± 11.63	1.007 ± 0.03	282.5 ± 9.01	289.5 ± 9.67	1.025 ± 0.022
O hyper [mOsm/L]	413.55 ± 14.14	418.82 ± 15.09	1.012 ± 0.016	418.21 ± 15	422.91 ± 16.26	1.008 ± 0.019	424.83 ± 9.97 #	425.5 ± 11.21	1.002 ± 0.015
Area	143.98 ± 5.94	142.97 ± 5.11	0.994 ± 0.034	150.2 ± 6.63 #	147.91 ± 5.91 #	0.98 ± 0.024 #	150.51 ± 4.71 #	153.95 ± 3.54 #,+	1.024 ± 0.038 +

Means ± S.D.; * $p < 0.05$ vs. before, # $p < 0.05$ vs. untrained control group, and + $p < 0.05$ vs. professional soccer players.

2.4. Red Blood Cell Aggregation

Table 4 summarizes the various erythrocyte aggregation parameters. Using light-transmission aggregometry M 5 s, M1 5 s, M 10 s, and M1 10 s index parameters were determined. In the untrained group, both M index values (at 0 s^{-1} shear rate) decreased, while M1 index values (at 3 s^{-1} shear rate) increased significantly (M 5 s: $p = 0.045$, M1 5 s: $p = 0.001$, and M1 10 s: $p < 0.001$). Similar changes were observed in the sportsmen groups (Soccer players' M1 5 s: $p < 0.001$, M1 10 s: $p < 0.001$; Ice hockey players' M 5 s: $p = 0.033$, M1 5 s: $p < 0.001$, and M1 10 s: $p = 0.022$), however, the largest increase was seen in the untrained control group.

Table 4. Red blood cell aggregation parameters in untrained control group and groups of professional soccer players and ice hockey players before and after the standardized physical exercise load by spiroergometry.

Variable	UnTrained Control Group			Professional Soccer Players			Professional Ice Hockey Players		
	Before	After	Before/After Ratio	Before	After	Before/After Ratio	Before	After	Before/After Ratio
M 5 s	3.38 ± 1.13	2.81 ± 1.07 *	0.89 ± 0.36	2.97 ± 1.07	2.65 ± 0.96	0.99 ± 0.64	3.39 ± 0.88	3.00 ± 1.41 *	0.88 ± 0.32
M1 5 s	2.75 ± 1.23	4.07 ± 1.15 *	1.80 ± 1.02	2.78 ± 1.03	3.61 ± 1.01 *,#	1.53 ± 0.94	3.14 ± 1.30	4.02 ± 1.14 *	1.34 ± 0.44
M 10 s	9.43 ± 3.97	9.00 ± 3.08	1.07 ± 0.41	7.62 ± 3.34 #	8.00 ± 2.55	1.27 ± 0.69	8.89 ± 3.36	8.36 ± 3.12	1.03 ± 0.50
M1 10 s	7.27 ± 2.57	10.53 ± 3.41 *	1.61 ± 0.68	6.20 ± 3.07	8.94 ± 2.86 *,#	1.74 ± 0.99	7.98 ± 3.80 +	9.81 ± 3.89 *	1.49 ± 0.97 #
AI [%]	74.83 ± 18.44	86.45 ± 10.12 *	1.24 ± 0.44	63.41 ± 4.90 #	68.31 ± 3.59 *,#	1.09 ± 0.06	78.43 ± 15.65 +	79.19 ± 9.59 #,+	1.03 ± 0.15 #,+
Amp [au]	7.05 ± 4.80	4.55 ± 4.20	0.79 ± 0.29	16.80 ± 2.66 #	15.54 ± 2.26 #	0.96 ± 0.18	1.39 ± 2.95 #,+	0.20 ± 0.27 #,+	0.85 ± 0.90
$t_{1/2}$ [s]	1.91 ± 2.04	0.97 ± 1.01	2.01 ± 3.26	2.25 ± 0.49	1.70 ± 0.29 *,#	0.75 ± 0.15	1.48 ± 1.29 +	1.36 ± 0.68 #,+	1.89 ± 1.96 +

Means ± S.D.; * $p < 0.05$ vs. before, # $p < 0.05$ vs. untrained control group, and + $p < 0.05$ vs. professional soccer players.

The aggregation values determined by the syllectometry, correlated with the changes of M1 index values, as the aggregation index (AI [%]) increased after the exercise, showing the largest rise in the untrained group ($p = 0.003$). The most stable values were seen in the ice hockey players. Amplitude values were very low in ice hockey players and were the highest in soccer players. After exercise values decreased in all groups, together with the half-time values ($t_{1/2}$ [s]) ($p < 0.001$ in Soccer players).

2.5. Maximal Oxygen Uptake, Respiratory Exchange Rate, and Lactate Concentration

The maximal oxygen consumption (VO_2 max) values were 42.89 ± 4.56 mL/min/kg in the untrained Control group, 58.91 ± 5.67 mL/min/kg in Soccer players ($p < 0.001$ vs. Control group), and 51.66 ± 3.36 mL/min/kg in Ice hockey players ($p < 0.001$ vs. Control group and vs. Soccer players).

The respiratory exchange rate (RER) showed the highest values in the untrained group and was lower in soccer players and the lowest in ice hockey players. The maximal lactate concentration and lactate concentration 5 min after the exercise did not differ significantly; however, when the exercise was finished, the values slightly decreased. The lowest values were seen in the untrained individuals, the highest in soccer players, and in between values were in ice hockey players. The correlation coefficient values between the RER and $lactate_{max}$ were the highest in untrained men and the second highest in ice hockey players (Table 5).

Table 5. Respiratory exchange rate (RER) and lactate concentration (maximal and 5 min and their ratio) and their correlation coefficients in untrained control group and groups of professional soccer players and ice hockey players before and after the standardized physical exercise load by spiroergometry.

Group	RER	$Lactate_{max}$ [mmol/L]	$Lactate_{5'}$ [mmol/L]	$Lactate_{max}/Lactate_{5'}$	R^2 of RER and $Lactate_{max}$	R^2 of RER and $Lactate_{5'}$
Untrained control group	1.22 ± 0.08	12.71 ± 1.91	12.29 ± 2	1.04 ± 0.13	0.4176	0.1319
Professional soccer players	1.18 ± 0.05	14.94 ± 3.01 #	14.58 ± 2.96 #	1.03 ± 0.16	0.0029	0.0287
Professional ice hockey players	1.13 ± 0.04 #,+	13.26 ± 1.96	12.53 ± 2.16	1.07 ± 0.14	0.2027	0.1439

Means ± S.D.; # $p < 0.05$ vs. untrained control group and + $p < 0.05$ vs. professional soccer players.

3. Discussion

Hemorheological changes induced by exercise is an intensively studied field that is still full of controversies. However, some general conclusions can be taken, thanks to the pioneer groups working on this field in the past two to three decades (e.g., L. Dintenfass, E. Ernst, J.F. Brun, P. Connes, and M. El-Sayed, among others) [16,17,19,21].

Many authors support that regular, low-intensity exercise activity has value for "hemorheological fitness". As a result of training, the expansion of blood volume, particularly plasma volume, results in better fluidity of the athletes' blood [15–17,21].

Microrheological parameters of red blood cells are influenced by oxidative stress, mechanical stress, metabolic alterations (e.g., accumulation of lactate and decrease in pH), and changes in oxygenation levels [5,6,9–12,26–28]; All are factors present during exercise in various magnitudes. When blood viscosity acutely increases during exercise (as short-term, exercise-induced hyperviscosity), in the case of a healthy vasculature, it means a key modulator for vasodilatation via the endothelial mechanoreceptor-mediated nitric oxide production process, as part of the adaptation [16,17,23,29]. Concerning the effect of lactate concentration increase and RBC deformability impairment [30], it has been demonstrated that red blood cell deformability does not impair under a lactate concentration threshold of 4 mmol/L. Above that, decreased red blood cell deformability can be found [16,31,32]. In athletes with high-level endurance, this correlation is not obvious, probably due to a kind of adaptation to exercise-induced hypoxemia with an improved lactate transfer through the cell membrane [16,31–38]. These findings correlate well with our study. However, the physical fitness level of trained and untrained people can be different, as well as the amount of accumulated lactate that affects lipid peroxidation [31], which influences rheological parameters, such as RBC deformability [5,6]. A wider investigative scope of metabolomics [36] would be important to better reveal the background of these alterations.

In our study, the age and BMI values of the healthy male volunteers were comparable. Therefore, we assumed that the observed changes can be dominantly related to the differences in physical activity level and profession. An important difference between the players of the two sports from the point of view of training physiology is that while soccer players play at medium intensity for a longer period of time, performing a higher proportion of aerobic activity, ice hockey players train at higher intensities at shorter time intervals. This is also supported by the higher VO_2 max values obtained by spiroergometry in the case of soccer players (better endurance). It is supposed that the differences in the muscle mass and its capillarity and thus metabolic responses might contribute to the explanation of the differences we found [39–41]. Hemorheological results (higher EI max values for hockey players under exercise) correlate with the results of maximum lactate levels after exercise and lactate levels measured 5 min after exercise. Presumably, due to the better red blood cell deformability, we may see a better lactate elimination in the case of hockey players as a result of the favorable microcirculation compared to the other two groups. Furthermore, the lower respiratory exchange rate values measured in hockey players can be explained by the positive effect of the high-intensity exercise on lipid and carbohydrate oxidation [33,38].

We have found that professional athletes experience more favorable hemorheological changes as a result of intense exercise than untrained individuals. While hematocrit normalized whole blood and plasma viscosity, and RBC deformability and aggregation deteriorated in the control group, in the case of athletes, these parameters remained unchanged or changed only to a lesser extent.

In our study, hematocrit increased almost equally in all three groups under intense exercise. When viscosity was normalized to 40% hematocrit, a significant increase in whole blood viscosity was observed only in the untreated group, with no increase in athletes relative to preload values. As whole blood viscosity depends primarily on hematocrit, plasma viscosity, and RBC aggregation and deformability [1–3], it was hypothesized that a smaller increase in viscosity was more likely to be due to a more favorable change in RBC aggregation and deformability in athletes.

Examining the change in plasma viscosity, a significant increase under load occurred only in the untrained group. An increase was also observed in athletes only with a smaller rate. Alis et al., during high-intensity interval training, observed an increase in serum protein, cholesterol, and triglyceride levels in healthy individuals in addition to an increase in hematocrit, explaining the increase in plasma viscosity [42]. The increase of hematocrit levels in athletes was slightly smaller than in the untrained men in our study, suggesting a lower hemoconcentration, which may explain the more moderate deterioration in plasma viscosity. It should also be emphasized that in our case, the total load time was much shorter than in the above-mentioned study, so no significant fluid loss could develop.

The intense exercise did not change the deformability of RBCs, only in untrained individuals. There are few data in the literature on changes in deformability under short-term intense loading. Kilic-Toprak et al. similarly analyzed hemorheological parameters on female volleyball players before and after the Yo-Yo intermittent recovery test level 1 (Yo-YoIR1) [43]. As a result of the test, an increase in whole blood viscosity and red blood cell deformability was also observed. This contradicts our own results, but may be explained by the different nature of the load, especially the interval type of the load. It is known that the deformability of RBCs improves in healthy adults as a result of regular training. Bizjak et al. also found an increase in the theoretical maximal elongation index at infinite shear stress during six weeks of moderately intense exercise [44].

It was an interesting observation that the value of the RBC aggregation index at rest was slightly higher in hockey players than in soccer players; however, under maximum load, it increased only slightly in hockey players, as opposed to the significant increase observed in soccer players. This difference in the two sports can also be explained by different training and match loads. In hockey, the proportion of short but high intensity loads and the acceleration that requires concentric muscle work is much more common [45].

4. Materials and Methods

4.1. Volunteers

Thirty-seven male volunteers (ethical permission nr.: DE RKEB/IKEB:5410-2020) took part in the study, forming three different groups. The untrained control group included 11 medical students with a hobby level of physical activity. The second group had 14 professional soccer players (time spent in competitive sport level: 14.4 ± 3.4 years). Twelve professional ice hockey players formed the third group (time spent in competitive sport level: 15.9 ± 4.5 years.). Anthropometric data are presented in Table 6.

Table 6. Age, height, weight, body mass index (BMI), and further body composition data (tested with InBody 770 device; InBody USA Co., Ltd., Cerritos, CA, USA) of the participants.

Group	Age [year]	Height [cm]	Weight [kg]	BMI [kg/m^2]	Percent Body Fat [%]	Skeletal Muscle Mass [kg]
Untrained control group	25.09 ± 2.55	184.09 ± 5.82	89.55 ± 13.02	26.36 ± 3.41	20.41 ± 3.91	38.87 ± 1.94
Professional soccer players	22.71 ± 3.43	182.21 ± 5.63	77.79 ± 6.55 #	23.64 ± 1.08	10.03 ± 4.55 #	39.69 ± 3.54
Professional ice hockey players	24.25 ± 4.29	183.92 ± 5.6	86.17 ± 8.72 +	25.42 ± 1.68 +	16.46 ± 3.94 #,+	43.35 ± 8.83 #

Means ± S.D., # $p < 0.05$ vs. untrained control group and + $p < 0.05$ vs. professional soccer players.

4.2. Spiroergometry Tests and Collection of Blood Samples

Five minutes before the beginning of the spiroergometry test [25,46,47], we collected venous blood samples from the individuals (median cubital vein, 23 G needle, Vacutainer tubes K_3-EDTA). After that, all participants had to complete a ramp protocol (Vitamaxima 12) on a treadmill ergometer with increasing performance (W) [46,47]. The protocol started with a 2 min warm-up phase with 4 km/h velocity. In the test phase following the warm-up, the workload of the ergometer increased every minute with 45 W by increasing the speed and elevation of the treadmill. The maximal velocity was 12 km/h reached in the 5th minute and, additionally, the maximal speed gradient was still continuously increasing to a maximum of 17.5 degrees. The reachable maximal workload was 645 W.

During the test, the athletes were connected via face mask to a Vyntus CPX hardver (Vyaire medical, Mettawa, IL, USA), giving minute ventilation, breathing frequency, oxygen uptake, carbon dioxide production, and other specific data. The device has an inbuilt automatic calibration mechanism that we proceeded after every 5th test. The heart rate was measured by a Polar H9 (Polar Electro, Kempele, Finland) chest belt, and real time data were trackable using a Bluetooth connection. The test was performed at a temperature of 20 °C. When participants reached their maximal performance capacity, we stopped the protocol. This running time was 7:04 ± 0:46 min in the control group, 8:19 ± 0:33 min in soccer players, and 6:59 ± 0:43 min in hockey players.

Recorded main parameters were: load [W] and time [min], heart rate [1/min], ventilation volume [L/min], breathing frequency [1/min], oxygen uptake as volume (VO_2 [mL/min]), carbon dioxide output as volume (VCO_2, [mL/min]), respiratory exchange ratio (RER, VCO_2/VO_2), systolic and diastolic blood pressure [mmHg]. In this paper, the RER values were correlated with the laboratory parameters.

Right after the exercise, we collected venous blood samples again and measured heart rate and blood pressure for 5 additional minutes. In this 5 min resting period, we also collected capillary blood samples from the individuals' fingertip (right hand, index finger with Accu-Chek Safe T-Pro lancets) to determine the blood lactate concentration in the first and fifth minutes of resting. To test lactate concentration [mmol/L], a Nova Lactate Plus device was used (Nova Biomedical, Waltham, MA, USA).

4.3. Laboratory Methods

4.3.1. Hematological Parameters

A Sysmex K-4500 automate (TOA Medical Electronics Co., Ltd., Kobe, Japan) was used to determine red blood cell count (RBC [$10^{12}/\mu$L]), white blood cell count (WBC [$10^9/\mu$L]), lymphocyte percent (Lymph [%]), granulocyte and monocyte percent (Gr+Mo [%]),]), platelet count (Plt [$10^9/\mu$L]), hematocrit (Hct [%]), hemoglobin concentration (Hgb [g/dL]), mean corpuscular volume (MCV [fL]), mean corpuscular hemoglobin (MCH [pg]), mean corpuscular hemoglobin concentration (MCHC [g/L], and mean platelet volume (MPV [fL]) values.

4.3.2. Hemorheological Parameters

Whole blood and plasma viscosity were determined by a Hevimet-40 capillary viscometer (Hemorex Ltd., Budapest, Hungary) at 90 s^{-1} shear rates. To calculate the whole blood viscosity, the hematocrit count was normalized to 40% according to the Matrai formula [48]: $WBV_{40\%}/PV = (WBV_{Hct}/PV)^{40\%/Hct}$ where WBV_{Hct} is the whole blood viscosity [mPas] measured at a 90 s^{-1} shear rate and at the actual hematocrit (Hct [%]); PV is the plasma viscosity [mPas], and Hct is the hematocrit of the sample.

RBC deformability was tested with a LoRRca MaxSis Osmoscan ektacytometer. For the conventional deformability tests, 10 μL of blood was diluted in 2 mL of a polyvinylpyrrolidone (PVP)/phosphate-buffered saline (PBS) solution (viscosity: 29.6 mPas, osmolarity: 293, mOsm/kg, and pH: 7.2). The elongation index (EI) values of red blood cells were determined in the function of shear stress (SS [Pa] at a range of 0.3–30 Pa) [49,50]. The individual EI-SS curves were compared using the EI values at 3 Pa, the maximal elongation index (EI_{max}) and the shear stress belonging to the half of it ($SS_{1/2}$, [Pa]) and their ratio ($EI_{max}/SS_{1/2}$) and calculated with the help of the Lineweaver–Burk equation [51].

RBC osmotic gradient deformability (osmoscan) parameters were determined using 250 μL of a blood sample that was suspended to 5 mL of a PVP–PBS solution. In this measurement, the SS was constant (30 Pa), while the osmolality (O [mOsm/kg]) of the suspending medium changed gradually from 0 to 500 mOsm/kg. The descriptive parameters of the EI-O curves are the followings: the minimal elongation index values measured in a low osmotic environment (EI min), the maximal elongation index values (EI max, not equal to EI_{max} above), the half EI max at high osmolality range (EI hyper), the belonging osmolality vales (O min and O (EI max) and O hyper, [mOsm/kg]), and the area calculated from the individual elongation index-osmolality curves [49,50,52].

RBC aggregation was tested using two devices operating with different methods. A Myrenne MA-1 erythrocyte aggregometer (Myrenne GmbH, Germany) was used to test erythrocyte aggregation by the light transmission method. After disaggregation (shear rate: 600 s^{-1}) of a 20 μL blood sample, the aggregation index values were determined at the 5th and the 10th second of the aggregation process in M mode (shear rate: 0 s^{-1}) and in M1 mode (shear rate: 3 s^{-1}) [49,50]. Accordingly, the parameters were M 5 s, M 10 s, M1 5 s, and M1 10 s index values. In the LoRRca MaxSis Osmoscan ektacytometer (Mechatronics BV, The Netherlands), based on the laser backscattering method, further parameters could be determined. In the Couette system, the blood sample (1 mL) was disaggregated by rotation, and as the rotor stopped the aggregation process started, while a syllectogarm of the reflected laser beam (intensity) was analyzed [49,50]. The obtained parameters were: amplitude (Amp [au]), aggregation index (AI [%]), and half-amplitude time ($t_{1/2}$ [s]).

4.4. Statistical Analysis

Data are expressed as means ± standard deviation (S.D.). A SigmaStat for Windows (Systat Software Inc., San Jose, USA) software was used for statistical analysis. The D'Agostino–Pearson normality test was used to determine Gaussian distribution. Intra- and inter-group differences were analyzed by ANOVA test followed by post hoc the Bonferroni test or Dunn's method, depending on the result of normality test. Before/after relations were also analyzed by paired t-test or the Wilcoxon Signed-Rank test, depending on the normality of data distribution. Probability values (p) less than 0.05 were considered as statistically significantly different.

5. Conclusions

Overall, exercise (physical exercise by spiroergometry) in the untrained group showed more significant metabolic and hemorheological changes. Macro and microrheological differences were found not only in the comparison of the untrained group and the athletes but also in the two sports (ice hockey vs. soccer: shorter and high intensity vs. longer and moderate intensity training and games). Regular, professional sport activity may result in a beneficial hemorheological status that improves tissue perfusion, together leading to a better performance (Figure 3). An examination of the hemorheological parameters can be a useful adjunct in assessing the health status of athletes. The results of our tests may help to monitor individual condition during their team's regular semiannual diagnostic surveys. In addition, it may provide an opportunity for the coaching staff to modify the training program and training methods, if necessary, taking into account any deviations that can be corrected. However, based on these promising results, a more detailed and more sophisticated metabolomics study will be important.

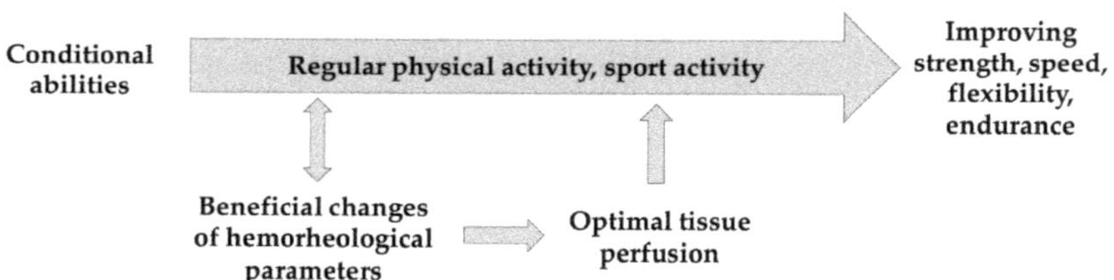

Figure 3. Supposed role in beneficial hemorheological changes supporting regular physical/sport activity leading to better performance.

Author Contributions: Conceptualization, S.S. and N.N.; physical examinations and spiroergometry test, T.M. and Z.G.; sample preparation and laboratory investigations, V.S., B.B., A.V., A.A.M., and L.B.B.; data analysis, S.S., T.M., A.V., A.A.M., and N.N.; writing—original draft preparation, S.S., T.M., Z.G., and N.N.; writing—review and editing, S.S. and N.N.; supervision, S.S.; funding acquisition, S.S. and N.N. All authors have read and agreed to the published version of the manuscript.

Funding: The work was supported by the Bridging Fund of the Faculty of Medicine, University of Debrecen and the National Research, Development and Innovation Office (NKFI-1 "OTKA" K-139184).

Institutional Review Board Statement: The study was conducted according to the guidelines of the Declaration of Helsinki and approved by the Regional and Institutional Ethics Committee. University of Debrecen, Clinical Center (permission nr.: DE RKEB/IKEB:5410-2020, 13 January 2020).

Informed Consent Statement: Informed consent was obtained from all subjects involved in the study. The medical students were not in direct teacher–student relation with any of the authors at the time of the study.

Data Availability Statement: Because of the participant consent obtained as part of the recruitment process, it is not possible to make these data publicly available. The data presented in this study are available on request from the corresponding author.

Acknowledgments: Authors are grateful to the staff of the Department of Sports Medicine and the Department of Operative Techniques and Surgical Research, Faculty of Medicine, University of Debrecen.

Conflicts of Interest: The authors declare no conflict of interest.

References

1. Baskurt, O.K.; Meiselman, H.J. Blood rheology and hemodynamics. *Semin. Thromb. Hemost.* **2003**, *29*, 435–450.
2. Cokelet, G.R.; Meiselman, H.J. Basic aspects of hemorheology. In *Handbook of Hemorheology and Hemodynamics*; Baskurt, O.K., Hardeman, M.R., Rampling, M.W., Meiselman, H.J., Eds.; IOS Press: Amsterdam, The Netherlands, 2007; pp. 21–33.
3. Cokelet, G.R.; Meiselman, H.J. Macro- and micro-rheological properties of blood. In *Handbook of Hemorheology and Hemodynamics*; Baskurt, O.K., Hardeman, M.R., Rampling, M.W., Meiselman, H.J., Eds.; IOS Press: Amsterdam, The Netherlands, 2007; pp. 45–71.
4. Baskurt, O.K.; Neu, B.; Meiselman, H.J. Determinants of red blood cell aggregation. In *Red Blood Cell Aggregation*; Baskurt, O.K., Neu, B., Meiselman, H.J., Eds.; CRC Press: Boca Raton, FL, USA, 2012; pp. 9–29.
5. Baskurt, O.K. Mechanisms of blood rheology alterations. In *Handbook of Hemorheology and Hemodynamics*; Baskurt, O.K., Hardeman, M.R., Rampling, M.W., Meiselman, H.J., Eds.; IOS Press: Amsterdam, The Netherlands, 2007; pp. 170–190.
6. Nemeth, N.; Peto, K.; Magyar, Z.; Klarik, Z.; Varga, G.; Oltean, M.; Mantas, A.; Czigany, Z.; Tolba, R.H. Hemorheological and microcirculatory factors in liver ischemia-reperfusion injury—An update on pathophysiology, molecular mechanisms and protective strategies. *Int J. Mol. Sci.* **2021**, *22*, 1864. [CrossRef] [PubMed]
7. Krüger-Genge, A.; Sternitzky, R.; Pindur, G.; Rampling, M.; Franke, R.P.; Jung, F. Erythrocyte aggregation in relation to plasma proteins and lipids. *J. Cell. Biotech.* **2019**, *5*, 65–70. [CrossRef]
8. Weed, R.I.; La Celle, P.L.; Merrill, E.W. Metabolic dependence of red blood cell deformability. *J. Clin. Investig.* **1969**, *48*, 795–809. [CrossRef]
9. Brun, J. Hormones, metabolism and body composition as major determinants of blood rheology: Potential pathophysiological meaning. *Clin. Hemorheol. Microcirc.* **2002**, *26*, 63–79.
10. Reinhart, W.H.; Gaudenz, R.; Walter, R. Acidosis induced by lactate, pyruvate, or HCl increases blood viscosity. *Crit. Care* **2002**, *17*, 38–42. [CrossRef] [PubMed]
11. Cicha, I.; Suzuki, Y.; Tateishi, N.; Maeda, N. Changes of RBC aggregation in oxygenation-deoxygenation: pH dependency and cell morphology. *Am. J. Physiol. Heart Circ. Physiol.* **2003**, *284*, H2335–H2342. [CrossRef] [PubMed]
12. Uyuklu, M.; Meiselman, H.J.; Baskurt, O.K. Effect of hemoglobin oxygenation level on red blood cell deformability and aggregation parameters. *Clin. Hemorheol. Microcirc.* **2009**, *41*, 179–188. [CrossRef] [PubMed]
13. Lipowsky, H.H. Microvascular rheology and hemodynamics. *Microcirculation* **2005**, *12*, 5–15. [CrossRef]
14. Adachi, H.; Sakurai, S.; Tanehata, M.; Oshima, S.; Taniguchi, K. Effect of long-term exercise training on blood viscosity during endurance exercise at an anaerobic threshold intensity. *Jpn. Circ. J.* **2000**, *64*, 848–850. [CrossRef]
15. Brun, J.F.; Varlet-Marie, E.; Connes, P.; Aloulou, I. Hemorheological alterations related to training and overtraining. *Biorheology* **2010**, *47*, 95–115. [CrossRef] [PubMed]
16. Brun, J.F.; Varlet-Marie, E.; Romain, A.J.; Guiraudou, M.; Raynaud de Mauverger, E. Exercise hemorheology: Moving from old simplistic paradigms to a more complex picture. *Clin. Hemorheol. Microcirc.* **2013**, *55*, 15–27. [CrossRef] [PubMed]
17. Connes, P.; Simmonds, M.; Brun, J.F.; Baskurt, O.K. Exercise hemorheology: Classical data, recent findings and unresolved issues. *Clin. Hemorheol. Microcirc.* **2013**, *53*, 187–199. [CrossRef]
18. Romain, A.J.; Brun, J.F.; Varlet-Marie, E.; Raynaud de Mauverger, E. Effects of exercise training on blood rheology: A meta-analysis. *Clin. Hemorheol. Microcirc.* **2011**, *49*, 199–205. [CrossRef]
19. Ernst, E. Influence of regular physical activity on blood rheology. *Eur. Heart J.* **1987**, *8*, 59–62. [CrossRef] [PubMed]
20. Varlet-Marie, E.; Maso, F.; Lac, G.; Brun, J.F. Hemorheological disturbances in the overtraining syndrome. *Clin. Hemorheol. Microcirc.* **2004**, *30*, 211–218.
21. El-Sayed, M.S.; Ali, N.; El-Sayed Ali, Z. Haemorheology in exercise and training. *Sports Med.* **2005**, *35*, 649–670. [CrossRef] [PubMed]
22. Sandor, B.; Nagy, A.; Toth, A.; Rabai, M.; Mezey, B.; Csatho, A.; Czuriga, I.; Toth, K.; Szabados, E. Effects of moderate aerobic exercise training on hemorheological and laboratory parameters in ischemic heart disease patients. *PLoS ONE* **2014**, *9*, e110751.
23. Nader, E.; Skinner, S.; Romana, M.; Fort, R.; Lemonne, N.; Guillot, N.; Gauthier, A.; Antoine-Jonville, S.; Renoux, C.; Hardy-Dessources, M.D.; et al. Blood rheology: Key parameters, impact on blood flow, role in sickle cell disease and effects of exercise. *Front. Physiol.* **2019**, *10*, 1329. [CrossRef] [PubMed]
24. Broich, H.; Sperlich, B.; Buitrago, S.; Mathes, S.; Mester, J. Performance assessment in elite football players: Field level test versus spiroergometry. *J. Hum. Sport Exerc.* **2012**, *7*, 287–295. [CrossRef]
25. Löllgen, H.; Leyk, D. Exercise testing in sports medicine. *Dtsch. Arztebl. Int.* **2018**, *115*, 409–416. [CrossRef]
26. Johnson, R.M. pH effects on red blood cell deformability. *Blood Cells* **1985**, *11*, 317–321. [PubMed]

27. Senturk, U.K.; Gunduz, F.; Kuru, O.; Kocer, G.; Ozkaya, Y.G.; Yesilkaya, A.; Bor-Kucukatay, M.; Uyuklu, M.; Yalcin, O.; Baskurt, O.K. Exercise-Induced oxidative stress leads hemolysis in sedentary but not trained humans. *J. Appl. Physiol.* **2005**, *99*, 1434–1441. [CrossRef] [PubMed]
28. Reinhart, W.H. The optimum hematocrit. *Clin. Hemorheol. Microcirc.* **2016**, *64*, 575–585. [CrossRef] [PubMed]
29. Krüger-Genge, A.; Blocki, A.; Franke, R.P.; Jung, F. Vascular endothelial cell biology: An update. *Int. J. Mol. Sci.* **2019**, *20*, 4411. [CrossRef] [PubMed]
30. Smith, J.A.; Telford, R.D.; Kolbuch-Braddon, M.; Weidemann, M.J. Lactate/H+ uptake by red blood cells during exercise alters their physical properties. *Eur. J. Appl. Physiol. Occup. Physiol.* **1997**, *75*, 54–61. [CrossRef] [PubMed]
31. Skelton, M.S.; Kremer, D.E.; Smith, E.W.; Gladden, L.B. Lactate influx into red blood cells from trained and untrained human subject. *Med. Sci. Sports Exerc.* **1998**, *30*, 536–542. [CrossRef]
32. Connes, P.; Bouix, D.; Py, G.; Prefaut, C.; Mercier, J.; Brun, J.F.; Caillaud, C. Opposite effects of In Vitro lactate on erythrocyte deformability in athletes and untrained subjects. *Clin. Hemorheol. Microcirc.* **2004**, *31*, 311–318.
33. Messonnier, L.; Freund, H.; Denis, C.; Féasson, L.; Lacour, J.R. Effects of training on lactate kinetics parameters and their influence on short high-intensity exercise performance. *Int. J. Sports Med.* **2006**, *27*, 60–66. [CrossRef]
34. Connes, P.; Caillaud, C.; Py, G.; Mercier, J.; Hue, O.; Brun, J.F. Maximal exercise and lactate do not change red blood cell aggregation in well trained athletes. *Clin. Hemorheol. Microcirc.* **2007**, *36*, 319–326. [PubMed]
35. Connes, P.; Tripette, J.; Mukisi-Mukaza, M.; Baskurt, O.K.; Toth, K.; Meiselman, H.J.; Hue, O.; Antoine-Jonville, S. Relationships between hemodynamic, hemorheological and metabolic responses during exercise. *Biorheology* **2009**, *46*, 133–143. [CrossRef] [PubMed]
36. Nemkov, T.; Skinner, S.C.; Nader, E.; Stefanoni, D.; Robert, M.; Cendali, F.; Stauffer, E.; Cibiel, A.; Boisson, C.; Connes, P.; et al. Acute cycling exercise induces changes in red blood cell deformability and membrane lipid remodeling. *Int. J. Mol. Sci.* **2021**, *22*, 896. [CrossRef] [PubMed]
37. Quittmann, O.J.; Abel, T.; Zeller, S.; Foitschik, T.; Strüder, H.K. Lactate kinetics in handcycling under various exercise modalities and their relationship to performance measures in able-bodied participants. *Eur. J. Appl. Physiol.* **2018**, *118*, 1493–1505. [CrossRef] [PubMed]
38. Quittmann, O.J.; Abel, T.; Vafa, R.; Mester, J.; Schwarz, Y.M.; Strüder, H.K. Maximal lactate accumulation rate and post-exercise lactate kinetics in handcycling and cycling. *Eur. J. Sport Sci.* **2021**, *21*, 539–551. [CrossRef] [PubMed]
39. Mitchell, E.A.; Martin, N.R.W.; Bailey, S.J.; Ferguson, R.A. Critical power is positively related to skeletal muscle capillarity and type I muscle fibers in endurance-trained individuals. *J. Appl. Physiol.* **2018**, *125*, 737–745. [CrossRef]
40. Hendrickse, P.; Degens, H. The role of the microcirculation in muscle function and plasticity. *J. Muscle Res. Cell. Motil.* **2019**, *40*, 127–140. [CrossRef] [PubMed]
41. Tesch, P.A.; Thorsson, A.; Kaiser, P. Muscle capillary supply and fiber type characteristics in weight and power lifters. *J. Appl. Physiol. Respir. Environ. Exerc. Physiol.* **1984**, *56*, 35–38. [CrossRef]
42. Alis, R.; Ibañez-Sania, S.; Basterra, J.; Sanchis-Gomar, F.; Romagnoli, M. Effects of an acute high-intensity interval training protocol on plasma viscosity. *J. Sports Med. Phys. Fitness.* **2015**, *55*, 647–653. [PubMed]
43. Kilic-Toprak, E.; Yapici, A.; Kilic-Erkek, O.; Koklu, Y.; Tekin, V.; Alemdaroglu, U.; Bor-Kucukatay, M. Acute effects of Yo-Yo intermittent recovery test level 1 (Yo-YoIR1) on hemorheological parameters in female volleyball players. *Clin. Hemorheol. Microcirc.* **2015**, *60*, 191–199. [CrossRef] [PubMed]
44. Bizjak, D.A.; Tomschi, F.; Bales, G.; Nader, E.; Romana, M.; Connes, P.; Bloch, W.; Grau, M. Does endurance training improve red blood cell aging and hemorheology in moderate-trained healthy individuals? *J. Sport Health Sci.* **2020**, *9*, 595–603. [CrossRef]
45. Harper, D.J.; Carling, C.; Kiely, J. High-Intensity acceleration and deceleration demands in elite team sports competitive match play: A systematic review and meta-analysis of observational studies. *Sports Med.* **2019**, *49*, 1923–1947. [CrossRef] [PubMed]
46. Maeder, M.; Wolber, T.; Atefy, R.; Gadza, M.; Ammann, P.; Myers, J.; Rickli, H. A nomogram to select the optimal treadmill ramp protocol in subjects with high exercise capacity: Validation and comparison with the Bruce protocol. *J. Cardiopulm. Rehabil.* **2006**, *26*, 16–23. [CrossRef]
47. Arena, R.; Myers, J.; Williams, M.A.; Gulati, M.; Kligfield, P.; Balady, G.J.; Collins, E.; Fletcher, G.; American Heart Association Committee on Exercise, Rehabilitation, and Prevention of the Council on Clinical Cardiology; American Heart Association Council on Cardiovascular Nursing. Assessment of functional capacity in clinical and research settings: A scientific statement from the American Heart Association Committee on Exercise, Rehabilitation, and Prevention of the Council on Clinical Cardiology and the Council on Cardiovascular Nursing. *Circulation* **2007**, *116*, 329–343. [PubMed]
48. Matrai, A.; Whittington, R.B.; Ernst, E. A simple method of estimating whole blood viscosity at standardized hematocrit. *Clin. Hemorheol.* **1987**, *7*, 261–265. [CrossRef]
49. Hardeman, M.; Goedhart, P.; Shin, S. Methods in hemorheology. In *Handbook of Hemorheology and Hemodynamics*; Baskurt, O.K., Hardeman, M.R., Rampling, M.W., Meiselman, H.J., Eds.; IOS Press: Amsterdam, The Netherlands, 2007; pp. 242–266.
50. Baskurt, O.K.; Boynard, M.; Cokelet, G.C.; Connes, P.; Cooke, B.M.; Forconi, S.; Liao, F.; Hardeman, M.R.; Jung, F.; Meiselman, H.J.; et al. New guidelines for hemorheological laboratory techniques. *Clin. Hemorheol. Microcirc.* **2009**, *42*, 75–97. [CrossRef] [PubMed]

51. Baskurt, O.K.; Hardeman, M.R.; Uyuklu, M.; Ulker, P.; Cengiz, M.; Nemeth, N.; Shin, S.; Alexy, T.; Meiselman, H.J. Parameterization of red blood cell elongation index–shear stress curves obtained by ektacytometry. *Scand. J. Clin. Lab. Investig.* **2009**, *69*, 777–788. [CrossRef] [PubMed]
52. Nemeth, N.; Kiss, F.; Miszti-Blasius, K. Interpretation of osmotic gradient ektacytometry (osmoscan) data: A comparative study for methodological standards. *Scand. J. Clin. Lab. Investig.* **2015**, *75*, 213–222. [CrossRef]

MDPI

Review

Effect of Exercise on Brain Health: The Potential Role of Lactate as a Myokine

Takeshi Hashimoto [1], Hayato Tsukamoto [1], Soichi Ando [2] and Shigehiko Ogoh [3,*]

1 Faculty of Sport and Health Science, Ritsumeikan University, Osaka 525-8577, Japan; thashimo@fc.ritsumei.ac.jp (T.H.); h-tsuka@fc.ritsumei.ac.jp (H.T.)
2 Graduate School of Informatics and Engineering, The University of Electro-Communications, Tokyo 182-8585, Japan; soichi.ando@uec.ac.jp
3 Department of Biomedical Engineering, Toyo University, Saitama 350-8585, Japan
* Correspondence: ogoh@toyo.jp

Abstract: It has been well established in epidemiological studies and randomized controlled trials that habitual exercise is beneficial for brain health, such as cognition and mental health. Generally, it may be reasonable to say that the physiological benefits of acute exercise can prevent brain disorders in late life if such exercise is habitually/chronically conducted. However, the mechanisms of improvement in brain function via chronic exercise remain incompletely understood because such mechanisms are assumed to be multifactorial, such as the adaptation of repeated acute exercise. This review postulates that cerebral metabolism may be an important physiological factor that determines brain function. Among metabolites, the provision of lactate to meet elevated neural activity and regulate the cerebrovascular system and redox states in response to exercise may be responsible for exercise-enhanced brain health. Here, we summarize the current knowledge regarding the influence of exercise on brain health, particularly cognitive performance, with the underlying mechanisms by means of lactate. Regarding the influence of chronic exercise on brain function, the relevance of exercise intensity and modality, particularly high-intensity interval exercise, is acknowledged to induce "metabolic myokine" (i.e., lactate) for brain health.

Keywords: executive function; mental health; brain-derived neurotrophic factor; insulin-like growth factor-1; vascular endothelial growth factor; neurogenesis; angiogenesis; cerebral blood flow; nicotinamide adenine dinucleotide hydrate

Citation: Hashimoto, T.; Tsukamoto, H.; Ando, S.; Ogoh, S. Effect of Exercise on Brain Health: The Potential Role of Lactate as a Myokine. *Metabolites* **2021**, *11*, 813. https://doi.org/10.3390/metabo11120813

Academic Editor: Norbert Nemeth

Received: 29 October 2021
Accepted: 27 November 2021
Published: 29 November 2021

1. Introduction

It has been well established that habitual exercise is beneficial for the cognition and brain health of most individuals, including older adults [1,2]. This view is not surprising because it is said that "exercise is the real polypill" based on organ-induced peripheral factors [3]. In general, it has been considered that the effects of habitual exercise on the human body are the result of repeated exercise and thus may be associated with cumulative acute responses to exercise.

Similarly, it may be reasonable to say that acute exercise favorable for improving brain function, although this is a transient response, is also beneficial for brain health with continuous repetition via chronic exercise training. However, the mechanisms of chronic exercise-improved brain function, especially how the effect of acute exercise on brain function determines that of chronic exercise, remain incompletely unknown. For instance, chronic exercise effects can be modified using the same acute exercise by changing exercise strength, duration, and frequency. Hence, the proper exercise prescription for chronic brain health may be difficult to build from results on the effect of acute exercise on brain function. Nonetheless, it is important to explore and organize the underlying mechanisms of acute exercise for brain health to provide insight into proper exercise prescriptions.

Among the acute responses to exercise, a growing body of evidence is accumulating to suggest that the myokine (i.e., muscle-induced peripheral factors) cathepsin B and irisin pass through the blood–brain barrier to enhance brain-derived neurotrophic factor (BDNF) production and hence improve neurogenesis, memory, and learning [4]. On the other hand, lactate, as an exercise-induced myokine favorable to the brain, was not investigated to identify the mechanism of exercise-induced improvement in brain function, although the production of lactate has been widely used as a biomarker to reflect exercise mode, strength, and duration [5–8].

In this minireview, we summarized the possibility of lactate as one of the underlying mechanisms linking brain health outcomes, particularly cognitive performance and mental health, to exercise regimens.

2. Exercise Intensity and Modality for Brain Health Regarding Chronic Exercise Adaptation (Implication of Lactate)

To promote and maintain health, the American College of Sports Medicine (ACSM) and American Heart Association (AHA) recommends that healthy adults aged 18–65 years perform sufficient volumes of exercise, such as moderate-intensity exercise for at least 30 min for 5 days/week or vigorous-intensity exercise for 20 min for 3 days/week [9]. Importantly, compared to habitual lower-intensity exercise, higher-intensity exercise can effectively improve cardiovascular and metabolic health [10–12]. In particular, long-term/chronic high-intensity interval exercise (HIIE) training (i.e., HIIT) is more effective than long-term/chronic moderate-intensity continuous exercise (MCE) because it increases exercise capacity in addition to cardiovascular and metabolic health in healthy individuals [13–15]. The effectiveness of HIIT over MCE training is also relevant for brain health. Recently, Mekari et al. demonstrated that HIIT was more effective for the improvement of executive function (EF) than MCE training in young adults [16]. A recent meta-analysis indicated that HIIT might be more effective for improving severe mental illness (e.g., cognition, negative and positive symptoms of schizophrenia, and depressive mood) than moderate-intensity exercise [17]. Given that HIIE produces more lactate than general exercise modalities, such as MCE, some beneficial effects of lactate on health, including brain health, can be implicated. For instance, based on the notion that acute exercise that is favorable for improving brain function is also beneficial for brain health with continuous repetition via chronic exercise training, our previous study demonstrated that HIIE could improve EF rather than MCE and was accompanied by more lactate production (Figure 1) [7], which may imply a potential benefit of lactate on increased cognitive performance by HIIE and subsequent HIIT.

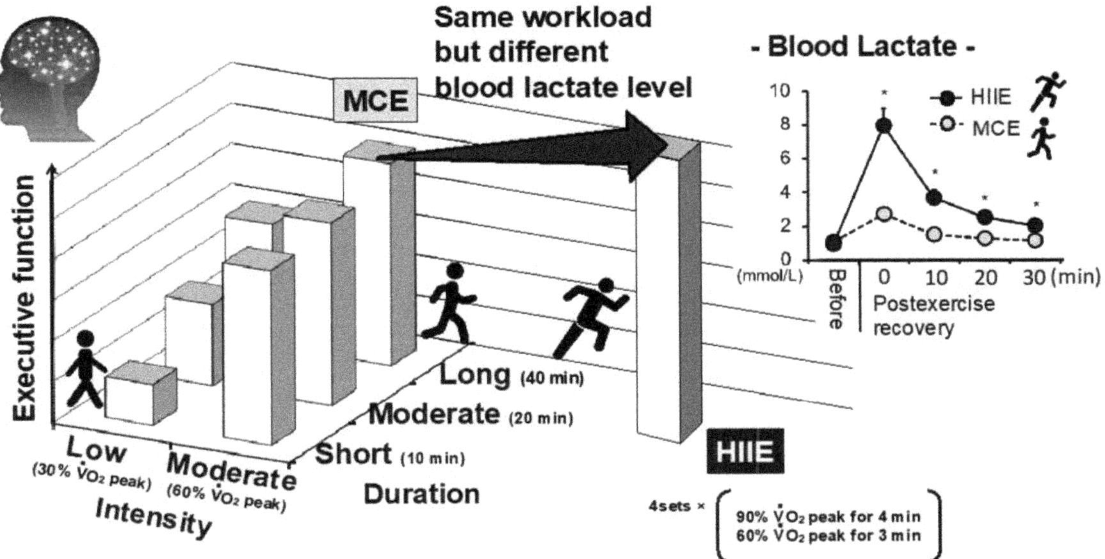

Figure 1. Impact of exercise intensity, duration, and modality on acute enhancement of executive function. The graph is illustrated by the authors based on previous studies [7,8,18]. HIIE could improve EF rather than volume-matched (i.e., same workload) MCE with more lactate production during postexercise recovery period [7]. * $p < 0.05$ vs. MCE.

3. Chronic Cognitive and Mental Alterations with Regular Exercise and Its Potential Link to Chronic Exercise-Induced Anatomical and Cerebral Microvasculature Alterations

The potential mechanisms of habitual exercise/physical activity-induced improvement as well as aging-induced impairments in cognitive performance and mental health remain unclear but are assumed to be associated with several physiological factors. For instance, the deleterious effects of aging on the brain comprise negative physiological and anatomical alterations, e.g., hemodynamic activity, synaptic plasticity, decreased brain volume and neurogenesis, while physical activity prevents the deleterious effects on the brain and, in contrast, induces brain neural alterations, including the formation of new neurons, the proliferation of neural cells, and integrated functional neural networks [19,20]. In particular, structural alterations, such as increased neurogenesis, synaptogenesis, angiogenesis, and brain volume, seem to be characteristics of the beneficial effects of chronic exercise on cognitive performance and mental health [2].

Regular aerobic exercise can increase or preserve the regional brain volume in areas associated with cognitive decline and portions of mental health [8,21,22]. It has been reported that aerobic exercise (i.e., 6 to 12 months of a walking program) increases spatial memory as well as gray and white matter volumes in both temporal (including the hippocampi) and prefrontal regions in healthy older adults (without dementia) [23]. In addition, Jonasson et al. demonstrated that following a 6-month exercise training period, the change in "cognitive score" determined by episodic memory, updating, processing speed, and EF was positively related to the thickness of the dorsolateral prefrontal cortex [24]. Regarding mental health, patients with major depressive disorder or schizophrenia show decreased hippocampal or gray matter volume [25,26], while an exercise-induced increase in hippocampal volume can be related to cognitive performance even in patients with schizophrenia [21]. However, whether brain structure is associated with psychiatric and neurological disorders is controversial [27], and whether the positive effects of aerobic exercise can be extended to psychiatric disorders is still unclear [8]. Further studies are needed to uncover the pathophysiology of mental disorders and improve the effect of exercise or physical activity.

In addition to brain structural/anatomical alterations, changes in cerebral microvasculature function can be a physiological factor that may elicit exercise-enhanced brain function. Since the energy reserve of the brain is relatively small, a continuous supply of glucose and oxygen from the cerebral circulation to the brain is required to maintain its function, e.g., cognitive performance. Thus, especially in the brain, synaptic activity suddenly increases the demand for energy for maintaining brain function and consequently might cause a relative lack of oxygen and glucose. However, in the brain, the neural activity causes neurovascular coupling with accordingly transient and adequate increases in regional cerebral blood flow (CBF) and consequently partially maintains brain function [28]. Indeed, the onset of cognitive impairment often occurs following cerebrovascular dysfunction, suggesting that dysfunction of CBF regulation is one of the mechanisms of the onset of dementia [29]. Furthermore, a decrease in the response of regional CBF to a simple motor task occurs when either intracranial carotid arteries or one vertebral artery is occluded in asymptomatic patients [30]. In addition, neural coupling to several physiological stimuli and resting CBF are reduced in patients with Alzheimer's disease [31–35]. These findings indicate that brain function via neurovascular coupling is attenuated by inadequate global or focal CBF regulation; thus, the regulation of global CBF is important to maintain adequate neural coupling [28] and thus brain function.

4. Can Acute Alterations in CBF to Exercise Affect Cognitive Performance?

As mentioned above, it is expected that maintaining brain function requires adequate CBF regulation as an important physiological factor. However, no study has examined whether alterations in CBF directly modify cognitive performance because CBF cannot be isolated from the many physiological factors that affect cognitive performance in patients with cerebral disease, vascular disease, or dementia, as well as in healthy older adults.

Basically, augmented cerebral metabolism or cerebral neural activity [36–38] are accompanied by transient increases in CBF [39–41] as well as cognitive performance [42,43] during and/or following mild- to moderate-intensity aerobic exercise. In contrast, similar to the decrease in CBF associated with hyperventilation during prolonged or heavy aerobic exercise [40], the exercise-induced facilitation of cognitive performance disappears during such prolonged exercise [44]. From this background, we previously examined for the first time whether manipulation of CBF alteration affects cognitive performance in young, healthy participants [45]. In contrast to our hypothesis, however, cognitive performance improved in response to the decrease in CBF during prolonged heavy exercise, and unexpectedly, an isolated change (i.e., hypercapnia-induced increase) in CBF did not affect cognitive performance at rest or during exercise [45]. Furthermore, several studies reported that increases in CBF during exercise were not directly related to changes in cognitive performance [46,47]. These findings suggest that acute exercise-induced cognitive improvement may not have the same narrative as that of chronic exercise in terms of the cerebrovascular system; thus, it is not simply due to an increase in global CBF, implying that another factor modified by exercise, rather than a change in CBF, affects cognitive performance.

5. Cerebral Lactate Metabolism and Cognitive Performance

A decrease in cerebral oxygenation is induced by prolonged exercise [45,48] or exercise under mild or severe hypoxia [49,50], while impaired cognitive performance is not evident in healthy young participants, suggesting a dissociation between an alteration in CBF and subsequent change in oxygen delivery to the brain and cerebral metabolism or cognitive performance. Indeed, albeit with a reduction in CBF during heavy exercise, the elevation of brain neural activity and metabolism might be accompanied by compensatory increases in the uptake of lactate, glucose, and oxygen support for the brain (arterial-jugular venous difference) [36]. Given that augmented brain neural activity and metabolism are independent of increases in CBF [51], extensive activation of motor and sensory systems due to the higher-order function of the prefrontal cortex may affect cognitive performance rather than cerebral perfusion in response to exercise.

Regarding metabolism, although the brain relies mainly on glucose at rest, during high-intensity exercise, the brain becomes dependent on lactate delivery [52,53] and repeated HIIE, which attenuates the increase in systemic blood lactate, resulting in impaired maintenance of HIIE-enhanced cognitive performance (i.e., EF) [18]. In particular, HIIE may facilitate neuronal activation and excitation levels to the extent that summation is facilitated to improve cognitive performance [7,54,55]. Neuronal activation is associated with an increase in energy requirements due to the transport of neurotransmitters and ions [56], and neurons preferentially utilize lactate as a fuel in vivo [57]. Sustained elevation of arterial/systemic lactate in response to intense exercise promotes the supply of lactate as an energy substrate to meet acute neuronal energy requirements [58–60]. In addition, intravenous infusion of 100 mM L-lactate into rats promoted cognitive recovery by preserving cerebral ATP generation following traumatic brain injury [61]. Furthermore, Skriver et al. found a correlation between systemic lactate concentration and the acquisition and retention of motor skills [62]. In addition, lactate supports synaptic activity [63], long-term potentiation and memory formation [64], and neuronal plasticity [65]. These findings suggest that brain function as expressed by cognitive performance depends on the provision of lactate. Indeed, we manipulated blood lactate during exercise at a given intensity by repeated HIIE and evaluated whether such manipulation of peripheral lactate metabolism affects brain lactate uptake (i.e., the arterial–jugular venous difference in lactate (a-v $diff_{lactate}$)) and EF [66]. We found that brain lactate uptake is associated with the arterial lactate concentration, and inadequate lactate provision to the brain might attenuate exercise (i.e., HIIE)-enhanced EF [66], irrespective of increased BDNF and catecholamine, both of which are supposed to relate to cognitive performance [55,67,68] (Figure 2). Given the reliance on lactate as a fuel for the brain, variations in blood lactate could affect cognitive performance during and after exercise and account for the significance of exercise (i.e., muscle contraction) for brain function.

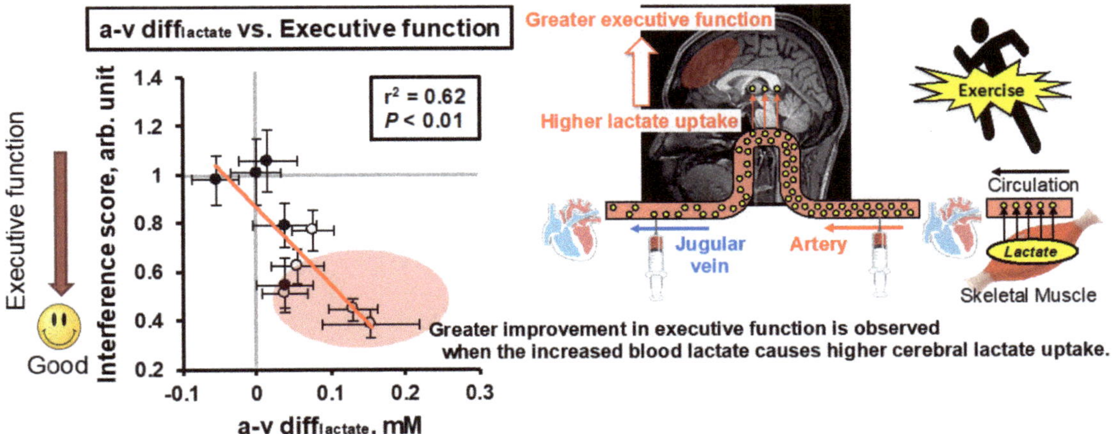

Figure 2. Relationship between a-v $diff_{lactate}$ and Δinterference score (i.e., executive function) during postexercise recovery. The open circles indicate the average of each time point during the post-first bout of HIIE recovery, and the solid circles indicate the average of each time point during the post-second bout of HIIE recovery in which a lower systemic lactate concentration is observed. This result suggests that brain lactate uptake is associated with better executive function. Values are the means ± SEM. Modified/adopted from Hashimoto et al. [66].

On the other hand, a recent study demonstrated that chronic lactate administration to mice promotes hippocampal neurogenesis but does not affect cognitive performance [69]. In addition, Sudo et al. found that recovery of prefrontal oxygenation affected cognitive performance after exhaustive exercise, irrespective of the blood lactate concentration [70]. Further studies are warranted to understand the role of lactate in brain function in acute and chronic exercise.

6. Exercise-Induced Improvement in Brain Health Based on Chronic Anatomical and Cerebral Microvasculature Alterations and Its Potential Link to Exercise-Produced Lactate in Active Muscle

As described above, brain structure may determine CBF regulation and volume that affects brain function. Of note, physical activity is useful to upregulate neurotrophins and growth factors, such as BDNF, insulin-like growth factor-1 (IGF-1), and vascular endothelial growth factor (VEGF), which are necessary to maintain existing neurons and neurogenesis for continued brain development [20]. The increases in BDNF, VEGF, and IGF-1 levels are positively related to augmented hippocampal volume, neurogenesis, and angiogenesis, thereby increasing cognitive performance, such as spatial memory, in older adults [8,20,23,71].

Among the growth factors, BDNF might be a key factor involved in cognitive performance improvement, at least regarding memory function and mental health, by means of promoting neurogenesis, synaptic plasticity, and cell survival, particularly in the cerebral cortex and hippocampus [8,72]. Indeed, poor cognitive function and mental health are associated with low circulating BDNF levels in both young and elderly persons and patients with a major depressive disorder [68,73,74]. On the other hand, Griffin et al. (2011) suggested that postexercise improvement in short-term memory performance was related to an acute increase in BDNF [68]. Additionally, to maintain a higher level of short-term memory for brain health, it is important that the acute increase in systemic BDNF is repeated [68]. In this connection, the effect of chronic exercise on cognitive function may be determined by repeated single exercise bout-induced physiological effects, as seen in muscle hypertrophy by resistance exercise training, and changes in some physiological and biological factors (e.g., BDNF) during single bouts of exercise may partially link such determination.

In line with this, the indirect effects of lactate should be a focus. Again, general structural alterations of the brain via chronic (i.e., repeated/habitual) exercise training/physical activity may be responsible for brain health, at least partly by growth factors. Interestingly, lactate infusion at rest induced an increase in blood BDNF in young male sports students [75]. Additionally, an increase in blood lactate concentration in response to acute graded exercise was correlated with an increase in serum BDNF in young, healthy subjects [76]. In this regard, it is not surprising that HIIE, which produces more lactate than MCE, increased serum BDNF more than MCE in young obese subjects [77]. Furthermore, acute sprint interval exercise-induced elevation in blood lactate concentration was associated with increased blood BDNF, IGF-1, and VEGF and improved cognitive performance in young subjects [78]. In addition, Hayek et al. suggested that exercise-produced lactate is transported through the circulation to the brain, whereby it induces BDNF expression via a signaling cascade between silent information regulator 1 (SIRT1), peroxisome proliferator-activated receptor gamma coactivator 1-alpha (PGC-1α), and fibronectin type III domain containing 5 (FNDC5) in the mouse hippocampus [79]. Importantly, the study also showed that such peripheral delivery of exercise-produced lactate promotes cognitive performance, such as learning and memory. These results suggest that either exercise-induced or exogenously administered lactate can be a trigger to augment BDNF expression (see [80]) and subsequent structural adaptations and hence may contribute to the improvement of cognitive performance.

Regarding VEGF, Morland et al. demonstrated that HIIE training and/or sodium lactate injections for 7 weeks promoted cerebral VEGF and angiogenesis via the lactate receptor hydroxycarboxylic acid receptor 1 (HCAR1) in an animal model [81]. These findings suggest that exercise-induced elevation of blood lactate can be an activator of neurogenesis and angiogenesis, which are favorable for brain health and should be considered an underlying molecular mechanism of HIIT benefits for the brain [82].

Interestingly, previous studies demonstrated that peripheral administration of lactate reduced behavioral despair and anhedonia-like behavior and reversed social avoidance [83,84]. It was suggested that the lactate-induced expression of genes/proteins related to neuronal plasticity, memory, neurogenesis, and neuroprotection, such as BDNF, VEGF, early growth response 1 (Egr1), CCAAT/enhancer-binding protein (C/EBP), Hes5, p11, and proto-oncogene c-Fos (c-Fos), as well as activity-regulated cytoskeletal-associated protein (Arc), might be associated with the antidepressant actions of lactate [65,83,85,86]. Recently, Carrard et al. suggested that hippocampal neurogenesis is important in the antidepressant actions of lactate [83]. In this study, chronic administration of corticosterone induced depression-like states with decreased hippocampal neurogenesis, while coadministration of lactate maintained hippocampal neurogenesis to the control level with suppression of oxidative stress. Importantly, this action was not induced by the administration of pyruvate but was elicited by β-hydroxybutyrate, which can be oxidized to acetoacetate with the production of nicotinamide adenine dinucleotide hydrate (NADH), suggesting that the antidepressant effect of lactate is associated with lactate oxidation-induced NADH rather than an energy substrate [83]. Indeed, NADH suppressed corticosterone-induced oxidative stress and a subsequent reduction in adult hippocampal stem/progenitor cell proliferation in an in vitro study [83]. Although physical activity/exercise-induced physiological strain that elicits brain functional adaptation may be multifactorial [82], we should recognize that muscle contraction-produced lactate might be a pivotal mediator of brain adaptation as a myokine for brain structure (Figure 3).

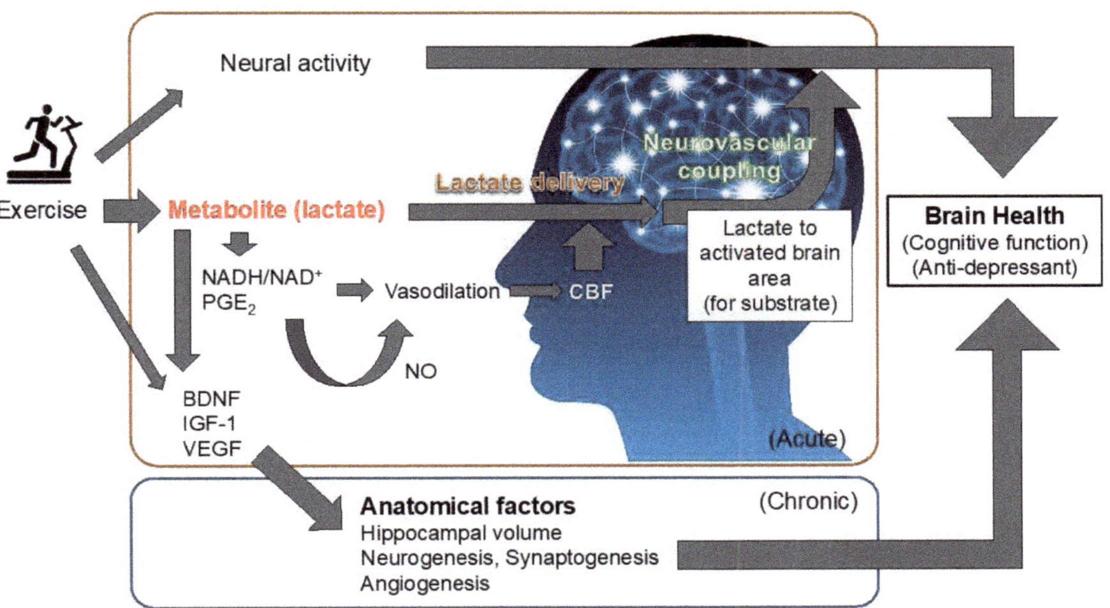

Figure 3. Potential acute and chronic effects of exercise-induced lactate on brain health. Scheme illustrating the potential acute and chronic effects of exercise-induced lactate on brain health.

7. Can Cerebral Blood Flow Regulation That Determines Brain Function Be Modified by Lactate?

Biochemical regulation of the cerebrovascular system by lactate is also evident in an acute setting. Gordon et al. demonstrated in rat brain slices that low oxygen levels facilitated lactate; hence, prostaglandin E_2 (PGE_2) elicited vasodilation [87]. In humans, the CBF response to physiological activation induced by visual stimulation was increased with lactate injection and plasma lactate/pyruvate ratio and subsequently augmented the NADH/NAD^+ ratio [88]. This increase in lactate/pyruvate and NADH/NAD^+ ratios may be related to the increase in CBF, probably through nitric oxide (NO) production [89]. In a clinical setting, hypertonic lactate injection increased cerebral perfusion and brain glucose availability and decreased the pulsatility index after acute brain injury [90]. In addition, the brain-injured person is hypermetabolic, and lactate has a pivotal role in supplying energy to bypass the restriction in glycolytic flux and spare limited glucose reserves for other cerebral metabolisms (e.g., pentose phosphate pathway for neuroprotection) (see [91]). Indeed, acute lactate infusion into mild traumatic brain injury patients improved their cognitive function as evaluated by the Mini Mental State Examination (MMSE), with several possible mechanisms, such as the energy substrate effect, the prevention of hyperchloremia, and the reduction in brain cell edema, by restoring impaired brain homeostasis and synapse function after brain injury [92].

8. Therapeutic Example of Exercise Modification to Consider the Interaction of Lactate

Given that resistance exercise is associated with several health benefits, such as a reduced risk for sarcopenia, osteoporosis, and metabolic dysfunction [93], this type of exercise is also attractive for improving quality of life. We found that an acute bout of localized resistance exercise could enhance cognitive performance immediately after exercise in a dose-dependent manner [94], whereby generally, high-intensity resistance exercise produces more lactate. Recently, we also found that resistance exercise with slow movement and tonic force generation improved EF more effectively than normal velocity movement exercise, accompanied by a considerable amount of lactate production even though the exercise intensity was low [95]. Interestingly, despite the application of a lower exercise load, resistance exercise with slow movement and tonic force generation improved postexercise EF similarly to high-intensity resistance exercise, which may be due to the equivalent blood lactate response between the two protocols in healthy young adults [96]. Therefore, it may be relevant to focus on exercise-induced lactate to predict the proper chronic exercise prescription for brain health.

9. Summary and Future Perspective

The potential mechanisms underlying the favorable effects of habitual exercise/physical activity on brain function are assumed to be multidimensional. In particular, structural alterations of the brain, such as increased neurogenesis, synaptogenesis, angiogenesis, and brain volume, might be characteristics of chronic exercise benefits because they cannot be achieved with only a single bout of acute exercise, although the cumulative effects of acute exercise-induced physiological stress are needed. It may be reasonable to say that acute exercise, if it is favorable for improvement of brain function, although it is a transient response, is also beneficial for brain health, including cognitive performance, with its continuous repetition via chronic exercise training. In this regard, it may be useful to understand the impact and mechanisms behind the favorable effects of acute exercise on brain function to develop a proper exercise prescription for brain health. Although such mechanisms are assumed to be multifactorial, cerebral metabolism may be an important physiological factor that determines brain function. Among metabolites, the provision of lactate to meet elevated neural activity and to regulate the cerebrovascular system and redox states in response to exercise may be responsible for exercise-enhanced brain health (Figure 3).

For this connection, the regulation of peripheral and cerebral lactate metabolism through exercise may be important for brain function. Furthermore, exercise intensity, duration, and modality also affect brain function possibly through the "metabolic myokine" (i.e., lactate). Particularly, HIIE might be practically relevant for brain health. Nonetheless, population (i.e., young and old) and gender (i.e., male and female) differences must be considered in future studies.

Funding: This research was funded by Grant-in-Aid for Scientific Research from the Japanese Ministry of Education, Culture, Sports, Science, and Technology (to T.H.), grant numbers #21H03384 and #20K21774.

Conflicts of Interest: The authors declare no conflict of interest.

References

1. Gomez-Pinilla, F.; Hillman, C. The influence of exercise on cognitive abilities. *Compr. Physiol.* **2013**, *3*, 403–428. [CrossRef]
2. Mandolesi, L.; Polverino, A.; Montuori, S.; Foti, F.; Ferraioli, G.; Sorrentino, P.; Sorrentino, G. Effects of Physical Exercise on Cognitive Functioning and Wellbeing: Biological and Psychological Benefits. *Front. Psychol.* **2018**, *9*, 509. [CrossRef]
3. Fiuza-Luces, C.; Garatachea, N.; Berger, N.A.; Lucia, A. Exercise is the real polypill. *Physiology* **2013**, *28*, 330–358. [CrossRef]
4. Pedersen, B.K. Physical activity and muscle-brain crosstalk. *Nat. Rev. Endocrinol.* **2019**, *15*, 383–392. [CrossRef]
5. Goodwin, M.L.; Harris, J.E.; Hernández, A.; Gladden, L.B. Blood lactate measurements and analysis during exercise: A guide for clinicians. *J. Diabetes Sci. Technol.* **2007**, *1*, 558–569. [CrossRef]
6. Krustrup, P.; Mohr, M.; Steensberg, A.; Bencke, J.; Kjaer, M.; Bangsbo, J. Muscle and blood metabolites during a soccer game: Implications for sprint performance. *Med. Sci. Sports Exerc.* **2006**, *38*, 1165–1174. [CrossRef]
7. Tsukamoto, H.; Suga, T.; Takenaka, S.; Tanaka, D.; Takeuchi, T.; Hamaoka, T.; Isaka, T.; Hashimoto, T. Greater impact of acute high-intensity interval exercise on post-exercise executive function compared to moderate-intensity continuous exercise. *Physiol. Behav.* **2016**, *155*, 224–230. [CrossRef]
8. Tsukamoto, H.; Takenaka, S.; Suga, T.; Tanaka, D.; Takeuchi, T.; Hamaoka, T.; Isaka, T.; Hashimoto, T. Impact of Exercise Intensity and Duration on Postexercise Executive Function. *Med. Sci. Sports Exerc.* **2016**, *49*, 774–784, Erratum in **2017**, *49*, 774–784. [CrossRef]
9. Haskell, W.L.; Lee, I.M.; Pate, R.R.; Powell, K.E.; Blair, S.N.; Franklin, B.A.; Macera, C.A.; Heath, G.W.; Thompson, P.D.; Bauman, A. Physical activity and public health: Updated recommendation for adults from the American College of Sports Medicine and the American Heart Association. *Med. Sci. Sports Exerc.* **2007**, *39*, 1423–1434. [CrossRef]
10. ACSM. American College of Sports Medicine Position Stand. The recommended quantity and quality of exercise for developing and maintaining cardiorespiratory and muscular fitness, and flexibility in healthy adults. *Med. Sci. Sports Exerc.* **1998**, *30*, 975–991.
11. Gormley, S.E.; Swain, D.P.; High, R.; Spina, R.J.; Dowling, E.A.; Kotipalli, U.S.; Gandrakota, R. Effect of intensity of aerobic training on VO2max. *Med. Sci. Sports Exerc.* **2008**, *40*, 1336–1343. [CrossRef]
12. Swain, D.P.; Franklin, B.A. Comparison of cardioprotective benefits of vigorous versus moderate intensity aerobic exercise. *Am. J. Cardiol.* **2006**, *97*, 141–147. [CrossRef]
13. Helgerud, J.; Hoydal, K.; Wang, E.; Karlsen, T.; Berg, P.; Bjerkaas, M.; Simonsen, T.; Helgesen, C.; Hjorth, N.; Bach, R.; et al. Aerobic high-intensity intervals improve VO2max more than moderate training. *Med. Sci. Sports Exerc.* **2007**, *39*, 665–671. [CrossRef]
14. Hood, M.S.; Little, J.P.; Tarnopolsky, M.A.; Myslik, F.; Gibala, M.J. Low-volume interval training improves muscle oxidative capacity in sedentary adults. *Med. Sci. Sports Exerc.* **2011**, *43*, 1849–1856. [CrossRef]
15. Talanian, J.L.; Galloway, S.D.; Heigenhauser, G.J.; Bonen, A.; Spriet, L.L. Two weeks of high-intensity aerobic interval training increases the capacity for fat oxidation during exercise in women. *J. Appl. Physiol. (1985)* **2007**, *102*, 1439–1447. [CrossRef]
16. Mekari, S.; Earle, M.; Martins, R.; Drisdelle, S.; Killen, M.; Bouffard-Levasseur, V.; Dupuy, O. Effect of High Intensity Interval Training Compared to Continuous Training on Cognitive Performance in Young Healthy Adults: A Pilot Study. *Brain Sci.* **2020**, *10*, 81. [CrossRef]
17. Korman, N.; Armour, M.; Chapman, J.; Rosenbaum, S.; Kisely, S.; Suetani, S.; Firth, J.; Siskind, D. High Intensity Interval training (HIIT) for people with severe mental illness: A systematic review & meta-analysis of intervention studies- considering diverse approaches for mental and physical recovery. *Psychiatry Res.* **2020**, *284*, 112601. [CrossRef]
18. Matura, S.; Fleckenstein, J.; Deichmann, R.; Engeroff, T.; Fuzeki, E.; Hattingen, E.; Hellweg, R.; Lienerth, B.; Pilatus, U.; Schwarz, S.; et al. Effects of aerobic exercise on brain metabolism and grey matter volume in older adults: Results of the randomised controlled SMART trial. *Transl. Psychiatry* **2017**, *7*, e1172. [CrossRef]
19. Tyndall, A.V.; Clark, C.M.; Anderson, T.J.; Hogan, D.B.; Hill, M.D.; Longman, R.S.; Poulin, M.J. Protective Effects of Exercise on Cognition and Brain Health in Older Adults. *Exerc. Sport Sci. Rev.* **2018**, *46*, 215–223. [CrossRef]
20. Tsukamoto, H.; Suga, T.; Takenaka, S.; Tanaka, D.; Takeuchi, T.; Hamaoka, T.; Isaka, T.; Ogoh, S.; Hashimoto, T. Repeated high-intensity interval exercise shortens the positive effect on executive function during post-exercise recovery in healthy young males. *Physiol. Behav.* **2016**, *160*, 26–34. [CrossRef]

21. Kandola, A.; Hendrikse, J.; Lucassen, P.J.; Yücel, M. Aerobic Exercise as a Tool to Improve Hippocampal Plasticity and Function in Humans: Practical Implications for Mental Health Treatment. *Front. Hum. Neurosci.* **2016**, *10*, 373. [CrossRef]
22. Pajonk, F.G.; Wobrock, T.; Gruber, O.; Scherk, H.; Berner, D.; Kaizl, I.; Kierer, A.; Müller, S.; Oest, M.; Meyer, T.; et al. Hippocampal plasticity in response to exercise in schizophrenia. *Arch. Gen. Psychiatry* **2010**, *67*, 133–143. [CrossRef] [PubMed]
23. Tarumi, T.; Zhang, R. The Role of Exercise-Induced Cardiovascular Adaptation in Brain Health. *Exerc. Sport Sci. Rev.* **2015**, *43*, 181–189. [CrossRef]
24. Erickson, K.I.; Voss, M.W.; Prakash, R.S.; Basak, C.; Szabo, A.; Chaddock, L.; Kim, J.S.; Heo, S.; Alves, H.; White, S.M.; et al. Exercise training increases size of hippocampus and improves memory. *Proc. Natl. Acad. Sci. USA* **2011**, *108*, 3017–3022. [CrossRef]
25. Schmaal, L.; Veltman, D.J.; van Erp, T.G.; Sämann, P.G.; Frodl, T.; Jahanshad, N.; Loehrer, E.; Tiemeier, H.; Hofman, A.; Niessen, W.J.; et al. Subcortical brain alterations in major depressive disorder: Findings from the ENIGMA Major Depressive Disorder working group. *Mol. Psychiatry* **2016**, *21*, 806–812. [CrossRef]
26. Ellison-Wright, I.; Bullmore, E. Anatomy of bipolar disorder and schizophrenia: A meta-analysis. *Schizophr. Res.* **2010**, *117*, 1–12. [CrossRef]
27. Besteher, B.; Gaser, C.; Nenadić, I. Brain Structure and Subclinical Symptoms: A Dimensional Perspective of Psychopathology in the Depression and Anxiety Spectrum. *Neuropsychobiology* **2020**, *79*, 270–283. [CrossRef]
28. Ogoh, S. Relationship between cognitive function and regulation of cerebral blood flow. *J. Physiol. Sci. JPS* **2017**, *67*, 345–351. [CrossRef]
29. Iadecola, C. Neurovascular regulation in the normal brain and in Alzheimer's disease. *Nat. Rev. Neurosci.* **2004**, *5*, 347–360. [CrossRef]
30. Rother, J.; Knab, R.; Hamzei, F.; Fiehler, J.; Reichenbach, J.R.; Buchel, C.; Weiller, C. Negative dip in BOLD fMRI is caused by blood flow–oxygen consumption uncoupling in humans. *NeuroImage* **2002**, *15*, 98–102. [CrossRef]
31. Hock, C.; Villringer, K.; Muller-Spahn, F.; Wenzel, R.; Heekeren, H.; Schuh-Hofer, S.; Hofmann, M.; Minoshima, S.; Schwaiger, M.; Dirnagl, U.; et al. Decrease in parietal cerebral hemoglobin oxygenation during performance of a verbal fluency task in patients with Alzheimer's disease monitored by means of near-infrared spectroscopy (NIRS)–correlation with simultaneous rCBF-PET measurements. *Brain Res* **1997**, *755*, 293–303. [CrossRef]
32. Kisler, K.; Nelson, A.R.; Montagne, A.; Zlokovic, B.V. Cerebral blood flow regulation and neurovascular dysfunction in Alzheimer disease. *Nat. Rev. Neurosci.* **2017**, *18*, 419–434. [CrossRef]
33. Korte, N.; Nortley, R.; Attwell, D. Cerebral blood flow decrease as an early pathological mechanism in Alzheimer's disease. *Acta Neuropathol.* **2020**, *140*, 793–810. [CrossRef] [PubMed]
34. Mentis, M.J.; Horwitz, B.; Grady, C.L.; Alexander, G.E.; VanMeter, J.W.; Maisog, J.M.; Pietrini, P.; Schapiro, M.B.; Rapoport, S.I. Visual cortical dysfunction in Alzheimer's disease evaluated with a temporally graded "stress test" during PET. *Am. J. Psychiatry* **1996**, *153*, 32–40. [CrossRef] [PubMed]
35. Warkentin, S.; Passant, U. Functional imaging of the frontal lobes in organic dementia. Regional cerebral blood flow findings in normals, in patients with frontotemporal dementia and in patients with Alzheimer's disease, performing a word fluency test. *Dement. Geriatr. Cogn. Disord.* **1997**, *8*, 105–109. [CrossRef]
36. Ide, K.; Schmalbruch, I.K.; Quistorff, B.; Horn, A.; Secher, N.H. Lactate, glucose and O2 uptake in human brain during recovery from maximal exercise. *J. Physiol.* **2000**, *522*, 159–164. [CrossRef]
37. Ogoh, S.; Ainslie, P.N. Cerebral blood flow during exercise: Mechanisms of regulation. *J. Appl. Physiol. (1985)* **2009**, *107*, 1370–1380. [CrossRef]
38. Ogoh, S.; Ainslie, P.N. Regulatory mechanisms of cerebral blood flow during exercise: New concepts. *Exerc. Sport Sci. Rev.* **2009**, *37*, 123–129. [CrossRef] [PubMed]
39. Ogoh, S.; Brothers, R.M.; Barnes, Q.; Eubank, W.L.; Hawkins, M.N.; Purkayastha, S.; O-Yurvati, A.; Raven, P.B. The effect of changes in cardiac output on middle cerebral artery mean blood velocity at rest and during exercise. *J. Physiol.* **2005**, *569*, 697–704. [CrossRef] [PubMed]
40. Ogoh, S.; Dalsgaard, M.K.; Yoshiga, C.C.; Dawson, E.A.; Keller, D.M.; Raven, P.B.; Secher, N.H. Dynamic cerebral autoregulation during exhaustive exercise in humans. *Am. J. Physiol. Heart Circ. Physiol.* **2005**, *288*, H1461–H1467. [CrossRef] [PubMed]
41. Sato, K.; Ogoh, S.; Hirasawa, A.; Oue, A.; Sadamoto, T. The distribution of blood flow in the carotid and vertebral arteries during dynamic exercise in humans. *J. Physiol.* **2011**, *589*, 2847–2856. [CrossRef]
42. Brisswalter, J.; Collardeau, M.; Rene, A. Effects of acute physical exercise characteristics on cognitive performance. *Sports Med.* **2002**, *32*, 555–566. [CrossRef]
43. McMorris, T.; Sproule, J.; Turner, A.; Hale, B.J. Acute, intermediate intensity exercise, and speed and accuracy in working memory tasks: A meta-analytical comparison of effects. *Physiol. Behav.* **2011**, *102*, 421–428. [CrossRef] [PubMed]
44. Grego, F.; Vallier, J.M.; Collardeau, M.; Rousseu, C.; Cremieux, J.; Brisswalter, J. Influence of exercise duration and hydration status on cognitive function during prolonged cycling exercise. *Int. J. Sports Med.* **2005**, *26*, 27–33. [CrossRef] [PubMed]
45. Ogoh, S.; Tsukamoto, H.; Hirasawa, A.; Hasegawa, H.; Hirose, N.; Hashimoto, T. The effect of changes in cerebral blood flow on cognitive function during exercise. *Physiol. Rep.* **2014**, *2*, e12163. [CrossRef]
46. Lucas, S.J.; Ainslie, P.N.; Murrell, C.J.; Thomas, K.N.; Franz, E.A.; Cotter, J.D. Effect of age on exercise-induced alterations in cognitive executive function: Relationship to cerebral perfusion. *Exp. Gerontol.* **2012**, *47*, 541–551. [CrossRef] [PubMed]

47. Smale, B.A.; Northey, J.M.; Smee, D.J.; Versey, N.G.; Rattray, B. Compression garments and cerebral blood flow: Influence on cognitive and exercise performance. *Eur. J. Sport Sci.* **2018**, *18*, 315–322. [CrossRef]
48. Ogoh, S.; Sato, K.; Okazaki, K.; Miyamoto, T.; Hirasawa, A.; Shibasaki, M. Hyperthermia modulates regional differences in cerebral blood flow to changes in CO2. *J. Appl. Physiol. (1985)* **2014**, *117*, 46–52. [CrossRef]
49. Ando, S.; Hatamoto, Y.; Sudo, M.; Kiyonaga, A.; Tanaka, H.; Higaki, Y. The effects of exercise under hypoxia on cognitive function. *PLoS ONE* **2013**, *8*, e63630. [CrossRef]
50. Komiyama, T.; Katayama, K.; Sudo, M.; Ishida, K.; Higaki, Y.; Ando, S. Cognitive function during exercise under severe hypoxia. *Sci. Rep.* **2017**, *7*, 10000. [CrossRef]
51. Miyazawa, T.; Horiuchi, M.; Ichikawa, D.; Sato, K.; Tanaka, N.; Bailey, D.M.; Ogoh, S. Kinetics of exercise-induced neural activation; interpretive dilemma of altered cerebral perfusion. *Exp. Physiol.* **2012**, *97*, 219–227. [CrossRef] [PubMed]
52. Quistorff, B.; Secher, N.H.; Van Lieshout, J.J. Lactate fuels the human brain during exercise. *FASEB J. Off. Publ. Fed. Am. Soc. Exp. Biol.* **2008**, *22*, 3443–3449. [CrossRef] [PubMed]
53. van Hall, G.; Stromstad, M.; Rasmussen, P.; Jans, O.; Zaar, M.; Gam, C.; Quistorff, B.; Secher, N.H.; Nielsen, H.B. Blood lactate is an important energy source for the human brain. *J. Cereb. Blood Flow Metab.* **2009**, *29*, 1121–1129. [CrossRef] [PubMed]
54. Egner, T.; Hirsch, J. The neural correlates and functional integration of cognitive control in a Stroop task. *NeuroImage* **2005**, *24*, 539–547. [CrossRef]
55. McMorris, T. Developing the catecholamines hypothesis for the acute exercise-cognition interaction in humans: Lessons from animal studies. *Physiol. Behav.* **2016**, *165*, 291–299. [CrossRef]
56. Dalsgaard, M.K.; Ide, K.; Cai, Y.; Quistorff, B.; Secher, N.H. The intent to exercise influences the cerebral O(2)/carbohydrate uptake ratio in humans. *J. Physiol.* **2002**, *540*, 681–689. [CrossRef]
57. Kemppainen, J.; Aalto, S.; Fujimoto, T.; Kalliokoski, K.K.; Langsjo, J.; Oikonen, V.; Rinne, J.; Nuutila, P.; Knuuti, J. High intensity exercise decreases global brain glucose uptake in humans. *J. Physiol.* **2005**, *568*, 323–332. [CrossRef]
58. Barros, L.F. Metabolic signaling by lactate in the brain. *Trends Neurosci.* **2013**, *36*, 396–404. [CrossRef]
59. Hu, Y.; Wilson, G.S. A temporary local energy pool coupled to neuronal activity: Fluctuations of extracellular lactate levels in rat brain monitored with rapid-response enzyme-based sensor. *J. Neurochem.* **1997**, *69*, 1484–1490. [CrossRef]
60. Smith, D.; Pernet, A.; Hallett, W.A.; Bingham, E.; Marsden, P.K.; Amiel, S.A. Lactate: A preferred fuel for human brain metabolism in vivo. *J. Cereb. Blood Flow Metab.* **2003**, *23*, 658–664. [CrossRef]
61. Holloway, R.; Zhou, Z.; Harvey, H.B.; Levasseur, J.E.; Rice, A.C.; Sun, D.; Hamm, R.J.; Bullock, M.R. Effect of lactate therapy upon cognitive deficits after traumatic brain injury in the rat. *Acta Neurochir.* **2007**, *149*, 919–927; discussion 927. [CrossRef]
62. Skriver, K.; Roig, M.; Lundbye-Jensen, J.; Pingel, J.; Helge, J.W.; Kiens, B.; Nielsen, J.B. Acute exercise improves motor memory: Exploring potential biomarkers. *Neurobiol. Learn. Mem.* **2014**, *116*, 46–58. [CrossRef]
63. Schurr, A.; West, C.A.; Rigor, B.M. Lactate-supported synaptic function in the rat hippocampal slice preparation. *Science* **1988**, *240*, 1326–1328. [CrossRef] [PubMed]
64. Suzuki, A.; Stern, S.A.; Bozdagi, O.; Huntley, G.W.; Walker, R.H.; Magistretti, P.J.; Alberini, C.M. Astrocyte-neuron lactate transport is required for long-term memory formation. *Cell* **2011**, *144*, 810–823. [CrossRef] [PubMed]
65. Yang, J.; Ruchti, E.; Petit, J.M.; Jourdain, P.; Grenningloh, G.; Allaman, I.; Magistretti, P.J. Lactate promotes plasticity gene expression by potentiating NMDA signaling in neurons. *Proc. Natl. Acad. Sci. USA* **2014**, *111*, 12228–12233. [CrossRef]
66. Hashimoto, T.; Tsukamoto, H.; Takenaka, S.; Olesen, N.D.; Petersen, L.G.; Sorensen, H.; Nielsen, H.B.; Secher, N.H.; Ogoh, S. Maintained exercise-enhanced brain executive function related to cerebral lactate metabolism in men. *FASEB J. Off. Publ. Fed. Am. Soc. Exp. Biol.* **2018**, *32*, 1417–1427. [CrossRef]
67. Chmura, J.; Nazar, K.; Kaciuba-Uscilko, H. Choice reaction time during graded exercise in relation to blood lactate and plasma catecholamine thresholds. *Int. J. Sports Med.* **1994**, *15*, 172–176. [CrossRef] [PubMed]
68. Griffin, E.W.; Mullally, S.; Foley, C.; Warmington, S.A.; O'Mara, S.M.; Kelly, A.M. Aerobic exercise improves hippocampal function and increases BDNF in the serum of young adult males. *Physiol. Behav.* **2011**, *104*, 934–941. [CrossRef]
69. Lev-Vachnish, Y.; Cadury, S.; Rotter-Maskowitz, A.; Feldman, N.; Roichman, A.; Illouz, T.; Varvak, A.; Nicola, R.; Madar, R.; Okun, E. L-Lactate Promotes Adult Hippocampal Neurogenesis. *Front. Neurosci.* **2019**, *13*, 403. [CrossRef]
70. Sudo, M.; Komiyama, T.; Aoyagi, R.; Nagamatsu, T.; Higaki, Y.; Ando, S. Executive function after exhaustive exercise. *Eur. J. Appl. Physiol.* **2017**, *117*, 2029–2038. [CrossRef]
71. Cotman, C.W.; Berchtold, N.C.; Christie, L.A. Exercise builds brain health: Key roles of growth factor cascades and inflammation. *Trends Neurosci.* **2007**, *30*, 464–472. [CrossRef]
72. Mattson, M.P.; Maudsley, S.; Martin, B. BDNF and 5-HT: A dynamic duo in age-related neuronal plasticity and neurodegenerative disorders. *Trends Neurosci.* **2004**, *27*, 589–594. [CrossRef] [PubMed]
73. Laske, C.; Banschbach, S.; Stransky, E.; Bosch, S.; Straten, G.; Machann, J.; Fritsche, A.; Hipp, A.; Niess, A.; Eschweiler, G.W. Exercise-induced normalization of decreased BDNF serum concentration in elderly women with remitted major depression. *Int. J. Neuropsychopharmacol.* **2010**, *13*, 595–602. [CrossRef]
74. Shimada, H.; Makizako, H.; Doi, T.; Yoshida, D.; Tsutsumimoto, K.; Anan, Y.; Uemura, K.; Lee, S.; Park, H.; Suzuki, T. A large, cross-sectional observational study of serum BDNF, cognitive function, and mild cognitive impairment in the elderly. *Front. Aging Neurosci.* **2014**, *6*, 69. [CrossRef] [PubMed]

75. Schiffer, T.; Schulte, S.; Sperlich, B.; Achtzehn, S.; Fricke, H.; Struder, H.K. Lactate infusion at rest increases BDNF blood concentration in humans. *Neurosci. Lett.* **2011**, *488*, 234–237. [CrossRef] [PubMed]
76. Ferris, L.T.; Williams, J.S.; Shen, C.L. The effect of acute exercise on serum brain-derived neurotrophic factor levels and cognitive function. *Med. Sci. Sports Exerc.* **2007**, *39*, 728–734. [CrossRef] [PubMed]
77. Rodriguez, A.L.; Whitehurst, M.; Fico, B.G.; Dodge, K.M.; Ferrandi, P.J.; Pena, G.; Adelman, A.; Huang, C.J. Acute high-intensity interval exercise induces greater levels of serum brain-derived neurotrophic factor in obese individuals. *Exp. Biol. Med. (Maywood N.J.)* **2018**, *243*, 1153–1160. [CrossRef]
78. Kujach, S.; Olek, R.A.; Byun, K.; Suwabe, K.; Sitek, E.J.; Ziemann, E.; Laskowski, R.; Soya, H. Acute Sprint Interval Exercise Increases Both Cognitive Functions and Peripheral Neurotrophic Factors in Humans: The Possible Involvement of Lactate. *Front Neurosci.* **2019**, *13*, 1455. [CrossRef]
79. El Hayek, L.; Khalifeh, M.; Zibara, V.; Abi Assaad, R.; Emmanuel, N.; Karnib, N.; El-Ghandour, R.; Nasrallah, P.; Bilen, M.; Ibrahim, P.; et al. Lactate Mediates the Effects of Exercise on Learning and Memory through SIRT1-Dependent Activation of Hippocampal Brain-Derived Neurotrophic Factor (BDNF). *J. Neurosci. Off. J. Soc. Neurosci.* **2019**, *39*, 2369–2382. [CrossRef]
80. Müller, P.; Duderstadt, Y.; Lessmann, V.; Müller, N.G. Lactate and BDNF: Key Mediators of Exercise Induced Neuroplasticity? *J. Clin. Med.* **2020**, *9*, 1136. [CrossRef]
81. Morland, C.; Andersson, K.A.; Haugen, O.P.; Hadzic, A.; Kleppa, L.; Gille, A.; Rinholm, J.E.; Palibrk, V.; Diget, E.H.; Kennedy, L.H.; et al. Exercise induces cerebral VEGF and angiogenesis via the lactate receptor HCAR1. *Nat. Commun.* **2017**, *8*, 15557. [CrossRef]
82. Lucas, S.J.; Cotter, J.D.; Brassard, P.; Bailey, D.M. High-intensity interval exercise and cerebrovascular health: Curiosity, cause, and consequence. *J. Cereb. Blood Flow Metab.* **2015**, *35*, 902–911. [CrossRef] [PubMed]
83. Carrard, A.; Cassé, F.; Carron, C.; Burlet-Godinot, S.; Toni, N.; Magistretti, P.J.; Martin, J.L. Role of adult hippocampal neurogenesis in the antidepressant actions of lactate. *Mol. Psychiatry* **2021**. [CrossRef] [PubMed]
84. Karnib, N.; El-Ghandour, R.; El Hayek, L.; Nasrallah, P.; Khalifeh, M.; Barmo, N.; Jabre, V.; Ibrahim, P.; Bilen, M.; Stephan, J.S.; et al. Lactate is an antidepressant that mediates resilience to stress by modulating the hippocampal levels and activity of histone deacetylases. *Neuropsychopharmacol. Off. Publ. Am. Coll. Neuropsychopharmacol.* **2019**, *44*, 1152–1162. [CrossRef] [PubMed]
85. Carrard, A.; Elsayed, M.; Margineanu, M.; Boury-Jamot, B.; Fragnière, L.; Meylan, E.M.; Petit, J.M.; Fiumelli, H.; Magistretti, P.J.; Martin, J.L. Peripheral administration of lactate produces antidepressant-like effects. *Mol. Psychiatry* **2018**, *23*, 392–399. [CrossRef]
86. Margineanu, M.B.; Mahmood, H.; Fiumelli, H.; Magistretti, P.J. L-Lactate Regulates the Expression of Synaptic Plasticity and Neuroprotection Genes in Cortical Neurons: A Transcriptome Analysis. *Front. Mol. Neurosci.* **2018**, *11*, 375. [CrossRef]
87. Gordon, G.R.; Choi, H.B.; Rungta, R.L.; Ellis-Davies, G.C.; MacVicar, B.A. Brain metabolism dictates the polarity of astrocyte control over arterioles. *Nature* **2008**, *456*, 745–749. [CrossRef]
88. Mintun, M.A.; Vlassenko, A.G.; Rundle, M.M.; Raichle, M.E. Increased lactate/pyruvate ratio augments blood flow in physiologically activated human brain. *Proc. Natl. Acad. Sci. USA* **2004**, *101*, 659–664. [CrossRef]
89. Hollyer, T.R.; Bordoni, L.; Kousholt, B.S.; van Luijk, J.; Ritskes-Hoitinga, M.; Østergaard, L. The evidence for the physiological effects of lactate on the cerebral microcirculation: A systematic review. *J. Neurochem.* **2019**, *148*, 712–730. [CrossRef]
90. Carteron, L.; Solari, D.; Patet, C.; Quintard, H.; Miroz, J.P.; Bloch, J.; Daniel, R.T.; Hirt, L.; Eckert, P.; Magistretti, P.J.; et al. Hypertonic Lactate to Improve Cerebral Perfusion and Glucose Availability After Acute Brain Injury. *Crit. Care Med.* **2018**, *46*, 1649–1655. [CrossRef] [PubMed]
91. Brooks, G.A.; Martin, N.A. Cerebral metabolism following traumatic brain injury: New discoveries with implications for treatment. *Front. Neurosci.* **2014**, *8*, 408. [CrossRef]
92. Bisri, T.; Utomo, B.A.; Fuadi, I. Exogenous lactate infusion improved neurocognitive function of patients with mild traumatic brain injury. *Asian J. Neurosurg.* **2016**, *11*, 151–159. [CrossRef]
93. Winett, R.A.; Carpinelli, R.N. Potential health-related benefits of resistance training. *Prev. Med.* **2001**, *33*, 503–513. [CrossRef] [PubMed]
94. Tsukamoto, H.; Suga, T.; Takenaka, S.; Takeuchi, T.; Tanaka, D.; Hamaoka, T.; Hashimoto, T.; Isaka, T. An acute bout of localized resistance exercise can rapidly improve inhibitory control. *PLoS ONE* **2017**, *12*, e0184075. [CrossRef] [PubMed]
95. Dora, K.; Suga, T.; Tomoo, K.; Sugimoto, T.; Mok, E.; Tsukamoto, H.; Takada, S.; Hashimoto, T.; Isaka, T. Effect of very low-intensity resistance exercise with slow movement and tonic force generation on post-exercise inhibitory control. *Heliyon* **2021**, *7*, e06261. [CrossRef] [PubMed]
96. Dora, K.; Suga, T.; Tomoo, K.; Sugimoto, T.; Mok, E.; Tsukamoto, H.; Takada, S.; Hashimoto, T.; Isaka, T. Similar improvements in cognitive inhibitory control following low-intensity resistance exercise with slow movement and tonic force generation and high-intensity resistance exercise in healthy young adults: A preliminary study. *J. Physiol. Sci. JPS* **2021**, *71*, 22. [CrossRef] [PubMed]

Article

Central and Peripheral Oxygen Distribution in Two Different Modes of Interval Training

Korbinian Sebastian Hermann Ksoll [1,2,*], Alexander Mühlberger [3] and Fabian Stöcker [3]

[1] Institute of Sport Sciences, Department of Human Sciences, Universität der Bundeswehr Munich, 85579 Neubiberg, Germany
[2] Professorship of Biomechanics in Sports, Department of Sport and Health Sciences, Technical University of Munich, 80992 Munich, Germany
[3] Prevention Center, Department of Sport and Health Sciences, Technical University of Munich, 80992 Munich, Germany; alex.muehli@freenet.de (A.M.); fabian.stoecker@tum.de (F.S.)
* Correspondence: korbinian.ksoll@unibw.de

Abstract: In high-intensity interval training the interval duration can be adjusted to optimize training results in oxygen uptake, cardiac output, and local oxygen supply. This study aimed to compare these variables in two interval trainings (long intervals HIIT3m: 3 min work, 3 min active rest vs. short intervals HIIT30s: 30 s work, 30 s active rest) at the same overall work rate and training duration. 24 participants accomplished both protocols, (work: 80% power output at VO_2peak, relief: 85% power output at gas exchange threshold) in randomized order. Spirometry, impedance cardiography, and near-infrared spectroscopy were used to analyze the physiological stress of the cardiopulmonary system and muscle tissue. Although times above gas exchange threshold were shorter in HIIT3m (HIIT3m 1669.9 ± 310.9 s vs. HIIT30s 1769.5 ± 189.0 s, $p = 0.034$), both protocols evoked similar average fractional utilization of VO_2peak (HIIT3m 65.23 ± 4.68% VO_2peak vs. HIIT30s 64.39 ± 6.78% VO_2peak, $p = 0.261$). However, HIIT3m resulted in higher cardiovascular responses during the loaded phases (VO_2 $p < 0.001$, cardiac output $p < 0.001$). Local hemodynamics were not different between both protocols. Average physiological responses were not different in both protocols owning to incomplete rests in HIIT30s and large response amplitudes in HIIT3m. Despite lower acute cardiovascular stress in HIIT30s, short submaximal intervals may also trigger microvascular and metabolic adaptions similar to HIIT3m. Therefore, the adaption of interval duration is an important tool to adjust the goals of interval training to the needs of the athlete or patient.

Keywords: interval exercise; oxygen uptake (VO_2); cardiac output (CO); oxygen availability (HHb/VO_2); near-infrared spectroscopy (NIRS)

Citation: Ksoll, K.S.H.; Mühlberger, A.; Stöcker, F. Central and Peripheral Oxygen Distribution in Two Different Modes of Interval Training. *Metabolites* **2021**, *11*, 790. https://doi.org/10.3390/metabo11110790

Academic Editor: Norbert Nemeth

Received: 28 October 2021
Accepted: 15 November 2021
Published: 18 November 2021

Publisher's Note: MDPI stays neutral with regard to jurisdictional claims in published maps and institutional affiliations.

1. Introduction

Interval training is an often-used training modality to improve endurance performance in athletes but also cardiorespiratory fitness in patients [1–3]. In contrast to continuous training, interval training consists of several alternating phases of high and low intensities. Buchheit & Laursen [1] defined multiple exercise variables used in the design of an interval exercise session including intensities and durations of work and relief phases, the work modality, and the combination of exercise series. The manipulation of these factors adjusts the interval training in order to meet the demands of the sport, the athlete's profile, or the patient's possibilities [1]. Sprint interval training or repeated sprint interval training, at maximum effort, highly affect the capability in maximal energy production by aerobic and anaerobic systems while short (<45 s) and long (2–4 min) high-intensity interval training (HIIT) is associated with higher emphasis on submaximal performance [1,4]. Due to intermittent exercise of work and relief, HIIT achieves longer times at high rates of oxygen uptake (VO_2) compared to long slow distance or moderate continuous training with the same training duration [1,5]. Accordingly, HIIT represents a greater stimulus on maximum aerobic energy production and hence is associated with a fast increase in peak oxygen uptake (VO_2peak) [5,6]. Both, HIIT and continuous training affect the cardiovascular system by increasing local perfusion [7]. Microvascular oxygen distribution and capillary perfusion are known as key determinants to promote oxidative metabolism [8]. Recent studies showed evidence for a higher impact of interval training on local muscle perfusion compared to continuous training [9,10]. However, there are several confounding variables in the investigation of interval training applications. One major issue in the research of intermittent training is the matching of interval intensity and duration. Using an isoeffort matching approach, Zafeiridis and colleagues compared a continuous training (70% VO_2max), HIIT with long intervals (2 min at 95% VO_2max, 2min passive rest), and HIIT with short intervals (30 s at 110% VO_2max, 30 s passive rest). Cardiovascular stress was highest in continuous and long interval training, while muscle oxygenation was equal in all protocols [11]. However, this study does not clarify the effect of different interval durations, as the work rate was not constant. Our study aimed to compare two interval regimens of equal overall work rate but the different composition of work- and relief-interval duration in respect of local and central cardiovascular effects. We hypothesized that long interval duration, i.e., 3 min work, has higher cardiometabolic demand compared to short interval duration, i.e., 30 s, at same overall work. Despite the different cardiometabolic demands, both interval protocols achieve similar effects in acute microvascular oxygen distribution.

2. Methods

24 male subjects (Table 1) participated in this study voluntarily. For this, informed consent was obtained from all subjects involved in the study. All test persons were healthy and performed recreational sport at least two times a week. This study was carried out in accordance with the Declaration of Helsinki and was approved by the local Ethics Committee of the Technical University of Munich (#67/14, 2014).

Table 1. Subject characteristics.

Parameter	Mean ± SD	N
Age [years]	24.3 ± 3.6	24
Height [cm]	181.4 ± 5.1	24
Weight [kg]	75.9 ± 7.6	24
Skinfold thickness at m. vastus lateralis [mm]	8.0 ± 3.2	24
Peak oxygen uptake (VO_{2peak}) [$L{*}min^{-1}$]	4.11 ± 0.53	24
Relative peak oxygen uptake (VO_{2peak}) [$mL{*}min^{-1}{*}kg^{-1}$]	54.1 ± 5.3	24
Gas Exchange Threshold (GET) [% VO_{2peak}]	52.9 ± 8.4	24
Respiratory Compensation Point (RCP) [% VO_{2peak}]	82.6 ± 6.9	24
peak heart rate (HR_{peak}) [bpm]	185.0 ± 7.7	24
peak cardiac output (CO_{peak}) [$L{*}min^{-1}$]	25.4 ± 3.4	17
peak stroke volume (SV_{peak}) [ml]	144.1 ± 19.4	17
peak power output (PO_{peak}) [W]	359.5 ± 44.8	24

On an electrically braked cycle ergometer (Lode Excalibur, Groningen, NL, USA) the participants performed three tests protocols which had to be separated at least 48 h from each other and executed within two weeks. During each measurement, an open spirometry device, Cortex Metalyzer 3b (Leipzig, GER), analyzed breath-by-breath the respiratory gas exchange. For this, the participants were not allowed to consume food and caffeinated drinks three hours before the measurements. The first test, a continuous ramp test to exhaustion (3 min baseline measurement, 3 min warm-up at 50 W, 20 W/min increase), was accomplished to determine the VO_2peak, the respiratory thresholds (gas exchange threshold, GET + respiratory compensation point, RCP), peak heart rate (HR_{peak}), cardiac output (CO_{peak}), stroke volume (SV_{peak}) and maximum power output (PO_{peak}). VO_2peak, HR_{peak}, CO_{peak}, $SV_{peak\ and}$ was defined as the highest value in the 30 s moving average of each parameter. Both, GET and RCP, were estimated using the modified V-slope method [12] by determining visual investigation of breakpoints in the plotted breath-by-breath data of carbon dioxide output (VCO_2) vs. oxygen uptake (VO_2) [12,13]. To increase the accuracy of GET and RCP, visual breakpoints in breath volume (VE) vs. time as well as in the equivalents of VE/VCO_2 and VE/VO_2 vs. time were also used for threshold detection. Additionally, VE vs. VCO_2 provided further information about the RCP. Without wearing special bike shoes, the participants were allowed to choose a comfortable cadence above 60 rpm which had to be maintained. The incremental ramp test was terminated by a drop of the pedal frequency below 60 rpm.

In randomized order, the participants performed two interval protocols that differed in the duration of the intervals. Each protocol consisted of five consecutive sections. In the long interval protocol (HIIT3m), one section was representing one complete interval including 3 min work phase and 3 min active rest. During the short interval protocol (HIIT30s), one section consisted of six repeated bouts of 30 s active recovery and 30 s work. The work-interval was set equal to the power output achieved at 80% of VO_2peak in the incremental ramp protocol for both, HIIT3m and HIIT30s work intervals. The intensity was then reduced by 5% to account for the delayed oxygen kinetics i.e., a mean response time of about 30 s [14]. The recovery intensity was set to the power output achieved at 85% GET. Due to the work to rest ratio of 1:1 and the same power output during work and relief phases, both interval protocols achieved the same amount of total work (Figure 1). Each interval session began with a 3 min baseline measurement of oxygen uptake, cardiac output, and muscle deoxygenation followed by a 5 min warm-up. The warm-up intensity was set equal to the active recovery intensity. The short intervals started with active rest to ensure that the last 30 s of a HIIT3m work interval was time-aligned with a work interval in the HIIT30s. After the last work phase of both protocols, 20 µL of blood was taken from the right earlobe for end-exercise lactate diagnosis. The blood-filled capillaries were stored in reaction cups and mixed with 1000 µL of hemolyzing solution. The calibrated Biosen S-Line system (EKF-diagnostic GmbH, Barleben, GER) analyzed the probes using the enzymatic-amperometric principle.

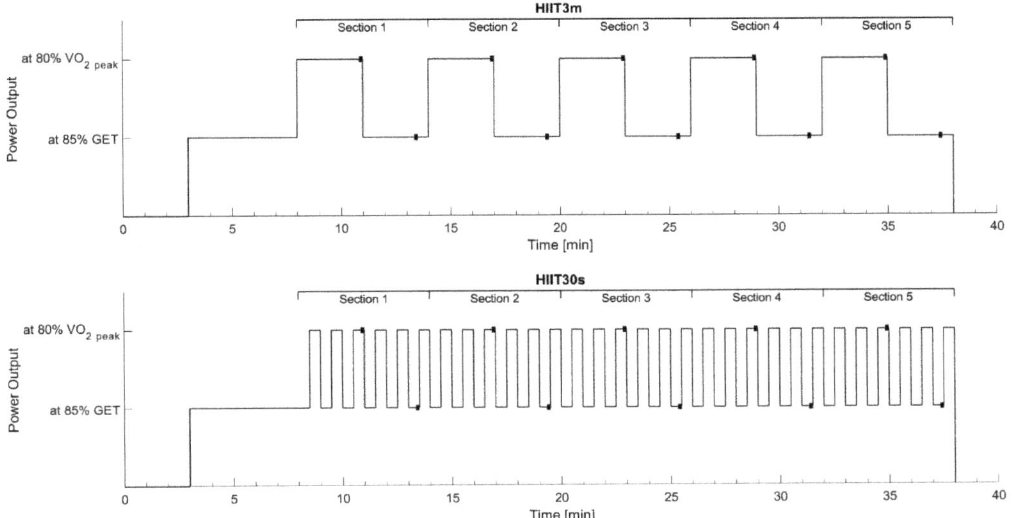

Figure 1. Scheme of HIIT3m and HIIT30s protocol regarding interval intensities and duration. The black squares demonstrate the 10 s period, where the means for work (loaded) and resting (unloaded) phases were calculated.

In addition to the heart rate (HR), stroke volume (SV) and cardiac output (CO) were analyzed non-invasively via impedance cardiography (PhysioFlow Enduro, Manatec Biomedical, Folschviller, France). Also known as transthoracic-electrical-bioimpedance, impedance cardiography detects changes of thoracic impedance caused by the alternating heart volume by sending a low amperage high-frequency current through the thorax [15,16]. The method of impedance cardiography, respectively the PhysioFlow System was validated against the direct Fick method and/or thermodilution in patients [15–17] and exercise tests [18]. For this method, six electrodes must be placed at the subject's neck and upper body. Previously, the skin was shaved and cleaned with skin preparation gel and alcohol to increase skin conductance and signal quality. After connecting the PhysioFlow, the blood pressure was measured and the device was calibrated. For calibration, the participants were asked to maintain a relaxed position on the ergometer. As soon as the signal stabilized, the software analyzed 30 heartbeats to calibrate the system. The PhysioFlow system was not available to the first seven participants during piloting. Hence, SV and CO were analyzed in only 17 subjects, whereas VO_2, HR, and muscle oxygenation were measured in all 24 test persons.

Muscle oxygen saturation was analyzed by near-infrared spectroscopy (NIRS). In general, a NIRS device consists of two optodes. The first emits near-infrared light into human tissue. On its way through the tissue, light, i.e., photons, can be absorbed or scattered by specific chromophores like oxygenated (O_2Hb) or deoxygenated hemoglobin (HHb). The second optode receives the remaining, attenuated light. According to the modified Lambert-Beer law, attenuation of light and therefore the concentration of the absorbing chromophores can be calculated by the logarithmic ratio between received and emitted light. Due to the unknown pathlength of the photons inside the tissue, a differential pathlength factor has to be included. In the used system, this factor was equal to 4 [19].

The near-infrared spectroscopy device (Portamon, Artinis medical Systems, Elst, NL) was placed at the shaved, right m. vastus lateralis similar to the SENIAM recommendations for EMG application [20]. The device emitted light (760 and 850 nm) continuously to the 35 mm distanced receiving optode, resulting in a penetration depth of approximately 17.5 mm [21]. To ensure that light penetrates muscle tissue, the skinfold at the NIRS position was measured by a caliper (Skinfold Caliper, Harpenden, Burgess Hill, UK). Limited by the continuous wave design of the device, only relative concentrations of O_2Hb [$\mu mol*L^{-1}$] and HHb [$\mu mol*L^{-1}$] can be measured. The NIRS software (Oxysoft, Artinis medical Systems, Elst, NL, USA) started recording with 10 Hz measuring rate and additionally calculated the total amount of hemoglobin (tHb = O_2Hb + HHb [$\mu mol*L^{-1}$]) as well as the tissue saturation index (TSI = $100*O_2Hb*tHb^{-1}$ [%]). Measured HHb was normalized from 0% (baseline) to 100% (maximum HHb response during ramp test) to be comparable between both interval protocols.

Microvascular oxygen distribution was estimated by ΔHHb per unit ΔVO_2 ($\Delta HHb/\Delta VO_2$). The ratio of normalized HHb to normalized VO_2 (100% = VO_2peak) reflects the continuous matching of O_2 distribution and its utilization. In general, a better microvascular O_2 distribution is indicated by a lower $\Delta HHb/\Delta VO_2$ ratio. This method was used previously in relation to microvascular O_2 adjustment during exercise on-transient [8,22] as well as a description of microvascular O_2 availability during and after exercise [11,23,24].

All data were collected and processed by customized Matlab scripts (Mathworks, Natick, MA, USA). The program interpolated the breath-by-breath data measured by the spirometry and PhysioFlow to second-by-second values. NIRS data were downsampled to 1 Hz. Data were smoothed using a moving average with a 30 s-time window. All data were normalized to the maximum value in the preceding maximal ramp-test. The section averages (1 long complete interval, 6 short complete intervals) as well as the end values of the work phase (loaded) and the resting phase (unloaded) were statistically analyzed. These end values were defined as the mean of the last 10 s of the loaded or unloaded phase (Figure 1) respectively. Repeated measures ANOVAs with two factors (interval training, section) were executed in SPSS 23 (IBM, Armonk, NY, USA). Results were corrected by the

Greenhouse–Geisser method if the assumption of sphericity, tested by Mauchly's test of sphericity, was violated. For the prevention of α-error accumulation, the Bonferroni–Holm method was used. Furthermore, time above GET, time above 80% VO_2peak, and average oxygen uptake were calculated to estimate the overall respiratory impact of the interval protocols. A paired t-test analyzed these parameters as well as end-exercise lactate for significant effects. The level of significance was set to $p = 0.05$.

3. Results

3.1. Aerobic Rate

Average oxygen uptake (Figure 2) increased progressively from Sections 1–5 in both protocols (HIIT3m: ANOVA $p \leq 0.001$, post hoc tests $p \leq 0.001$; HIIT30s: ANOVA ≤ 0.001, post hoc tests $p \leq 0.05$). Although there was no significant difference in average oxygen uptake in any of the five sections between both protocols, there was a significant interaction effect between the first and second as well as the fourth and fifth section (ANOVA $p \leq 0.001$, post hoc tests $p \leq 0.002$).

The loaded phases of all sections were significantly higher in the HIIT3m compared to HIIT30s (ANOVA $p \leq 0.001$, post hoc tests $p \leq 0.001$). During the unloaded phases, VO_2 was higher in HIIT30s (ANOVA $p \leq 0.001$, post hoc tests $p \leq 0.001$, Figure 3).

3.2. Heart Rate

The average heart rate (Figure 2) also increased progressively from Sections 1–5 in HIIT3m (ANOVA $p \leq 0.001$, post hoc tests $p \leq 0.001$). In HIIT30s, relative HR increased up to Section 3 and flattened afterward (ANOVA $p \leq 0.001$, post hoc tests $p \leq 0.05$).

The loaded phases of all sections were significantly higher in HIIT3m compared to HIIT30s (ANOVA $p \leq 0.001$, post hoc tests $p \leq 0.001$). In contrast, HIIT30s resulted in higher HR during the unloaded phases (ANOVA $p \leq 0.001$, post hoc tests $p \leq 0.001$) (Figure 3).

3.3. Cardiac Output

Cardiac output (Figure 2) increased progressively with increasing exercise duration in both protocols (ANOVA $p \leq 0.001$, post hoc tests $p \leq 0.002$), except between Sections 2 and 3 in HIIT3m (Figure 3). Furthermore, there was a significant difference between both protocols in the first two sections, where the HIIT3m evoked a higher cardiac output (ANOVA $p = 0.045$, post hoc tests $p \leq 0.025$).

Cardiac output during the loaded phases was significantly higher ($p \leq 0.001$) in HIIT3m for all sections. Contrary to the loaded phases, HIIT30s resulted in higher cardiac output in the unloaded phases ($p \leq 0.001$, Figure 3).

3.4. Stroke Volume

Stroke volume remained stable despite the increasing exercise duration, and section averages did not differ significantly between the two protocols. This also applied when data was separated into loaded and unloaded phases (Figures 2 and 3).

3.5. Muscle Hemodynamics

HHb/VO_2 increased from Sections 1–3 (ANOVA $p = 0.001$, post hoc tests $p \leq 0.05$) in both protocols and stayed stable thereafter (Figure 2). However, there was a significant interaction effect indicating a steeper increase in HIIT30s between Sections 1 and 2 ($p = 0.027$).

Loaded and unloaded phases did not differ significantly between both protocols. But in HIIT30s, there was a significant interaction effect of HHb/VO_2 between Sections 1 and 2 as well as Sections 2 and 3 in the loaded phases ($p \leq 0.05$, Figure 3).

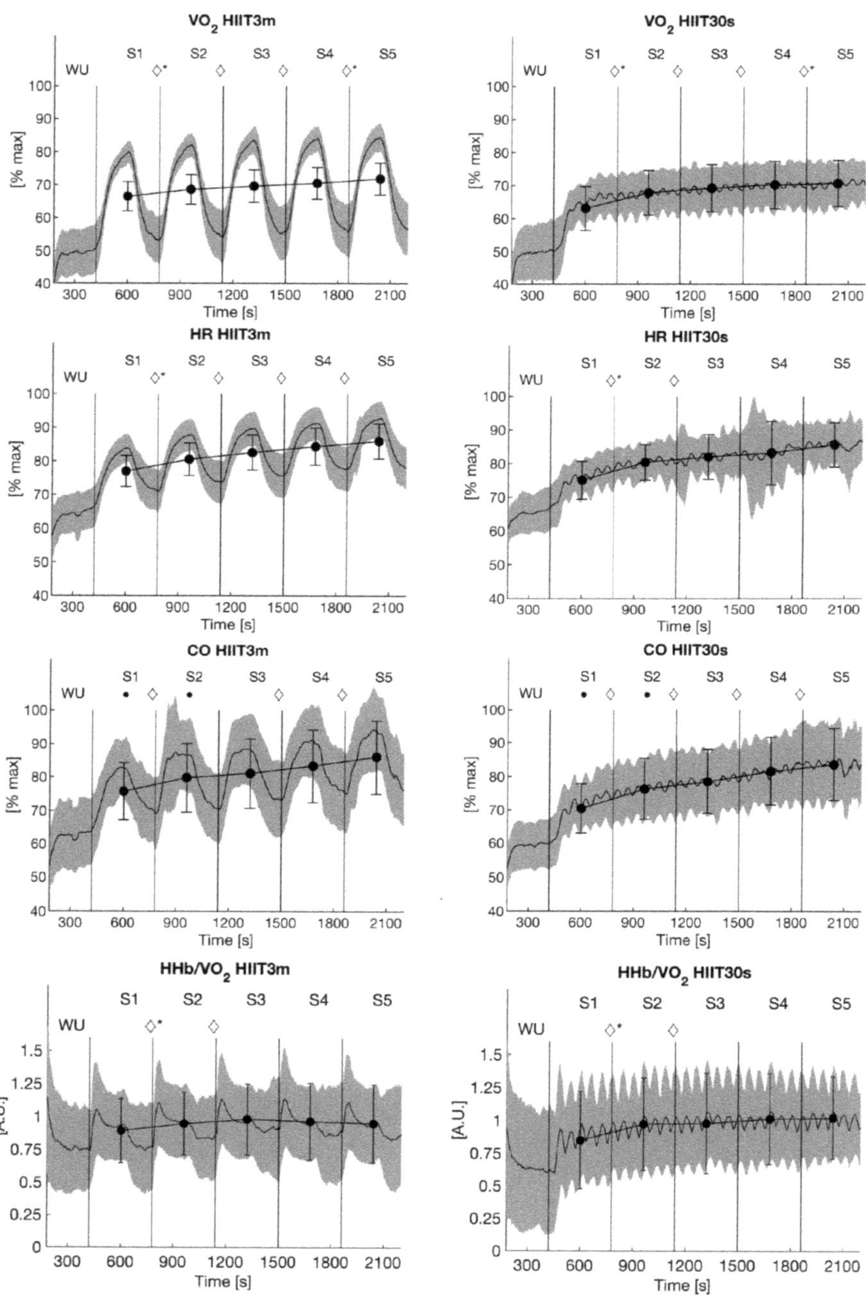

Figure 2. Averaged traces of relative oxygen uptake, heart rate, cardiac output and ΔHHb/VO_2 $\pm$ SD (gray area); averaged values of each 6 min-section are presented as mean $\pm$ SD; $\Diamond$ indicate significant differences between two successive sections; interaction effects are marked with * ($p \leq 0.05$).

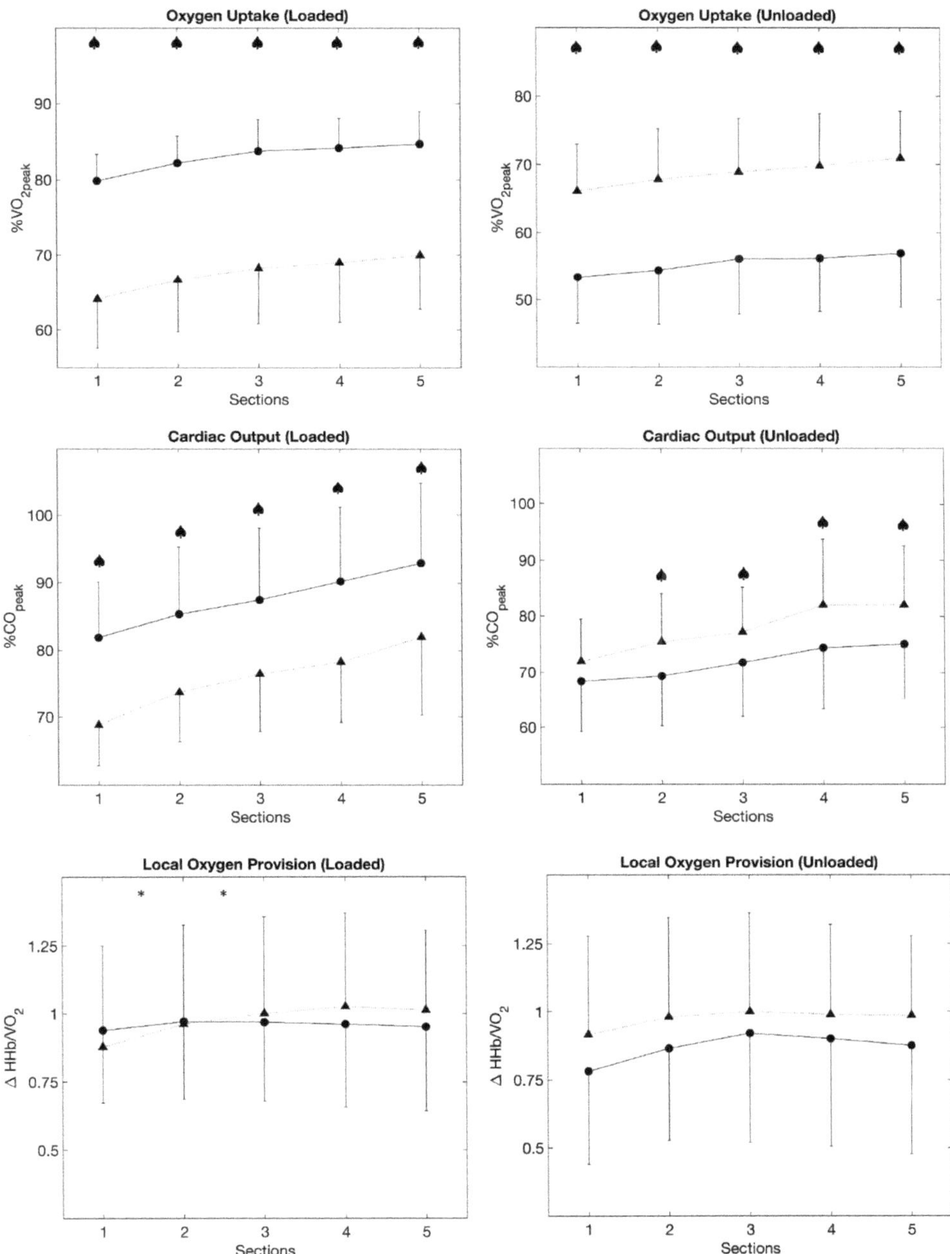

Figure 3. Averaged values for oxygen uptake, cardiac output and $HHbVO_2$ at the end of loaded and unloaded phases for HIIT3m (•, solid line) and HIIT30s (▲, dashed line); ♠ indicates significant differences between both protocols ($p \leq 0.05$); significant interaction effects are marked with * ($p \leq 0.05$).

In HIIT3m, TSI dropped significantly in response to the first interval ($p < 0.001$), decreased further after the second interval ($p = 0.012$), and remained stable thereafter. This drop was not significant in HIIT30s.

3.6. End-Exercise Lactate

End-exercise lactate was significantly ($p \leq 0.001$) higher in HIIT3m (6.2 ± 2.2 mmol/L) compared to HIIT30s (4.2 ± 2.0 mmol/L).

4. Discussion

In order to optimize the effects of training, it is important to know the impact of different training regimens on the desired performance-predicting physiological systems. Our results demonstrate that interval training with long intervals increases training time at high rates of oxygen uptake (i.e., higher aerobic rates) and cardiac output compared to a short interval training at an identical overall workload. This suggests that long-interval training approaches the central cardiovascular system more intensively. Although average responses (i.e., including low intervals) of oxygen uptake and cardiac output did not differ significantly, it agrees with recent findings that the exercise time at high intensities might be crucial for triggering positive effects [25]. Overall metabolism rises disproportionately [24,26], therefore higher aerobic rates represent considerably higher metabolic stimuli, despite the longer unloaded phases during the HIIT3m. This is also supported by much higher lactate values in the HIIT3m.

We have shown in previous studies that isolated 30 s work intervals might lead to lower cardiovascular responses compared to 3 min work intervals [23]. However, repeated 30 s work-interval bouts could nevertheless be effective in a similar way compared to longer interval durations. In the current study, results promote the 3 min intervals to evoke a higher impact on the cardiovascular system. Short-interval training is very often prescribed using short maximal or supramaximal intensities, interspersed by either passive breaks or longer, active relief intervals [1]. This is associated with having a greater effect on VO_2max development by higher rates of oxygen consumption during training. In the current study, we did not consider this type of interval training as we aimed to focus solely on the effect of interval duration without changing the work rate.

4.1. Oxygen Uptake and Central Responses

As mentioned before, long intervals had a higher impact on peak oxygen uptake and peak cardiac output during work intervals. This was associated with the long-lasting demand for oxygen, which is reflected by a plateau or only a minor increase in oxygen uptake at the end of the loaded phases. Work intervals were too short to affect a similar VO_2 response in HIIT30s and therefore VO_2 reached lower values without attaining a plateau. On the other hand, VO_2 dropped lower during resting phases in HIIT3m compared to HIIT30s, resulting in a high amplitude of VO_2 response to the training session. Contrary, there was only a small variation between loaded and unloaded phases in HIIT30s. Despite this, the average response of VO_2 to both protocols was similar. Nevertheless, the participants' aerobic rate did not recover to baseline values in both protocols. Thus, average oxygen uptake increased almost consistently with each interval section.

The high difference in VO_2 between the loaded and unloaded phases in HIIT3m as well as the small variation in HIIT30s resulted in an equal average response in all sections and protocols. In detail, the HIIT3m elicited shorter training times above GET than the short interval training (1669.9 ± 310.9 s vs. 1769.5 ± 189.0 s, $p = 0.034$) while the average fractional utilization of VO_2peak was similar (65.23 ± 4.68% VO_2peak vs. 64.39 ± 6.78% VO_2peak, $p = 0.261$). However, the long interval training caused a significantly longer time above the target intensity of 80% VO_2peak (377.3 ± 254.4 s vs. 121.7 ± 370.9 s, $p = 0.001$). Further, 15 subjects exceeded the fractional utilization of VO_2peak that is corresponding to the occurrence of RCP in the ramp test. In contrast, only 4 subjects reached that level during the short interval training. As oxygen uptake responses to constant load exercise above GET is not only dependent on exercise intensity but also on exercise time due to the oxygen uptake slow component [27], this result is not surprising. However, it indicates higher aerobic flux during the HIIT3m protocol. 8 subjects reached their RCP at intensities of 85% VO_2peak and higher and therefore did not reach RCP level in fractional utilization of VO_2peak despite the slow component. These findings indicate that the HIIT3m affects a longer training period in the severe intensity domain. which is typically one major reason for the application of interval regimens in endurance sports.

Directly linked to oxygen uptake, a similar behavior was shown in cardiac output. As a result of incomplete recovery, cardiac output increased in both protocols due to an increase in heart rate, while stroke volume remained constant. Again, the long intervals resulted in higher cardiovascular stress during the loaded phases compared to short intervals. But the section average values of cardiac output showed again a similar impact on the cardiovascular system. However, adjustment of the cardiac output took longer in the long-interval protocol (Figure 3).

4.2. Microvascular O_2 Provision

As previously mentioned, microvascular oxygen provision was estimated by ΔHHb per unit VO_2. The larger the ratio, the more O_2 is extracted in relation to VO_2 uptake. Lower values would indicate a smaller increase of ΔHHb at a given workload by higher local muscle oxygen provision [8,24]. In both protocols, the average ratio of local oxygen supply to systemic demand increased until Section 3, i.e., until the 18th minute of both interval training, and flattened afterward. Local vasodilation could be associated with the improvement of local O_2 provision after the first three sections. It is assumed that vasoactive substances, like lactate or acetylcholine, might promote tissue perfusion and therefore enhance oxygen uptake on-kinetics, e.g., the response to the work-intervals in this study [24,27,28]. Furthermore, acid substances are known to shift the O_2 dissociation curve and thus improve muscular O_2 utilization according to the Bohr Effect [29,30]. However, the Bohr Effect might be of smaller importance in the current study as exercise times in the severe intensity domain were quite short and therefore is not likely to lower local pH that much. Moreover, intermittent contractions during cycling might also enhance blood flow in terms of muscle pumps. Especially during the active rest, where small intensities had to be handled, mean arterial pressure might be maintained constant during the recovery phases compared to a passive rest [31]. However, ΔHHb/VO_2 increased significantly during the loaded phases in HIIT30s compared to HIIT3m (Figure 3). Although ΔHHb/VO_2 did not differ following Section 3, HIIT3m presumably promotes vascular effects faster than the HIIT30s.

These results were partially in line with the results of Zafeiridis' study [11]. Although different intensities and passive rests had been used, no significant differences in local muscle perfusion were detected, while cardiovascular stress was increased by a long interval training. Zafeiridis and colleagues speculated that both long and short intervals trigger microvascular and metabolic adaption due to equal average central responses and similar ΔHHb/VO_2 [11].

The adaption of interval duration at submaximal workloads is an important parameter to control the cardiorespiratory loading. Reductions in interval length should therefore be accompanied by an increase in workload to achieve a comparable training stimulus for the cardiorespiratory system. The equal work rates might limit the results of the study due to different intensity domains resulting in different metabolic demands. However, interval training affects muscle perfusion independent of interval duration. Thus, short submaximal interval training can be a part of rehabilitation programs to improve muscular perfusion with less cardiac stress. A training study would be necessary to validate the results and estimate long-term adaption effects in the muscle and cardiorespiratory systems.

5. Conclusions

In conclusion, the study aimed to detect possible differences in central and periphera hemodynamics in two interval protocols at the same work rate. Tested short and long interval durations increased average response in oxygen uptake, cardiac output, heart rate and local oxygen provision while stroke volume maintained almost constant. Differences between the protocols were detected regarding loaded and unloaded phases in central hemodynamic parameters and oxygen uptake, where the long-interval training regimen results in stronger cardiovascular and metabolic responses. Therefore, interval duration is a useful tool to adapt interval training to the needs and aims of the training process.

Author Contributions: K.S.H.K. and F.S. have given substantial contributions to the conceptualization and methodology of the project. K.S.H.K. and A.M. collected the data, while all authors were responsible for software, validation, investigation resources, and data curation. The original draft was prepared and edited after a critical review of all authors by K.S.H.K. and F.S. were responsible for the visualization. F.S. held the supervision and project administration. All authors have read and agreed to the published version of the manuscript.

Funding: This research received no external funding.

Institutional Review Board Statement: This study was carried out in accordance with the Declaration of Helsinki and was approved by the local Ethics Committee of the Technical University of Munich (#67/14, 2014).

Informed Consent Statement: Informed consent was obtained from all subjects involved in the study.

Data Availability Statement: All data, tables and figures presented in this manuscript are original. Further inquiries can be directed to the corresponding author.

Acknowledgments: We acknowledge financial support from Universität der Bundeswehr München.

Conflicts of Interest: The authors declare no conflict of interest.

References

1. Buchheit, M.; Laursen, P.B. High-intensity interval training, solutions to the programming puzzle: Part I: Cardiopulmonary emphasis. *Sports Med.* **2013**, *43*, 313–338. [CrossRef]
2. Guiraud, T.; Nigam, A.; Juneau, M.; Meyer, P.; Gayda, M.; Bosquet, L. Acute responses to high-intensity intermittent exercise in CHD patients. *Med. Sci. Sports Exerc.* **2011**, *43*, 211–217. [CrossRef] [PubMed]
3. Meyer, P.; Guiraud, T.; Gayda, M.; Juneau, M.; Bosquet, L.; Nigam, A. High-intensity aerobic interval training in a patient with stable angina pectoris. *Am. J. Phys. Med. Rehabil.* **2010**, *89*, 83–86. [CrossRef] [PubMed]
4. Buchheit, M.; Laursen, P.B. High-intensity interval training, solutions to the programming puzzle: Part II: Anaerobic energy, neuromuscular load and practical applications. *Sports Med.* **2013**, *43*, 927–954. [CrossRef] [PubMed]
5. Helgerud, J.; Høydal, K.; Wang, E.; Karlsen, T.; Berg, P.; Bjerkaas, M.; Simonsen, T.; Helgesen, C.; Hjorth, N.; Bach, R.; et al. Aerobic high-intensity intervals improve VO2max more than moderate training. *Med. Sci. Sports Exerc.* **2007**, *39*, 665–671. [CrossRef] [PubMed]
6. Milanović, Z.; Sporis, G.; Weston, M. Effectiveness of high-intensity interval training (HIT) and continuous endurance training for VO2max improvements: A systematic review and meta-analysis of controlled trials. *Sports Med.* **2015**, *45*, 1469–1481. [CrossRef] [PubMed]
7. Calbet, J.A.L.; Lundby, C. Skeletal muscle vasodilatation during maximal exercise in health and disease. *J. Physiol.* **2012**, *590*, 6285–6296. [CrossRef] [PubMed]
8. Murias, J.M.; Spencer, M.D.; DeLorey, D.S.; Gurd, B.J.; Kowalchuk, J.M.; Paterson, D.H.; Vissing, K.; Latchman, H.; Lamboley, C.; Lamb, G.D.; et al. Speeding of VO2 kinetics during moderate-intensity exercise subsequent to heavy-intensity exercise is associated with improved local O2 distribution. *J. Appl. Physiol.* **2011**, *111*, 1410–1415. [CrossRef]
9. Belfry, G.R.; Paterson, D.H.; Murias, J.M.; Thomas, S.G. The effects of short recovery duration on VO2 and muscle deoxygenation during intermittent exercise. *Eur. J. Appl. Physiol.* **2012**, *112*, 1907–1915. [CrossRef] [PubMed]
10. Fu, T.-C.; Wang, C.-H.; Lin, P.-S.; Hsu, C.-C.; Cherng, W.-J.; Huang, S.-C.; Liu, M.-H.; Chiang, C.-L.; Wang, J.-S. Aerobic interval training improves oxygen uptake efficiency by enhancing cerebral and muscular hemodynamics in patients with heart failure. *Int. J. Cardiol.* **2013**, *167*, 41–50. [CrossRef] [PubMed]
11. Zafeiridis, A.; Kounoupis, A.; Dipla, K.; Kyparos, A.; Nikolaidis, M.G.; Smilios, I.; Vrabas, I.S. Oxygen delivery and muscle deoxygenation during continuous, long- and short-interval exercise. *Int. J. Sports Med.* **2015**, *36*, 872–880. [CrossRef] [PubMed]
12. Sue, D.Y.; Wasserman, K.; Moricca, R.B.; Casaburi, R. Metabolic acidosis during exercise in patients with chronic obstructive pulmonary disease use of the V-slope method for anaerobic threshold determination. *Chest* **1988**, *94*, 931–938. [CrossRef]

13. Beaver, W.L.; Wasserman, K.; Whipp, B.J. A new method for detecting anaerobic threshold by gas exchange. *J. Appl. Physiol.* **1986**, *60*, 2020–2027. [CrossRef]
14. Iannetta, D.; Inglis, E.C.; Mattu, A.T.; Fontana, F.Y.; Pogliaghi, S.; Keir, D.A.; Murias, J.M. A critical evaluation of current methods for exercise prescription in women and men. *Med. Sci. Sports Exerc.* **2020**, *52*, 466–473. [CrossRef] [PubMed]
15. Dupuis, M.; Noël-Savina, E.; Prévot, G.; Têtu, L.; Pillard, F.; Riviere, D.; Didier, A. determination of cardiac output in pulmonary hypertension using impedance cardiography. *Respir. Int. Rev. Thorac. Dis.* **2018**, *96*, 500–506. [CrossRef] [PubMed]
16. Charloux, A.; Lonsdorfer-Wolf, E.; Richard, R.; Lampert, E.; Oswald-Mammosser, M.; Mettauer, B.; Geny, B.; Lonsdorfer, J. A new impedance cardiograph device for the non-invasive evaluation of cardiac output at rest and during exercise: Comparison with the "direct" Fick method. *Eur. J. Appl. Physiol.* **2000**, *82*, 313–320. [CrossRef]
17. Moshkovitz, Y.; Kaluski, E.; Milo, O.; Vered, Z.; Cotter, G. Recent developments in cardiac output determination by bioimpedance: Comparison with invasive cardiac output and potential cardiovascular applications. *Curr. Opin. Cardiol.* **2004**, *19*, 229–237. [CrossRef] [PubMed]
18. Richard, R.; Lonsdorfer-Wolf, E.; Charloux, A.; Doutreleau, S.; Buchheit, M.; Oswald-Mammosser, M.; Lampert, E.; Mettauer, B.; Geny, B.; Lonsdorfer, J. Non-invasive cardiac output evaluation during a maximal progressive exercise test, using a new impedance cardiograph device. *Eur. J. Appl. Physiol.* **2001**, *85*, 202–207. [CrossRef] [PubMed]
19. Delpy, D.T.; Cope, M.; Van Der Zee, P.; Arridge, S.; Wray, S.; Wyatt, J. Estimation of optical pathlength through tissue from direct time of flight measurement. *Phys. Med. Biol.* **1988**, *33*, 1433–1442. [CrossRef]
20. Hermens, H.J.; Freriks, B.; Disselhorst-Klug, C.; Rau, G. Development of recommendations for SEMG sensors and sensor placement procedures. *J. Electromyogr. Kinesiol.* **2000**, *10*, 361–374. [CrossRef]
21. Van Beekvelt, M.C.P.; Borghuis, M.S.; Van Engelen, B.G.M.; Wevers, R.A.; Colier, W.N.J.M. Adipose tissue thickness affects in vivo quantitative near-IR spectroscopy in human skeletal muscle. *Clin. Sci.* **2001**, *101*, 21–28. [CrossRef]
22. DeLorey, D.S.; Kowalchuk, J.M.; Paterson, D.H. Effect of age on O(2) uptake kinetics and the adaptation of muscle deoxygenation at the onset of moderate-intensity cycling exercise. *J. Appl. Physiol.* **2004**, *97*, 165–172. [CrossRef] [PubMed]
23. Stöcker, F.; Von Oldershausen, C.; Paternoster, F.K.; Schulz, T.; Oberhoffer, R. Relationship of post-exercise muscle oxygenation and duration of cycling exercise. *BMC Sports Sci. Med. Rehabil.* **2016**, *8*, 9. [CrossRef]
24. Stocker, F.; Von Oldershausen, C.; Paternoster, F.K.; Schulz, T.; Oberhoffer, R. End-exercise ΔHHb/ΔVO2and post-exercise local oxygen availability in relation to exercise intensity. *Clin. Physiol. Funct. Imaging* **2017**, *37*, 384–393. [CrossRef] [PubMed]
25. Romijn, J.A.; Coyle, E.F.; Sidossis, L.S.; Gastaldelli, A.; Horowitz, J.F.; Endert, E.; Wolfe, R.R. Regulation of endogenous fat and carbohydrate metabolism in relation to exercise intensity and duration. *Am. J. Physiol.* **1993**, *265*, E380–E391. [CrossRef] [PubMed]
26. Murias, J.M.; Kowalchuk, J.M.; Paterson, D.H. Speeding of Vo2 kinetics with endurance training in old and young men is associated with improved matching of local O_2 delivery to muscle O_2 utilization. *J. Appl. Physiol.* **2010**, *108*, 913–922. [CrossRef] [PubMed]
27. Burnley, M.; Jones, A.M. Oxygen uptake kinetics as a determinant of sports performance. *Eur. J. Sport Sci.* **2007**, *7*, 63–79. [CrossRef]
28. Saltin, B.; Rådegran, G.; Koskolou, M.D.; Roach, R. Skeletal muscle blood flow in humans and its regulation during exercise. *Acta Physiol. Scand.* **1998**, *162*, 421–436. [CrossRef]
29. Gerbino, A.; Ward, S.A.; Whipp, B.J. Effects of prior exercise on pulmonary gas-exchange kinetics during high-intensity exercise in humans. *J. Appl. Physiol.* **1996**, *80*, 99–107. [CrossRef]
30. Rossiter, H.; Ward, S.A.; Kowalchuk, J.M.; Howe, F.; Griffiths, J.R.; Whipp, B.J. Effects of prior exercise on oxygen uptake and phosphocreatine kinetics during high-intensity knee-extension exercise in humans. *J. Physiol.* **2001**, *537*, 291–303. [CrossRef]
31. Tschakovsky, M.E.; Sheriff, D.D. Immediate exercise hyperemia: Contributions of the muscle pump vs. rapid vasodilation. *J. Appl. Physiol.* **2004**, *97*, 739–747. [CrossRef] [PubMed]

Article

Plasma Fibrinogen Independently Predicts Hypofibrinolysis in Severe COVID-19

Diana Schrick [1], Margit Tőkés-Füzesi [2], Barbara Réger [2] and Tihamér Molnár [1,*]

[1] Anesthesiology and Intensive Therapy, Clinical Centre, Medical School, University of Pécs, 7624 Pecs, Hungary; schrickdiana1@gmail.com

[2] Department of Laboratory Medicine, Clinical Centre, Medical School, University of Pécs, 7624 Pecs, Hungary; tfgitta@gmail.com (M.T.-F.); reger.barbara@pte.hu (B.R.)

* Correspondence: tihamermolnar@yahoo.com

Abstract: High rates of thrombosis are present in patients with severe acute respiratory syndrome coronavirus 2 (SARS-CoV-2). Deeper insight into the prothrombotic state is essential to provide the best thromboprophylaxis care. Here, we aimed to explore associations among platelet indices, conventional hemostasis parameters, and viscoelastometry data. This pilot study included patients with severe COVID-19 (n = 21) and age-matched controls (n = 21). Each patient received 100 mg aspirin therapy at the time of blood sampling. Total platelet count, high immature platelet fraction (H-IPF), fibrinogen, D-dimer, Activated Partial Thromboplastin Time, von Willebrand factor antigen and von Willebrand factor ristocetin cofactor activity, plasminogen, and alpha2-antiplasmin were measured. To monitor the aspirin therapy, a platelet function test from hirudin anticoagulated whole blood was performed using the ASPI test by Multiplate analyser. High on-aspirin platelet reactivity (n = 8) was defined with an AUC > 40 cut-off value by ASPI tests. In addition, in vitro viscoelastometric tests were carried out using a ClotPro analyser in COVID-associated thromboembolic events (n = 8) ($p = 0.071$) nor the survival rate ($p = 0.854$) showed associations with high on-aspirin platelet reactivity status. The platelet count ($p = 0.03$), all subjects. COVID-19 patients presented with higher levels of inflammatory markers, compared with the controls, along with evidence of hypercoagulability by ClotPro. H-IPF (%) was significantly higher among non-survivors (n = 18) compared to survivors ($p = 0.011$), and a negative correlation ($p = 0.002$) was found between H-IPF and plasminogen level in the total population. The platelet count was significantly higher among patients with high on-aspirin platelet reactivity ($p = 0.03$). Neither the ECA-A10 ($p = 0.008$), and ECA-MCF ($p = 0.016$) were significantly higher, while the tPA-CFT ($p < 0.001$) was significantly lower among patients with high on-aspirin platelet reactivity. However, only fibrinogen proved to be an independent predictor of hypofibrinolysis in severe COVID-19 patients. In conclusion, a faster developing, more solid clot formation was observed in aspirin 'non-responder' COVID-19 patients. Therefore, an individually tailored thromboprophylaxis is needed to prevent thrombotic complications, particularly in the hypofibrinolytic cluster.

Keywords: COVID-19; platelet; IPF; hemostasis; aggregometry; viscoelastic test

Citation: Schrick, D.; Tőkés-Füzesi, M.; Réger, B.; Molnár, T. Plasma Fibrinogen Independently Predicts Hypofibrinolysis in Severe COVID-19. *Metabolites* **2021**, *11*, 826. https://doi.org/10.3390/metabo11120826

Academic Editor: Norbert Nemeth

Received: 2 November 2021
Accepted: 25 November 2021
Published: 30 November 2021

1. Introduction

The COVID-19 pandemic, caused by SARS-CoV-2, is a contagious potentially life-threatening disease that has caused more than five million deaths worldwide [1]. Clinical characteristics of the disease can range from mild upper tract respiratory infection to multiple organ disfunction (MODS), multiple organ failure (MOF), and fatal hypoxemic respiratory failure [2,3]. There is a great body of evidence that COVID-19 significantly affects the coagulation system, contributing to hypercoagulable states and thrombotic events. The reason for such alterations is multifactorial, including the activation of the thrombo-inflammatory cascade and endothelial dysfunction [4,5]. There is a great need to identify novel markers to stratify disease severity and predict the outcome of disease.

Such attempts can not only provide deeper insight into the pathological process, but also allow more accurate triage and faster therapeutic interventions. Based on this concept, a faster correction of abnormal coagulation parameters might be associated with improved prognosis in infected patients [6–8]. In addition to primary hemostasis, platelets play an important role in inflammatory and immune responses. According to a recent study, platelet count per se provide valuable data in the assessment of disease severity and outcome [9]. Zhao et al. reported that an early decrease in blood platelet count was associated with poor prognosis in COVID-19 patients [10]. Hypothetically, an infection-induced cytokine storm or the virus itself directly affects bone marrow via CD13/CD66a and destroys cells and inhibits hematopoiesis [9,11,12]. In contrast, SARS-CoV-2-associated thrombocythemia has been reported too [12,13]. Platelet indices, such as immature platelet fraction (IPF, %), are valid indicators of thrombopoiesis level [14]. IPF represents young cells that have recently been released into the circulation and contain a higher concentration of ribonucleic acid than mature platelets [14]. Recently, IPF was reported as a novel early predictive marker for disease severity in patients with COVID-19 [15]. Critically ill COVID-19 patients have impaired fibrinolysis. The hypofibrinolytic state due to decreased fibrinolytic response may contribute to COVID-associated thromboembolic events. The decreased fibrinolytic response was recently defined as a lysis time (LT) >393 s. The aim of this clinical study was to explore associations among platelet indices and conventional hemorheological parameters in patients with severe SARS-CoV-2 infection and their impact on the clinical outcome [16].

The aim of this clinical study was to explore associations among platelet indices and conventional hemorheological parameters in patients with severe SARS-CoV-2 infection. In addition, the changes of platelet reactivity and fibrinolytic response contributing to the etiology of an increased thrombotic risk associated with COVID-19 were also examined here.

2. Results

2.1. Patients and Healthy Subjects

A total number of 21 COVID-19 patients (male: 12) and 21 age-matched SARS-CoV-2 PCR-negative control subjects were enrolled into this prospective observational study. All patients had SARS-CoV-2 PCR positivity, and they required intensive care with oxygen therapy with or without ventilator support. The demography, past medical history (co-morbidities), and clinical data of the study population are summarized in Table 1. Patients were compared to an age-matched control group (69 years; IQR: 52–71 vs. 67 years; IQR 63–69; $p = 0.222$). The gender ratio was the same in both study groups. There was no significant difference in their body mass index (BMI). A total of 76% of enrolled patients had hypertension and 57% of them was treated for diabetes (T2DM). Not surprisingly, significantly higher erythrocyte sedimentation rates (ESR), D-dimer levels, von Willebrand factor antigen and von Willebrand factor ristocetin cofactor activities ($p < 0.001$) were observed on admission to the ICU. Furthermore, serum levels of IL-6 and ferritin also exceeded the normal laboratory range in patients, but these markers were not measured in the controls (not shown in the table).

We analyzed associations between erythrocyte sedimentation rate (ESR) and acute phase proteins, such as fibrinogen and hs-CRP. We found strong positive correlations between ESR and plasma fibrinogen levels, as well as serum hs-CRP concentration in patients, but not in healthy controls, reflecting the ongoing inflammatory response in severe SARS-CoV-2-infected patients ($r = 0.812$, $p < 0.001$ and $r = 0.666$, $p = 0.001$) (Figure 1).

Table 1. Demography, comorbidities, and admission laboratory parameters of the total study population.

	Patients n = 21	Controls n = 21	p
Age (y)	69 (52–71)	67 (63–69)	0.222
Male (n)	12 (50%)	11 (52%)	0.757
BMI	27 (26–33)	25 (24–26)	0.189
Hypertension	16 (76)		
Diabetes mellitus	12 (57)		
Thromboembolic event (stroke/TIA, DVT)	5 (24)		
Myocardial infarct	4 (19)		
Heart failure	1 (5)		
ESR (mm/h)	84 (36–107)	4 (4–10)	<0.001
Platelet (g/L)	214 (114–355)	261 (248–265)	0.950
IPF (%)	8.7 (6.1–12.5)	7.6 (6.7–8.2)	0.753
H-IPF (%)	2.2 (0.6–4.1)	0.9 (0.8–1.0)	0.308
Fibrinogen (g/L)	5.1 (3.5–5.4)	3.2 (2.8–3.2)	0.059
D-dimer (μg FEU/L)	2296 (1415–6260)	495 (363–575)	<0.001
APTT (s)	12.9 (12.4–14.0)	26.3 (24.9–27.8)	0.002
TT (s)	34.0 (31.2–36.7)	10.4 (10.2–10.8)	<0.001
vWF:Ag (%)	488 (412–605)	109 (97–109)	<0.001
vWF:RCo (%)	399 (353–568)	104 (97–109)	<0.001
Plasminogen (%)	86 (64–96)	104 (98–106)	0.028
Alpha-2-antiplasmin (%)	118 (107–126)	119 (116–121)	1.000
hs-CRP(mg/L)	90.5 (27.7–126.1)	1.2 (0.9–1.5)	<0.001

BMI: body mass index; TIA: transient ischemic attack; DVT: deep vein thrombosis; ESR: erythrocyte sedimentation rate; IPF: immature platelet fraction, H-IPF: high-immature platelet fraction, APTT: activated partial thromboplastin time, TT: thrombin time, vWF:Ag: von Willebrand factor antigen, vWF:RCo: von Willebrand factor ristocetin cofactor activity. hs-CRP: high-sensitivity C-reactive protein. Data are presented as count (%) or median (25th–75th percentiles).

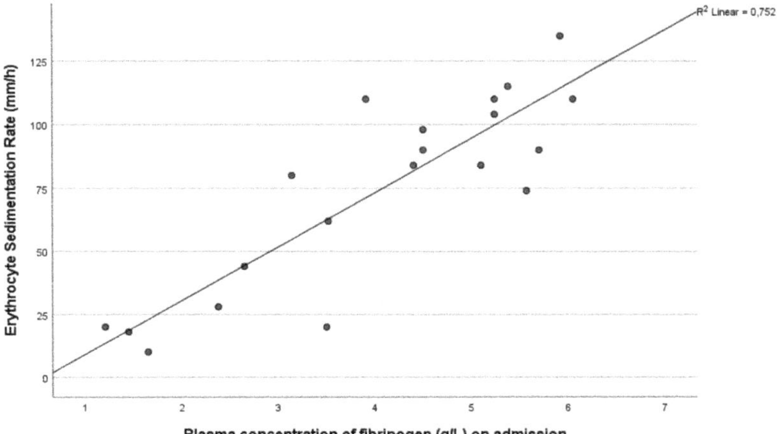

Figure 1. Correlation between erythrocyte sedimentation rate (ESR) and plasma fibrinogen level in COVID-19 patients on admission ($p < 0.001$).

A strong positive correlation was observed between von Willebrand factor antigen and plasma level of von Willebrand factor ristocetin cofactor activity in patients with severe COVID-19 (r = 0.966, $p < 0.001$). Furthermore, a positive correlation was seen between either von Willebrand factor antigen or von Willebrand factor ristocetin cofactor activity and plasma level of D-dimer (r = 0.683, $p < 0.001$; r = 0.675, $p < 0.001$, respectively—data not shown).

2.2. Non-Survivals vs. Survivals

A total of 18 patients died during intensive care, while 3 patients were discharged from hospital alive. The total platelet count showed no difference between non-survivors and survivors ($p = 0.08$). In contrast, H-IPF (%) showed significant differences when the non-survival vs. survival subgroups were compared (2.5, 1.0–4.2 vs. 0.5, 0.45–0.55; $p = 0.011$) (Figure 2). Interestingly, we detected that activated partial thromboplastin time (APTT, sec) was lower in those who died compared to survivors (13, 12.3–14.0 vs. 15.8, 14.95–26.35; $p = 0.024$).

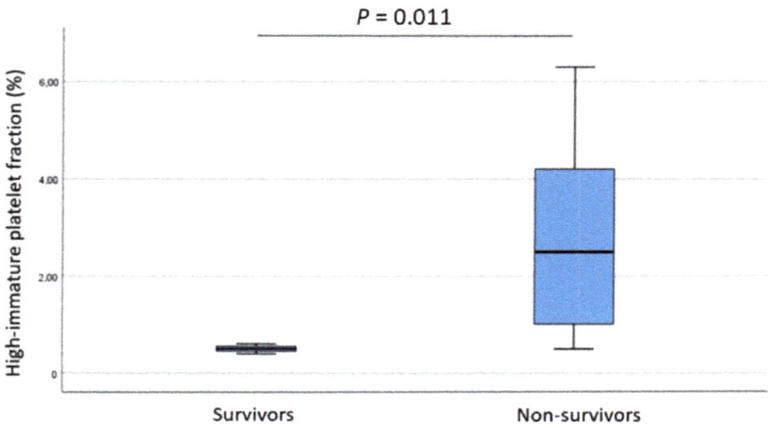

Figure 2. High-immature platelet fraction (%) in survivals and non-survivals.

A significant negative correlation was observed between H-IPF (%) and plasma plasminogen (%) among non-survivors (r = −0.572, $p = 0.002$), but not in survivors (Figure 3).

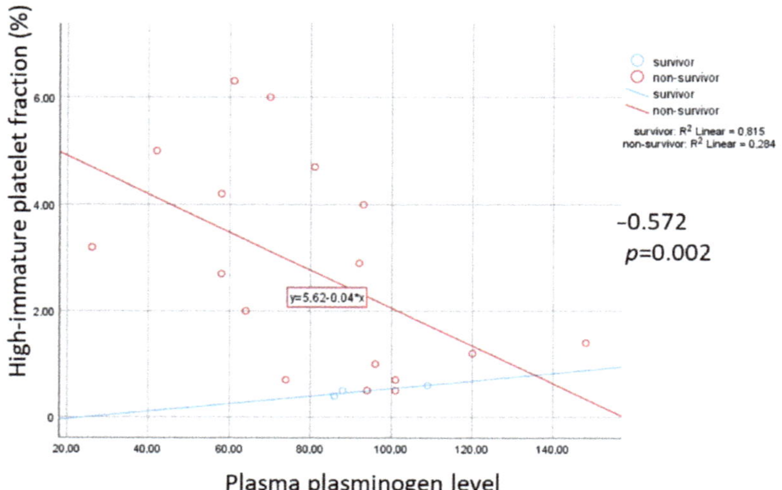

Figure 3. Correlation of high-immature platelet fraction (H-IPF, %) and plasma plasminogen level in survivals and non-survivals. Rho and *p* indicate negative correlation among non-survivals.

2.3. 'Responders' vs. 'Non-Responders'

Despite aspirin alone or in combination with enoxaparin, eight patients developed symptomatic thrombosis during their ICU stay. Patients were divided into two subgroups based on their ex vivo platelet reactivity measured by a Multiplate analyzer. High on-aspirin platelet reactivity was found in eight COVID patients using the AUC > 40 cut-off value by the ASPI test [17]. Next, COVID patients were dichotomized based on their fibrinolytic response; in patients with impaired fibrinolytic response, the AUC measured by ASPI, Risto, TRAP and ADP tests showed significant differences when aspirin responders and non-responders were compared (all $p = 0.024$, respectively) (Figure 4). Neither the thromboembolic events related to COVID-19 ($p = 0.071$), nor survival rate ($p = 0.854$) showed associations with high on-aspirin platelet reactivity status. The platelet count ($p = 0.03$), the ECA-A10 ($p = 0.008$), and ECA-MCF ($p = 0.016$) were significantly higher, while the tPA-CFT ($p < 0.001$) was significantly lower among patients with high on-aspirin platelet reactivity. In addition, the platelet count showed positive correlations with the AUC by Risto, ASPI, TRAP and ADP tests (0.500, $p = 0.021$; 0.500, $p = 0.021$; 0.760, $p < 0.001$; 0.621, $p = 0.003$, respectively), while H-IPF negatively correlated with the AUC by TRAP and ADP tests (−0.559, $p = 0.008$; −0.530, $p = 0.013$, respectively). The acute COVID-related thromboembolic events were acute coronary syndrome (n = 2), pulmonary embolism (n = 4), and ischemic stroke (n = 2). Both vWF antigen and vWF ristocetin cofactor activity showed positive correlations with H-IPF (both $p = 0.016$), and platelet count was significantly higher among patients with high on-aspirin platelet reactivity ($p = 0.03$) (data not shown).

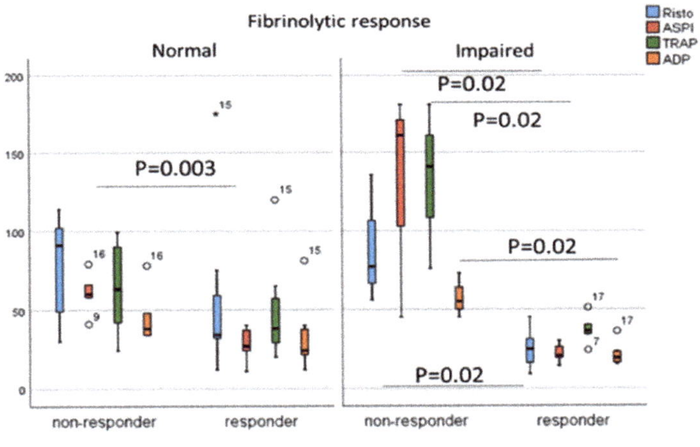

Figure 4. Comparisons of Risto, ASPI, TRAP and ADP tests (AUC) in patients with normal and impaired fibrinolytic response dichotomized based on their responsiveness to aspirin (responder vs. non-responder status). Mann-Whitney test (Asterix and white circles indicate extreme values).

Maximal clot firmness (MCF) was significantly higher measured by the ECA test ($p = 0.016$) in patients with high on-aspirin platelet reactivity (n = 8) compared to the 'responder' subgroup (n = 13), indicating larger and more solid clots despite 100 mg aspirin treatment. (Figure 5A,B). When the manufacturer's AUC < 71 cut-off value was used in comparison, the tPA lysis time (tPA LT) tended to increase among aspirin 'responder' COVID-19 patients ($p = 0.06$) compared to 'non-responders', and eight of these patients showed features of the 'fibrinolysis shut-down' phenomenon.

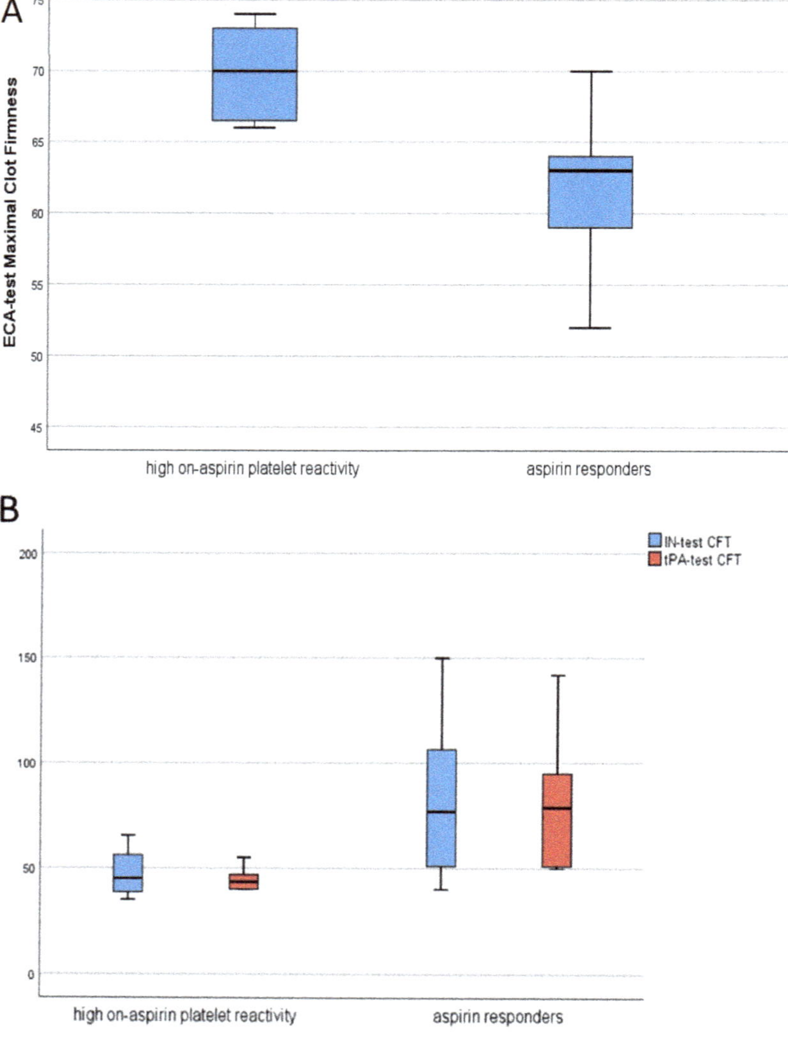

Figure 5. (**A**) Maximal clot firmness measured by ECA test in patients with high on-aspirin platelet reactivity vs. aspirin *'responders'* (Mann–Whitney test, $p = 0.016$); (**B**) Clot formation time measured by IN and tPA test in patients with high on-aspirin platelet reactivity vs. aspirin *'responders'* (Mann–Whitney test, $p = 0.039$ and $p < 0.001$, respectively).

2.4. Admission Platelet Count in COVID-19 Patients

Next, patients were divided into two subgroups based on their platelet count on admission (thrombocytopenia <150 g/L; and normocythemia >150 g/L). Both von Willebrand factor antigen (vWF:Ag) and von Willebrand factor ristocetin cofactor activities (vWF:RCo) were significantly higher in COVID-19 patients independently from their platelet count on admission compared to healthy subjects ($p < 0.001$) (data not shown). In contrast, the plasma level of plasminogen, but not alpha-2-antiplasmin, was significantly lower among COVID-19 patients with thrombocytopenia (<150 g/L) than in patients with normal platelet counts (>150 g/L) (Figure 6). In addition, significantly lower MCF values were observed in the IN and ECA tests among patients with lower platelet counts (both $p = 0.004$), while significantly higher FIB and tPA test values were detected in patients with normal platelet counts compared to healthy controls ($p = 0.014$ and $p = 0.03$, respectively) (Figure 7).

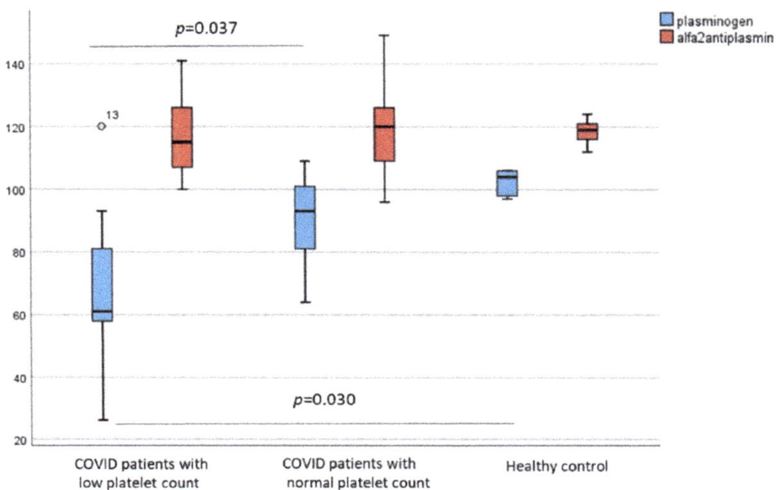

Figure 6. Plasminogen and alpha-2-antiplasmin levels in patients with low platelet counts (<150 g/L), with normal platelet counts (>150 g/L), and healthy controls.

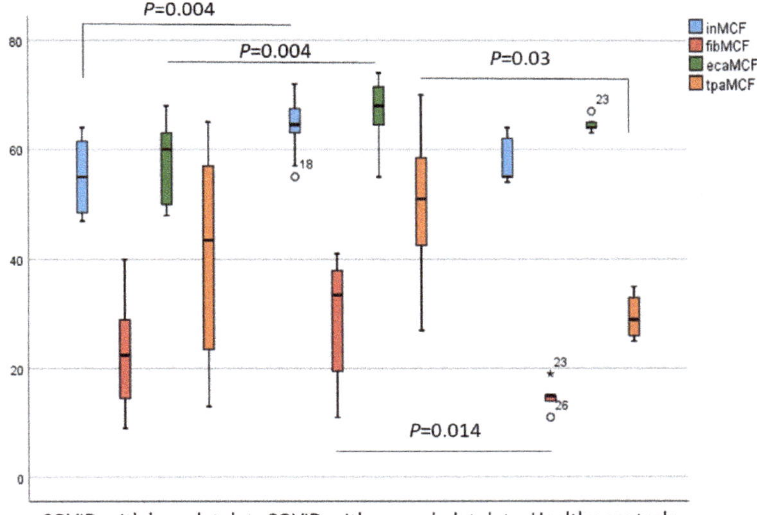

Figure 7. Maximal clot firmness MCF) by IN/FIB/ECA/tPA test respectively in patients with low platelet counts (<150 g/L), with normal platelet counts (>150 g/L), and healthy controls.

Next, COVID patients were dichotomized based on their fibrinolytic response again. Thereafter, the low platelet vs. normal platelet subgroups were compared. In patients with impaired fibrinolytic response, the AUCs measured by ASPI and TRAP tests ($p = 0.03$) were significantly higher in patients with normal platelet counts. In contrast, in the normal fibrinolytic group, only the ADP test showed significant differences ($p = 0.04$) (Figure 8).

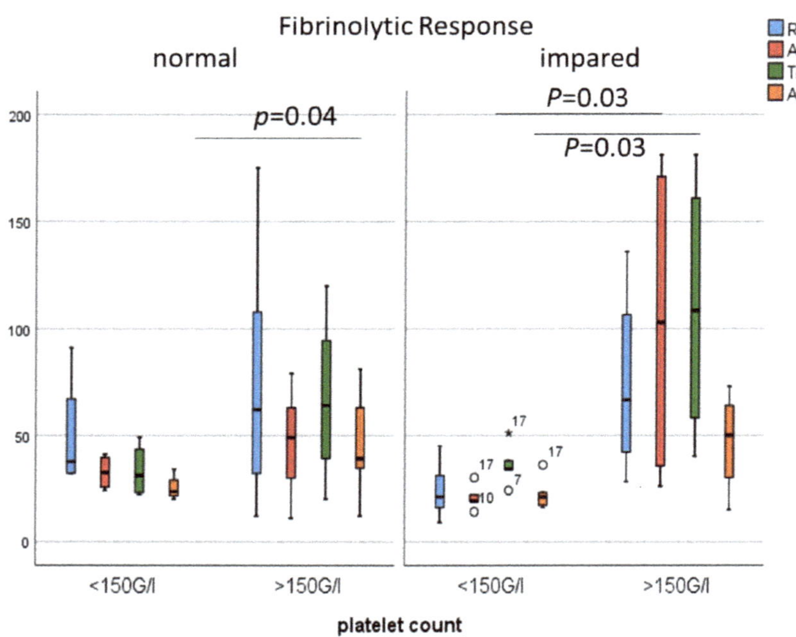

Figure 8. Comparisons of Risto, ASPI, TRAP and ADP tests (AUC) in patients with normal and impaired fibrinolytic response, dichotomized based on their platelet count (g/L).

2.5. Independent Predictors of Ompaired Fibrinolytic Response

A separate analysis was run with hypofibrinolysis as the outcome of interest. Based on binary logistic regression analysis including age, gender, D-dimer, fibrinogen, and aspirin responsiveness based on impedance electrode aggregometry by Multiplate, only fibrinogen (OR: 3.55, 95% CI: 1.33–9.47, $p = 0.01$) proved to be an independent predictor of hypofibrinolysis. The ROC analysis of plasma fibrinogen level as a predictor of hypofibrinolysis in severe COVID patients revealed a cut-off value of 3.86 g/L (AUC of 0.800, 95% CI: 0.623–0.976; $p = 0.006$) with a 78% sensitivity and 73% specificity.

3. Discussion

Despite the small number of included patients, an elevated level of H-IPF (%) was found to be predictive of the fatal outcome here. Welder et al. found that elevated percentages of IPF at presentation was predictive of the length of hospitalization, the need of ICU admission, and mechanical ventilation [15]. Importantly, a higher H-IPF level was associated with lower plasminogen levels in those COVID-19 patients who died. This finding is indirectly supported by Bertolin et al., who reported increased plasminogen activator inhibitor 1 (PAI-1) activity in COVID-19 patients, that may be due to the consumption of plasminogen in association with hypercoagulability [18]. Taken together, endothelial dysfunction (elevated vWF level), with the release of fibrinolysis inhibitor PAI-1, and hyperimmune response (increased ESR, CRP, ferritin, and IL-6) with younger (higher H-IPF) activated platelets seem to be significant contributors to thrombogenesis in COVID-19. Importantly, in our cohort, the aspirin non-responder patient group presented with not only higher platelet count, but also increased platelet reactivity based on either Risto, ASPI, TRAP or ADP tests in the hypofibrinolysis (LT > 393 s) group revealed by ClotPro (likewise by Bachler et al.) [16]. Bertolin et al. also observed lower platelet reactivity based on Multiplate aggregometry compared to healthy controls, despite having higher levels of D-dimer, fibrinogen, and PAI-1, and hypercoagulability by thromboelastometry [18]. In accordance, eight COVID-patients on aspirin exhibited an increased platelet reactivity via the ASPI test (referred to in this article as 'non-responders'). Significantly higher maximal clot firmness (MCF) was observed in ECA tests; meanwhile, significantly lower IN-, and tPA-CFT were found in 'non-responders' compared to 'responders', indicating faster developing, larger and more solid clots despite aspirin treatment. A large randomized clinical trial (*RECOVERY*) found that aspirin did not improve survival for patients hospitalized with COVID-19 [19,20]. Therefore, there is an urgent need to identify aspirin low/non-responders providing a modified antiplatelet regime or alternative strategies (e.g., activated protein C, PAI-1 antagonists, and tissue plasminogen activators) to combat thrombosis in this disease.

Based on our findings, besides platelet count itself, the MCF depends on several other factors such as plasminogen and the von Willebrand factor. Kruse et al. observed lower levels of plasminogen, suggesting that it was integrated into the clot, but unable to disintegrate it effectively, presumably by the inhibitory effect of alpha2-antiplasmin, which makes thrombi resistant to plasmin; meanwhile, plasminogen activator inhibitor (PAI-1) inhibited the activation of tissue plasminogen activator (tPA). The net effect of these may result in the 'fibrinolysis shut-down' phenomenon, leading to lysis-resistant microthrombi formation in different organs, particularly in the lungs [21]. Importantly, we aimed to explore predictors of impaired fibrinolytic response and found only fibrinogen with an OR: 3.55 as an independent predictor of hypofibrinolysis.

Moreover, ESR was found to be significantly higher in patients with severe SARS-CoV-2 infection and it showed strong positive correlation with fibrinogen concentration. Similar data were shown by Henry et al., who performed a meta-analysis involving 21 studies showing that inflammatory markers such as ESR, CRP, serum ferritin, IL-6, procalcitonin, and IL-2R were significantly elevated in patients with severe and fatal COVID-19 [22]. Another systematic review and meta-analysis detected that ESR positively correlated with COVID-19 severity [23]. CRP is an exquisitely sensitive systemic marker of acute-phase response in inflammation, infection, and tissue damage [24]. Elevation in serum CRP levels has been suggested in several studies as a reliable indicator of the presence and severity of SARS-CoV-2 infection [25–27]. D-dimer arises from the lysis of cross-linked fibrin and indicates the activation of coagulation and fibrinolysis [25,28]. Although the tPA lysis time (tPA LT) tended to be increased among aspirin 'non-responder' COVID-19 patients ($p = 0.06$), the D-dimer concentration was not different between 'responder' and 'non-responder' subgroups. There was no difference in D-dimer concentrations between survival and non-survival subgroups in our cohort as well (also see limitations of our study). In contrast, larger studies reported D-dimer level as a predictor for mortality. Zhang et al. stated that it is an independent factor of all-cause death in hospitalized patients with COVID-19 [29]. Regarding the kinetics, Corrado et al. found that non-survivals had rapidly increasing D-dimer levels [30]. Due to the very low survival rate of our cohort, independent predictors of mortality could not be analyzed here.

Despite the fact that we detected reduced activated partial thromboplastin time (APTT) in non-survivors, in a recent Dutch study evaluating ICU patients with COVID-19, prolongation of the prothrombin time >3 s and activated partial thromboplastin time >5 s were found to be independent predictors of thrombotic complications [31]. In our cohort, the thromboembolic complication was only associated with a reduced clotting time in the FIB test.

In fact, a high number of patients with COVID-19 die due to thromboembolic complications. Della-Morte et al. hypothesized plasminogen as the precursor for fibrinolysis [32]. Our finding was supported by them, because they found that low levels of plasminogen strongly correlated with mortality. Recently, plasminogen was suggested to play a pivotal role in controlling the complex mechanisms beyond COVID-19 complications, so it could be a useful prognostic marker and a potential therapeutic target [33].

Elevated vWF levels, as we also observed in our own cohort, imply activated or damaged endothelium [34]. It would be anticipated that damaged endothelium would result in the release of ultra-large vWF multimers capable of interacting with platelets, leading to platelet activation, microthrombi, and platelet consumption [5]. In accordance, we also found positive correlations between vWF antigen and activity and H-IPF (%) among patients with high on-aspirin platelet reactivity. Studies have shown that patients with COVID-19 have significantly elevated levels of vWF antigen and activity, likely contributing to an increased risk of thrombosis [35].

In summary, a faster-developing, larger and more solid clot formation was observed in aspirin 'non-responder' COVID-19 patients than 'responders' here. Based on ClotPro analysis, the clot seemed to be resistant to lysis in the 'non-responders' (longer lysis), suggesting that this cluster of patients belong to the 'hypofibrinolysis or fibrinolysis shut-down' group, but this requires further validation. Nevertheless, several physiological aspects should also be considered in viscoelastic studies because activation of the vessel wall, the endothelium, and platelets entering the clot is not present in vitro. Nevertheless, our results suggest the necessity of an individual approach regarding antiplatelet therapy, as was recently confirmed in other vascular diseases [36,37]. Our observations deserve further validation in a larger prospective cohort, as there is an urgent need for individually tailored thromboprophylaxis to prevent fatal complications such as symptomatic thrombosis in severe COVID-19 patients.

4. Limitations

First, this is a small single-center study. Results need to be confirmed on a larger sample size of COVID patients with different severity clusters. Secondly, sampling at multiple time points instead of a single time could clarify whether the kinetics of such variables differ in various outcome subgroups. Thirdly, the rigid inclusion/exclusion criteria limit the generalizability of this study.

5. Methods

This pilot study was approved by the Hungarian Medical Research Council (20783-5/2020/EÜIG). All procedures were performed in accordance with the ethical guidelines of the 1975 Declaration of Helsinki. Written informed consent was provided by all participants or relatives before enrolment in the present study. A total of 21 patients with severe SARS-CoV-2 infection (with the inclusion criteria: requirement of O_2 supplementation and signed informed consent) were retrospectively analyzed from a prospective database at the Coronavirus Crisis Centre of the Clinical Centre at the University of Pécs, Pécs, Hungary. Patients under 18 years old, with congenital hemostatic abnormalities, anamnestic/current malignancy, and pregnant women were excluded from the study. Patients hospitalized in the ICU were on 100 mg/d aspirin and prophylactic anticoagulation (with enoxaparin uniformly, 1x/d) based on our local therapeutic protocol. Patients who were enrolled into the study did not receive non-steroid anti-inflammatory drugs; only paracetamol was given occasionally (e.g., in case of fever) and basic analgosedation was conducted with opioids (sufentanil uniformly). Twenty-one SARS-CoV-2 PCR negative health care workers served as healthy controls. Blood samples for measurements were drawn into a closed system blood sampling tube with 3.2% Na_3-citrate (Becton Dickinson, Diagon Ltd., Budapest, Hungary), hirudin (Sarstedt S-Monovette® 1.6 mL Hirudin), and K_3-EDTA (Becton Dickinson, Diagon Ltd., Budapest, Hungary) as anticoagulant and serum separator tubes without anticoagulant. Samples were processed within a maximum of 1 h after collection. The blood collection from volunteers was carried out through vein puncture with a 21-gauge needle into a closed system.

5.1. Blood Count, Platelet Count, High Immature Platelet Fraction (H-IPF) Measurement

The total blood cell count from the whole blood and the absolute neutrophil count after 1 h of sedimentation from the upper and lower part of the blood were measured on a Sysmex XN 9000 integrated automated hematology analyzer (Sysmex Co., Kobe, Japan, 2017). The platelet number (PLT-F) was measured using the fluorescent platelet channel of the analyzer. In this channel, the platelets were specifically stained intracellularly with fluorescent dye and measured on the principle of flow cytometry, analyzing the forward scattered light (FSC), side scattered light (SSC) and side fluorescent light (SFL). The platelets were counted and additionally, the plots in the area with high fluorescence intensities were separated into the immature platelet fraction and the research parameter, the high immature platelet fraction (H-IPF).

5.2. The Erythrocyte Sedimentation Rate (ESR)

The ESR test measures how quickly red blood cells sedimentate in the test tube. The rate at which red blood cells settle is measured as the number of millimeters of clear plasma present at the top of the column after one hour (mm/h). For the manual determination of ESR according to Westergren, we used a BD seditainer stand with an adjustable zero mark. After swiveling the tube to mix the blood sample and preparation, the tubes were immediately placed in the stand to start the measurement. After 1 h of sedimentation, the results were read.

5.3. Hemostasis

Fibrinogen (quantitatively determined based on the Clauss method) and Activated Partial Thromboplastin Time (APTT) were measured as part of the routine hemostasis parameters on an ACL-TOP-750 analyzer (Werfen, Hungary) with Q.F.A. Thrombin (Bovine; HemosIL®, Werfen, Hungary) and APTT-SP (liquid; HemosIL®, Werfen, Hungary) reagent, respectively.

The special hemostasis tests were measured on an ACL-TOP-500 analyzer (Werfen, Hungary). The quantitative determination of von Willebrand factor antigen (vWF:Ag) and von Willebrand factor ristocetin cofactor activity (vWF:RCo) was performed with an automated latex enhanced immunoassay, both with HemosIL® reagent. For quantitative measurement of plasminogen we used an automated chromogenic assay (Plasminogen; HemosIL®). The quantitative determination of alpha2-antiplasmin as an important regulator of the fibrinolytic system was carried out using an automated chromogenic assay (Plasmin Inhibitor, HemosIL®).

To monitor the aspirin therapy, we performed a platelet function test from hirudin anticoagulated whole blood within 1 h after blood sampling on a Cobas® Multiplate® Analyzer (Roche Diagnostics, Mannheim, Germany) using the ASPI test (using arachidonic acid as an activator). The aggregation level was expressed as the area under the curve (AUC). The AUC was calculated by the analyzer using the product of aggregation unit (AU) $\times$ time (minutes). Given the lack of universal cut-off values, the normal aggregation range for the ASPI test was expected as AUC: 71-115U according to the manufacturer (laboratory cut off value). However, previous studies suggest that patients were considered as 'responders' to aspirin therapy with an AUC < 40; and 'non-responders' with an AUC ≥ 40. In our data set, (n = 13) were defined as 'responders' and (n = 8) 'non-responders', showing high on-aspirin platelet reactivity.

Viscoelastrometric testing was carried out on a ClotPro (DiaCare Solutions, Mumbai, India) in vitro *POCT* coagulation analyzer. It uses pipettes prefilled with starting reagents and 340 μL of citrated whole blood to initiate measurement. For measurement, it uses a stationary pin placed in a moving cup, from which the reduction in movement is detected and charted as the amplitude resulting in thrombelastometry curves. As standard tests in COVID-19 and control patients, we used the EX test (tissue factor-activated assay with polibrene), IN test (ellagic acid-activated assay), FIB test (tissue factor activated assay, without functional platelet), ECA test (ecarin-based assay), and tPA test (r-tPA within an extrinsic pathway-based assay). Of note, the EX test, tPA test, and FIB test contain polybrene to neutralize heparin. In each test, we recorded the next parameters which characterized the whole course of coagulation: clotting time (CT), clot formation time (CFT), α angle, "amplitude of the clot" at a given time x (A(x)), maximum clot firmness (MCF), maximum lysis (ML), and lysis time (LT). The critically ill COVID-19 patients were divided into two groups based on their fibrinolytic response. A decreased fibrinolytic response (n = 9) was defined as LT > 393 s [18,38].

5.4. Statistical Analysis

Statistical analysis of the collected data was evaluated by IBM SPSS Statistics® 27.0. To analyze demographic and clinical factors, the chi-square test was used for categorical data. The Kolmogorov–Smirnov test was applied to test for normality of continuous variables distribution. Comparisons of continuous non-normally distributed data between COVID vs. control groups were carried out using the Mann–Whitney U-test, while COVID vs. controls with or without ASA subgroups were tested using a one-way ANOVA test. A Student's *T*-test was used for the analysis of normally distributed continuous data. Continuous variables are reported as median and interquartile range or mean and standard error of mean (SEM). Correlation analysis was performed calculating Spearman's correlation coefficient (rho). Correlations between variables were analyzed with univariate and multivariate linear regression with corresponding beta values and 95% confidence intervals. Multivariable logistic regression was used to identify factors independently associated with decreased fibrinolytic response defined as hypofibrinolysis. A p value <0.05 was considered statistically significant.

Author Contributions: Conceptualization, T.M.; project administration, T.M.; methodology, M.T.-F. and B.R.; statistical analysis, T.M.; investigation, D.S.; data curation, D.S.; writing—original draft preparation, D.S.; visualization, D.S.; review and editing, D.S., M.T.-F., B.R., and T.M. All authors have read and agreed to the published version of the manuscript.

Funding: The study was supported by EFOP-3.6.3-VEKOP-16-2017-00009 at the University of Pécs.

Institutional Review Board Statement: This study was conducted according to the guidelines of the Declaration of Helsinki and approved by the Hungarian Medical Research Council (20783-5/2020/EÜIG).

Informed Consent Statement: Informed consent was obtained from all subjects (or from their relatives due to their critical medical conditions) involved in the study.

Data Availability Statement: The data that support the findings of this study are available on request from the corresponding author, due to high amount of raw data.

Acknowledgments: Our thanks go to all the patients and staff who participated in this research.

Conflicts of Interest: The authors declare no conflict of interest.

References

1. WHO Coronavirus (COVID-19). Dashboard | WHO Coronavirus (COVID-19) Dashboard with Vaccination Data. Available online: https://covid19.who.int/ (accessed on 20 July 2021).
2. Cohen, A.; Harari, E. Immature platelets in patients hospitalized with Covid-19. *J. Thromb. Thrombolysis* **2021**, *51*, 608–616. [CrossRef] [PubMed]
3. Inchingolo, A.D.; Inchingolo, A.M. SARS-CoV-2 Disease Adjuvant Therapies and Supplements Breakthrough for the Infection Prevention. *Microorganisms* **2021**, *9*, 525. [CrossRef] [PubMed]
4. Ulanowska, M.; Olas, B. Modulation of Hemostasis in COVID-19; Blood Platelets May Be Important Pieces in the COVID-19 Puzzle. *Pathogens* **2021**, *10*, 370. [CrossRef] [PubMed]
5. Zhen, W.M.; van Wijk, X. Role of von Willebrand Factor in COVID-19 Associated Coagulopathy. *J. Appl. Lab. Med.* **2021**, *6*, 1305–1315. [CrossRef]
6. Yusuf, S.; Hawken, S. Effect of potentially modifiable risk factors associated with myocardial infarction in 52 countries (the INTERHEART study): Case-control study. *Lancet* **2004**, *364*, 937–952. [CrossRef]
7. Arachchillage, D.; Laffan, M. Abnormal coagulation parameters are associated with poor prognosis in patients with novel coronavirus pneumonia. *J. Thromb. Haemost.* **2020**, *18*, 1233–1234. [CrossRef] [PubMed]
8. Tang, N.; Bai, H. Anticoagulant treatment is associated with decreased mortality in severe coronavirus disease 2019 patients with coagulopathy. *J. Thromb. Haemost.* **2020**, *18*, 1094–1099. [CrossRef]
9. Amgalan, A.; Othman, M. Hemostatic laboratory derangements in COVID-19 with a focus on platelet count. *Platelets* **2020**, *31*, 740–745. [CrossRef] [PubMed]

10. Zhao, X.; Wang, K. Early decrease in blood platelet count is associated with poor prognosis in COVID-19 patients—Indications for predictive, preventive, and personalized medical approach. *EPMA J.* **2020**, *11*, 139. [CrossRef]
11. Xu, P.; Zhou, Q. Mechanism of thrombocytopenia in COVID-19 patients. *Ann. Hematol.* **2020**, *99*, 1205–1208. [CrossRef]
12. Tchachil, J. What do monitoring platelet counts in COVID-19 teach us? *J. Thromb. Haemost.* **2020**, *18*, 2071–2072. [CrossRef] [PubMed]
13. Ouyang, S.M.; Zhu, H.Q. Temporal changes in laboratory markers of survivors and non-survivors of adult inpatients with COVID-19. *BMC Infect. Dis.* **2020**, *20*, 1–10. [CrossRef] [PubMed]
14. Abe, Y.; Wada, H. A simple technique to determine thrombopoiesis level using immature platelet fraction (IPF). *Thromb Res.* **2006**, *118*, 463–469. [CrossRef]
15. Welder, D.; Jeon-Slaughter, H. Immature platelets as a biomarker for disease severity and mortality in COVID-19 patients. *Br. J. Haematol.* **2021**, *194*, 530–536. [CrossRef] [PubMed]
16. Bachler, M.; Bosch, J. Impaired fibrinolysis in critically ill COVID-19 patients. *Br. J. Anaesth.* **2021**, *126*, 590–598. [CrossRef] [PubMed]
17. Zhang, J.W.; Liu, W.W. Predictors of high on-aspirin platelet reactivity in elderly patients with coronary artery disease. *Clin. Interv. Aging* **2017**, *12*, 1271–1279. [CrossRef] [PubMed]
18. Bertolin, A.J.; Dalcoquio, T.F. Platelet Reactivity and Coagulation Markers in Patients with COVID-19. *Adv. Ther.* **2021**, *38*, 3911–3923. [CrossRef]
19. Iacobucci, G. Covid-19: Aspirin does not improve survival for patients admitted to hospital, trial reports. *BMJ* **2021**, *373*. [CrossRef]
20. RECOVERY Collaborative Group. Aspirin in patients admitted to hospital with COVID-19 (RECOVERY): A randomised, controlled, open-label, platform trial. *Lancet* **2021**. [CrossRef]
21. Kruse, J.M.; Magomedow, A. Thromboembolic complications in critically ill COVID-19 patients are associated with impaired fibrinolysis. *Crit. Care* **2020**, *24*, 1–10. [CrossRef] [PubMed]
22. Henry, B.M.; Santos de Oliveira, M.H. Hematologic, biochemical and immune biomarker abnormalities associated with severe illness and mortality in coronavirus disease 2019 (COVID-19): A meta-analysis. *Clin. Chem. Lab. Med.* **2020**, *58*, 1021–1028. [CrossRef]
23. Ghahramani, S.; Tabrizi, R. Laboratory features of severe vs. non-severe COVID-19 patients in Asian populations: A systematic review and meta-analysis. *Eur. J. Med. Res.* **2020**, *25*, 1–10. [CrossRef] [PubMed]
24. Zeng, F.; Huang, Y. Association of inflammatory markers with the severity of COVID-19: A meta-analysis. *Int. J. Infect. Dis.* **2020**, *96*, 467–474. [CrossRef]
25. Kermali, M.; Khalsa, R.K. The role of biomarkers in diagnosis of COVID-19—A systematic review. *Life Sci.* **2020**, *254*, 117788. [CrossRef] [PubMed]
26. Liu, F.; Li, L. Prognostic value of interleukin-6, C-reactive protein, and procalcitonin in patients with COVID-19. *J. Clin. Virol.* **2020**, *127*, 104370. [CrossRef] [PubMed]
27. Wang, L. C-reactive protein levels in the early stage of COVID-19. *Médecine Mal. Infect.* **2020**, *50*, 332–334. [CrossRef]
28. Poudel, A.; Poudel, Y. D-dimer as a biomarker for assessment of COVID-19 prognosis: D-dimer levels on admission and its role in predicting disease outcome in hospitalized patients with COVID-19. *PLoS ONE* **2021**, *16*, e0256744. [CrossRef]
29. Zhang, L.; Yan, X. D-dimer levels on admission to predict in-hospital mortality in patients with Covid-19. *J. Thromb. Haemost.* **2020**, *18*, 1324–1329. [CrossRef] [PubMed]
30. Lodigiani, C.; Iapichino, G. Venous and arterial thromboembolic complications in COVID-19 patients admitted to an academic hospital in Milan, Italy. *Thromb. Res.* **2020**, *191*, 9–14. [CrossRef]
31. Klok, F.A.; Kruip, M.J.H.A. Incidence of thrombotic complications in critically ill ICU patients with COVID-19. *Thromb. Res.* **2020**, *191*, 145–147. [CrossRef]
32. Della-Morte, D.; Pacifici, F. Low level of plasminogen increases risk for mortality in COVID-19 patients. *Cell Death Dis.* **2021**, *12*, 1–8. [CrossRef]
33. Wool, G.D.; Miller, J.L. The Impact of COVID-19 Disease on Platelets and Coagulation. *Pathobiology* **2021**, *88*, 15–27. [CrossRef] [PubMed]
34. Ward, S.E.; Curley, G.F. Von Willebrand factor propeptide in severe coronavirus disease 2019 (COVID-19): Evidence of acute and sustained endothelial cell activation. *Br. J. Haematol.* **2021**, *192*, 714–719. [CrossRef] [PubMed]
35. Incir, S.; Komesli, Z. Immature platelet fraction: Is a novel early predictive marker for disease severity in patients with Covid-19 pneumonia? *Turkish J. Biochem.* **2021**. [CrossRef]
36. Schrick, D.; Ezer, E. Novel predictors of future vascular events in post-stroke patients—A pilot study. *Front. Neurol.* **2021**, *12*, 971. [CrossRef] [PubMed]
37. Ezer, E.; Schrick, D. A novel approach of platelet function test for prediction of attenuated response to clopidogrel. *Clin Hemorheol Microcirc.* **2019**, *73*, 359–369. [CrossRef]
38. Roh, D.J.; Eiseman, K. Hypercoagulable viscoelastic blood clot characteristics in critically ill coronavirus disease 2019 patients and associations with thrombotic complications. *J. Trauma Acute Care Surg.* **2021**, *90*, 7–12. [CrossRef]

MDPI

Review

Translational Application of Fluorescent Molecular Probes for the Detection of Reactive Oxygen and Nitrogen Species Associated with Intestinal Reperfusion Injury

Gustavo Sampaio de Holanda [1,*], Samuel dos Santos Valença [2], Amabile Maran Carra [1], Renata Cristina Lopes Lichtenberger [1], Bianca de Castilho [1], Olavo Borges Franco [1], João Alfredo de Moraes [2] and Alberto Schanaider [1]

1 Centre of Experimental Surgery, Post Graduate Program in Surgical Sciences, Department of Surgery, Faculty of Medicine, Federal University of Rio de Janeiro, Rio de Janeiro 219491-590, Brazil; amabilemcarra@gmail.com (A.M.C.); renata.2001berger@gmail.com (R.C.L.L.); biacaastilo@gmail.com (B.d.C.); olavobf1988@gmail.com (O.B.F.); albertoscha@gmail.com (A.S.)

2 Redox Biology Laboratory, Institute of Biomedical Sciences, Federal University of Rio de Janeiro, Rio de Janeiro 21941-901, Brazil; samuelv@icb.ufrj.br (S.d.S.V.); joaomoraes@icb.ufrj.br (J.A.d.M.)

* Correspondence: gsholanda@gmail.com; Tel.: +55-21-9657-13794

check for updates

Citation: de Holanda, G.S.; dos Santos Valença, S.; Carra, A.M.; Lichtenberger, R.C.L.; de Castilho, B.; Franco, O.B.; de Moraes, J.A.; Schanaider, A. Translational Application of Fluorescent Molecular Probes for the Detection of Reactive Oxygen and Nitrogen Species Associated with Intestinal Reperfusion Injury. *Metabolites* **2021**, *11*, 802. https://doi.org/10.3390/metabo11120802

Academic Editor: Norbert Nemeth

Received: 8 November 2021
Accepted: 23 November 2021
Published: 26 November 2021

Publisher's Note: MDPI stays neutral with regard to jurisdictional claims in published maps and institutional affiliations.

Abstract: Acute mesenteric ischemia, caused by an abrupt interruption of blood flow in the mesenteric vessels, is associated with high mortality. When treated with surgical interventions or drugs to re-open the vascular lumen, the reperfusion process itself can inflict damage to the intestinal wall. Ischemia and reperfusion injury comprise complex mechanisms involving disarrangement of the splanchnic microcirculatory flow and impairment of the mitochondrial respiratory chain due to initial hypoxemia and subsequent oxidative stress during the reperfusion phase. This pathophysiologic process results in the production of large amounts of reactive oxygen (ROS) and nitrogen (RNS) species, which damage deoxyribonucleic acid, protein, lipids, and carbohydrates by autophagy, mitoptosis, necrosis, necroptosis, and apoptosis. Fluorescence-based systems using molecular probes have emerged as highly effective tools to monitor the concentrations and locations of these often short-lived ROS and RNS. The timely and accurate detection of both ROS and RNS by such an approach would help to identify early injury events associated with ischemia and reperfusion and increase overall clinical diagnostic sensitivity. This abstract describes the pathophysiology of intestinal ischemia and reperfusion and the early biological laboratory diagnosis using fluorescent molecular probes anticipating clinical decisions in the face of an extremely morbid disease.

Keywords: ischemia-reperfusion injury; molecular probes; oxidative stress; mesenteric ischemia

1. Introduction

Acute mesenteric ischemia (AMI) is characterized by an abrupt interruption or reduction of the intestinal blood supply, either temporarily or permanently, and is a medical-surgical emergency that requires immediate attention. Despite advances in medical imaging techniques, the evolution of clinical knowledge, and the development of biotechnology beyond the adoption of less invasive treatments, AMI still represents a major diagnostic and therapeutic challenge, largely due to the absence of specific markers related to the severity of the condition [1–3]. The clinical course of the disease is correlated with a high mortality rate (70%) in patients with AMI admitted to the emergency room [4].

It should be clarified that in clinical practice mesenteric ischemia is more severe when the occlusion occurs more centrally in a mesenteric vessel. Mechanical obstruction is the most common etiology of ischemia, followed by intestinal reperfusion. It can be caused by an embolus or thrombus within the vascular lumen, but also by phenomena that directly affect the intestinal wall such as volvulus (loop twist), intestinal invagination, incarcerated hernia, or extrinsic compression (caused by tumors or adhesions among others). In addition to a mechanical obstruction, an intestinal transplant, blood pressure variations (hypovolemic, cardiogenic, or neurogenic shock), and non-occlusive phenomena (e.g., due to the use of drugs) represent other reported causal events [3].

The quicker appropriate treatment of AMI is provided, the greater the chances of successful visceral perfusion with the return of effective blood flow and reduced risk of any sequelae. However, reperfusion inevitably produces reactive oxygen (ROS) and nitrogen (RNS) species, indicative of oxidative and nitrosative stress, respectively. The severity of intestinal tissue damage due to increased ROS and RNS levels depends on the magnitude of exacerbated synthesis of these humoral mediators, which is determined by the persistence of the deleterious causal agent and the extent of depletion of the organic defensive capacity to scavenge such reactive species. Under conditions of sustained stress, degradation of energy sources, enzyme synthesis, and activation of nuclear transcriptional factors occur, resulting in a chain reaction with significant production of ROS and/or RNS and the formation of several oxidizing substances, including lipid peroxides and carbonyl proteins. Such oxidants can trigger extensive cell damage and aggravate ischemia-initiated injury in the intestinal loop [5]. Beyond their roles in processes like modulation of cell survival, differentiation, cell death, cell signaling, and inflammation-related factor production, some reactive species have clear beneficial actions, such as the containment of invading pathogens, when present in physiologically 'normal' levels. However, when overproduced they typically become harmful to cellular homeostasis and can cause local and distant tissue damage [6].

Currently, the diagnosis of mesenteric ischemia is based on patient history, main symptoms, physical examination findings, and the use of state-of-the-art imaging methods. Abdominal computed tomography angiography with three-dimensional imaging can identify features of acute splanchnic vascular obstruction and intestinal injury. However, it is inaccurate in assessing the extent and severity of parietal involvement [7]. In this context, there is a lack of serum biomarkers and/or molecular methods to identify AMI with satisfactorily specificity and sensitivity to enable a straightforward and rapid diagnosis when required (i.e., as soon as AMI is suspected) [8].

Fluorescent probes allow for accurate detection of complex biomolecular components, such as ROS and RNS. Taking advantage of the diagnostic potential of these probes in diseases that affect visceral perfusion can benefit clinical-surgical practices, especially when it comes to patients with early acute abdominal pain. The present review will explore fluorescence techniques that rely on molecular probes for the measurement of ROS and RNS and evaluate how they could serve as a useful complementary resource in the diagnosis of AMI.

2. Discussion

2.1. Historical Context

The first studies related to oxygen-derived free radicals date back to 1931 when Haber and Weiss described the monovalent reduction of molecular oxygen to superoxide anion (O_2^-), a free radical capable of oxidizing organic structures and enzymes [9]. The conversion of superoxide to the highly reactive hydroxyl radical ($OH^{\cdot}$) was described three years later [10].

In 1968, McCord and Fridovich referred to xanthine oxidase (XO) as a biological source of superoxide production and found it was present in various organic tissues. The same authors later described the discovery of the superoxide dismutase enzyme capable of inactivating the superoxide radical [11]. In the 1970s, N-acetylcysteine, a small molecule inclosing the thiol group, was identified as a ROS scavenger, acting as a potent antioxidant [12].

It was not until the early 1980s that the inexplicable worsening of patients after supposedly adequate treatment to relieve AMI was starting to be clarified in terms of pathophysiology, biochemistry, and molecular biology. The first reports on potentially harmful effects of intestinal reperfusion were related to cell damage. De novo oxygenation of ischemic tissues was found to produce ROS and cause deleterious effects on cell structures by lipid peroxidation, protein oxidation, and nucleotide involvement, including purine bases [13–16]. In 1981, Granger and colleagues characterized the presence of XO in the intestine of different animals and humans and determined its relationship with tissue damage following intestinal ischemia and reperfusion. These data were fundamental in sparking research focusing on the role of reactive species biology in the digestive system under physiological and pathological conditions [17–19].

During the 1980s, endovascular intervention using catheters in the vessel lumen to locally deliver drugs therapy or for revascularization of AMI emerged as an alternative approach. It is a less invasive procedure than surgically opening the abdominal cavity and, if successful, reduces the need for intestinal resection, incidence of postoperative complications, length of hospitalization and mortality [20,21].

In the mid-1980s, the first probes consisting of a single-stranded deoxyribonucleic acid (DNA) fragment conjugated to a product (radioisotope, biotin, fluorescent compound) were developed. Nevertheless, fluorescent molecular probes to study oxidative stress and detect specific intracellular components in complex biomolecular matrices, with applicability in various visceral ischemic conditions, were only introduced very recently [22].

2.2. Free Radical Synthesis and the Pathophysiology of Ischemia/Reperfusion

When (partial or total) occlusion of the superior mesenteric artery or its branches occurs, splanchnic perfusion is limited due to reduction or, more frequently, interruption of blood flow. Blockage of oxygen supply and an impediment to aerobic energy metabolism induce an acute pathophysiological changes in the affected tissue(s) [23]. The lack of oxygen supply causes tissue ischemia and, if not restored promptly, will result in cellular dysfunction and cell death, ultimately resulting in parietal necrosis [24,25] (Figure 1).

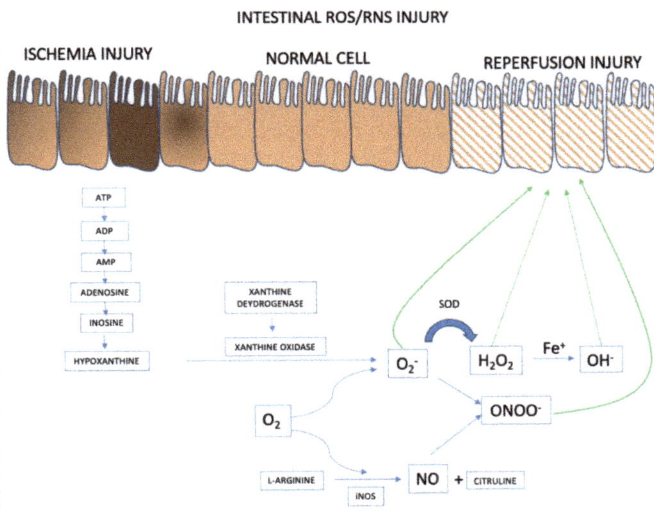

Figure 1. ROS and RNS formation mechanisms in the AMI setting. Adapted from [23].

Intestinal Epithelial Cells

ROS and RNS formation begins in the intestinal ischemia phase with adenosine triphosphate (ATP) accumulation generated in the anaerobic metabolism. There is degradation until the accumulation of hypoxanthine that at the beginning of the reperfusion phase, when there is reintroduction of oxygen to the intestinal tissue, interact with xanthine oxidase forming the superoxide anion (O_2^-), the first ROS formed. From there, the organism launches defenses such as superoxide dismutase (SOD), attenuating and forming ROS such as hydrogen peroxide (H_2O_2). However, if the response to reperfusion injury continues, the hydrogen peroxide is transformed into hydroxyl from the metal Iron (Fe^+) into hydroxyl (OH^-), in the so-called Fenton reaction. And in parallel, there may be the activation of RNS with the formation of nitric oxide (NO) from L-arginine, mediated by inducible nitric oxide synthase (iNOS). The combination of NO and superoxide anion forms the highly reactive species called nitrite peroxide (ONOO) which will further damage the intestinal cell's epithelium.

During the ischemic phase, mitochondrial oxidative phosphorylation is inhibited rendering a drop in the production and storage of adenosine triphosphate (ATP). ATP is successively degraded to adenosine diphosphate (ADP), adenosine monophosphate (AMP), adenosine, inosine, and finally hypoxanthine. Lack of cellular energy causes sodium-potassium (Na^+/K^+) pump failure resulting in intracellular Na^+ accumulation and K^+ out of cells, ultimately leading to cellular edema and organelle dysfunction. In addition, an influx of calcium (Ca^{2+}) and chloride (Cl^-) ions into the intracellular environment occurs and triggers the activation of calpain protease, which in turn promotes the breakdown of a peptide bridge of the enzyme xanthine dehydrogenase (XDH) and subsequent formation of XO.

Although essential for the rescue of morphofunctional integrity of the affected tissues, restoration of mesenteric blood flow and consequent ischemic tissue reoxygenation has a deleterious effect because, paradoxically, reperfusion itself aggravates the damage [26,27]. Oxygen together with hypoxanthine and XO, synthesized during ischemia, catalyze the formation of ROS [28,29]. Re-introducing oxygen into the visceral circulation via reperfusion leads to the formation of $O_2{}^-$ and hydrogen peroxide (H_2O_2) after successive monovalent reductions. In the presence of iron, copper, cobalt, chromium, or vanadium, the production of highly reactive hydroxyl radical (OH·) is promoted via the Haber-Weiss and Fenton reactions [30]. There is an activity burst of the oxidative process characterized by the abundant production of multiple ROS and RNS within a few minutes after the restoration of blood flow [27]. The events underlying the damage caused by ischemia/reperfusion produce an uncontrolled and excessive release of ROS and RNS that overcome the organic line of defense represented by free radical scavengers [31].

The mitochondrial respiratory electron transport chain is the main intracellular site of ROS production and polymorphonuclear leukocytes play an important role in several pathological conditions also generating free radicals and nitric oxide (NO) synthesis. Different forms of mitochondrial dysfunction and tissue inflammation can affect the organ undergoing ischemia and reperfusion and may even compromise other organs and systems with a paracrine or and endocrine effect. This phase can lead to the failure of multiple organs and systems [23,32].

Nitric oxide (NO) dynamics underpin changes involving RNS. NO is produced from L-arginine by three main isoforms of nitric oxide synthase (NOS): epithelial NOS (eNOS), related to vasodilation and vascular regulation; neuronal NOS (nNOS), linked to various intracellular signaling pathways; and inducible NOS (iNOS), which has been reported to have beneficial microbicidal, antiviral, antiparasitic and antitumoral actions, but has also been implicated in the pathophysiology of colitis [33]. While the production of NO by nNOS and eNOS is regulated by a Ca^{2+}/calmodulin-dependent mechanism, iNOS is activated in response to triggers such as endotoxins or cytokines, which can lead to rapid production of large amounts of NO. Several diseases have been associated with excessive levels of NO production, resulting in serious deleterious cell-physiological consequences [34–38]. All products formed by NO reactions are collectively called RNS. Despite the discovery of NO as an endothelium-derived relaxing factor, it plays a critical role in the pathophysiology of sepsis as an important mediator of endotoxin-induced arteriolar vasodilatation, hypotension, and shock [39]. At high concentrations, NO is importantly involved in inflammatory, infectious, and degenerative diseases [40]. Via reactions with other free radicals produced during oxidative stress, NO can be converted to nitrogen dioxide (NO_2), peroxynitrite ($ONOO^-$), and dinitrogen trioxide (N_2O_3). NO_2 is formed from NO autoxidation (reaction of NO with oxygen). $ONOO^-$ is a powerful electron oxidant and is formed through the diffusion-controlled reaction between O_2^- and NO; its most relevant targets are peroxiredoxins, glutathione peroxidase (GSH), CO_2, and metal centers. N_2O_3 can be formed from a reaction between NO_2 and NO and is considered an important intermediate in the autoxidation of NO. N_2O_3 is rapidly hydrolyzed to NO_2 [41]. All these compounds can subsequently react with various classes of biomolecules, including lipids, DNA, thiols, amino acids, and metals, leading to oxidation and nitration. If produced at high levels, RNS will detrimentally impact cell function, leading to changes in membrane integrity, loss of enzyme function, and DNA mutations [42].

It is noteworthy that, despite its typically beneficial antioxidant and vasodilatory functions, NO in high concentrations induces caspase-mediated apoptosis of epithelial cells in the intestinal tissue during ischemia and reperfusion. In addition, O_2^- rapidly reacts with NO to produce $ONOO^-$, which is another potent oxidant [43]. In the vasculature, the reaction of NO with O_2^- leads to the formation of $ONOO^-$ and decreases the vasorelaxant efficacy of NO. $ONOO^-$ is a strong oxidant that can hydroxylate aromatic amino acids, oxidize thiols and lipids, and nitrate-free and protein-bound tyrosine residues. The number of possible reactions leading to secondary RNS formation illustrates the strong potential of NO to contribute to oxidative damage. High concentrations of NO, particularly in combination with increased oxidant production, cause tissue damage and inflammation through the production of NO_2, $ONOO^-$ and other nitrating, nitrosating, and oxidizing intermediates, and via inhibition of metal-dependent enzymes [44,45].

Several enzymes, such as cytochrome P450, the enzyme complexes of the mitochondrial respiratory chain, XO [46], eNOS [47], heme oxygenase (HO) [48], myeloperoxidase (MPO) [49], lipoxygenase (LOX), cyclooxygenase (COX) [50], and NADPH oxidases (NOX) [51] generate ROS under pathological conditions leading to oxidative stress [52]. All these factors contribute to persistent oxidative stress in the cellular environment, which will result in progressive functional impairment of critical intracellular organelles and structures, including membranes, mitochondria, the endoplasmic reticulum, the cytoskeleton, and the nucleus. These deleterious effects occur mainly due to the oxidation of proteins, DNA, and lipids, ultimately culminating in cell death [53,54].

A balance between ROS levels and the activity of inactivating (antioxidant) enzymes is crucial for the maintenance of cellular homeostasis. Erythroid-related nuclear factor 2 (Nrf2) is a transcription factor that plays an important role in the response to oxidative stress to maintain redox balance. Under homeostatic conditions, Nrf2 is bound to its chaperone Keap1 (Kelch-like ECH association protein 1) in the cytoplasm. However, when oxidative stress occurs, Nrf2 dissociates from the inactive Keap1-Nrf2 complex and translocates to the nucleus, where it regulates specific gene expression to induce the synthesis of antioxidant enzymes [55]. O_2^- and H_2O_2 are inactivated by superoxide dismutase and catalase or the glutathione peroxidase system, respectively. OH· is typically more harmful than these ROS, as this oxygen-derived free radical does not have an intracellular inactivator. Its production intensifies the severity of injuries to cell structures, causing DNA damage caused by adducts of lipid peroxidation, and the production of other free radicals (such as malondialdehyde, hydroperoxide, and $ONOO^-$, among other substances capable of stimulating the adherence of granulocytes to the microvascular endothelium [55,56].

2.3. Molecular Probe Fundamentals

Oxidative and nitrosative stress biomarkers are important tools to assess the balance between reactive species and antioxidants, contributing to the understanding of the pathophysiology of diseases [57]. Direct measurement of your cellular levels is a challenge, as direct and accurate measurement is complex, due to its short productive life and fast reactivity with other REDOX regulators [58]. Fluorescent probes for ROS selectively assess cellular levels of ROS in a very simple way, but it is important to consider their limitations.

Fluorescent probes are able to monitor the behavior of a target biomolecule in live cells in real time [59].

The dihydrorhodamine 123 (DHR123) probe passively diffuses the cell membrane and concentrates in the intracellular space. In the presence of H_2O_2, hypochlorous acid (HOCl), or $ONOO^-$, it is oxidized to rhodamine (R123) which exhibits green fluorescence. DHR123 is considered an intracellular probe for general detection of ROS; however, it has a lower stability than several other commercially available probes.

The CM-H_2DCFDA (5-diacetate and 6-chloromethyl-2′,7′-dichlorodihydro-fluorescein) probe passively crosses the plasma membrane to enter the cell after which its acetate groups are cleaved by esterases to generate intracellular CM-H_2DCF; the thiol-reactive chloromethyl group reacts with intracellular glutathione and other thiols, and subsequent oxidation renders a fluorescent intracellular adduct. This probe is used to detect intracellular ROS and can react with H_2O_2, OH, $ONOO^-$ and other peroxide radicals. However, it is easily auto-oxidized resulting in a spontaneous increase in fluorescence, which must be corrected for at the time of the reading, discounting the value of a cell-free well containing the probe, as described by Hempel et al. Although this type of probe mainly detects H_2O_2, OH, and $ONOO^-$, it is not specific for any oxidant because it responds to a wide range of oxidizing reactions; the CM-H_2DCFDA probe is therefore considered a probe for general detection of ROS [60].

Fluorogenic complex probes containing boronate are used as a basis for detecting intracellular H_2O_2. Aromatic boronates react with H_2O_2, to generate a corresponding phenol, forming a highly fluorescent molecule in cells. Arylboronates also react with $ONOO^-$, six times faster than with H_2O_2, verified by flow kinetics technique and high performance liquid chromatography (HPLC) analysis [61]. One of the characteristics is its photophysical properties, such as high photostability and suitable high fluorescence. In addition, the iminocoumarin by-products have excitation and emission wavelengths that are longer, whereas rapid cyclization would generate the highly fluorescent benzothiazolyl iminocoumarin [62].

Amplex Red reagent is a colorless, highly sensitive, non-fluorescent compound used as a stable probe to detect the generation of H_2O_2. It is oxidized by horseradish peroxidase (HRP) to a fluorescent product, resorufin. One of the main complicating factors is photochemical oxidation in the presence of biological reducers (glutathione) that induce the formation of free radicals (O_2^- and H_2O_2), making the measurement of intracellular H_2O_2 a problem, even in the absence of HRP and H_2O_2. It is a highly sensitive method for detecting H_2O_2 and resorufin is stable for some time. However, it is impervious to cells and cannot be used to detect intracellular H_2O_2. Amplex Red is a very sensitive method for detecting ROS in organelles, as well as extracellular ROS, which is freely diffusible. The Amplex Red assay is also used to evaluate ROS formation in mitochondria [61,63].

CellRox represents another class of probes used for the general detection of ROS and comes in different models capable of emitting distinct fluorescence signals. In a reduced state, these cell-permeant dyes are non- or weakly fluorescent and become fluorescent upon oxidation by ROS. In general, CellRox can be oxidized by OH. and O_2^-, while CellRox orange is also capable of detecting H_2O_2, NO and ONOO-. These probes exhibit outstanding photostability compared to DCF [64] and it has also been shown that these probes can detect signals not detected by DCF [65]. Furthermore, depending on the model, can be used for in situ detection, allowing the assessment of real-time ROS dynamics in any given tissue [66].

The dihydroethidium (DHE) probe is capable of specifically detecting O_2^- radicals in intracellular and extracellular environment [67]. In addition, it can also be used to detect O_2^- in situ. The primary radical hydroethidine is derived from the loss of an aromatic amino hydrogen atom that, upon rearrangement, further reacts with another O_2^- anion to form DHE. Acetylation of the aromatic amino groups in hydroethidine inhibited its reaction with O_2^- [68]. MitoSox is the preferred probe for the specific analysis of mitochondrial O_2^-; this reagent selectively targets mitochondria where it is rapidly oxidized by O_2^-

(but not by other ROS or RNS) producing a red fluorescent signal, the oxidized product is highly fluorescent upon binding to nucleic acid [69].

DAF-FM (4-amino-5-methylamino-2′,7′-difluorofluorescein diacetate) is the leading molecular probe for the detection of NO. Like CM-H_2DCFDA, DAF-FM diacetate also passively diffuses the plasma membrane and is cleaved by esterases to generate intracellular DAF-FM. Subsequent oxidation by NO yields a triazole product accompanied by increased fluorescent recovery [70]. DAF-FM is not a reversible balance sensor, which limits its ability to track rapid target substance (NO) fluctuations in real time.

The aminophenyl fluorescein (APF) and hydroxyphenyl fluorescein (HPF) probes provide better selectivity and stability than CM-H_2DCFDA for specific detection of OH· and $ONOO^-$ with relatively high resistance to light-induced oxidation. In their initial (reduced) form, the APF and HPF molecular probes are not fluorescent until they react with $ONOO^-$ or OH, producing bright green fluorescence [45], resulting in cleavage of the aminophenyl ring from the fluorescein ring system, which is highly fluorescent. APF will also be transformed into the fluorescent form if exposed to a combination of H_2O_2 and horseradish peroxidase (HRP); HRP catalyzes the oxidation of APF by H_2O_2 [71].

The main fluorescent probes are widely used, mainly due to their simplicity, sensitivity, selectivity, execution speed and wide possibility of use in liquids and organic materials. There are limitations that should be known, such as autoxidation, but this could possibly be mitigated by the combined use of several fluorescent probes. Some regularly used probes are listed in Table 1.

Table 1. Probes for reactive species [72]. In the table are represented some fluorescent probes and possible reactive species identified in each reaction.

Probes	Reactive Species	Chemical Structure
DHR123	Hydrogen peroxide Hypochlorous acid Peroxynitrite anion	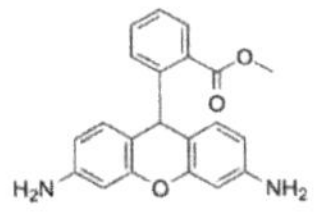 H_2N … NH_2
CM-H_2DCFDA	Hydrogen peroxide Hydroxyl radical Peroxynitrite anion Peroxyl radical	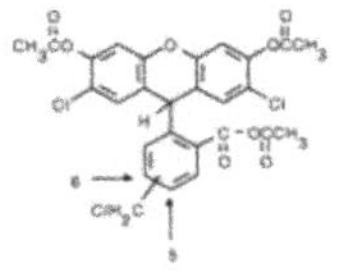 CH_3CO … $OCCH_3$, Cl, Cl, H, $C-OCCH_3$, ClH_2C
CellRox	Hydrogen peroxide Hydroxyl radical Nitric oxide Peroxynitrite anion Superoxide anion	
Dihydroethidium	Superoxide anion	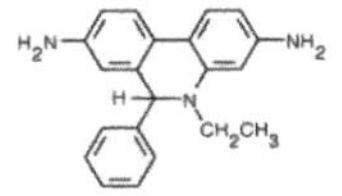 H_2N … NH_2, H, N, CH_2CH_3
MitoSox	Superoxide anion	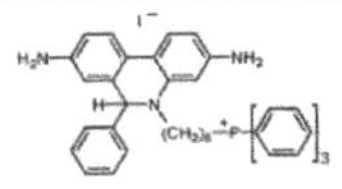 I^-, H_2N … NH_2, H, N, $(CH_2)_6-P^+[C_6H_5]_3$

Table 1. *Cont.*

DAF-FM	Nitric oxide	
APF and HPF	Hypochlorous acid Peroxynitrite anion Hydroxyl radical	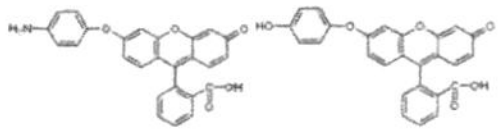
Boronate	Hydrogen peroxide Peroxynitrite anion	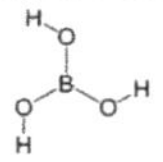
Amplex red	Hydrogen peroxide	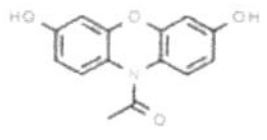

2.4. Translational Studies

Molecular fluorescent probes for the detection of free radicals have been increasingly used in experimental animal studies and clinical trials, with proof of diagnostic efficacy in injuries resulting from visceral ischemia and reperfusion in a range of diseases. Childs and co-workers (2002) conducted a study with a fluorescent probe sensitive to hydroperoxides (DHR123) in Sprague-Dawley rats submitted to hemorrhagic shock. They evaluated the production of ROS in real-time and demonstrated an 80% elevation 5 min into the reperfusion phase, followed by an increase in leukocyte adherence between 5 and 10 min of reperfusion after volume replacement [73]. Others recently reported the attenuation of oxidative damage as measured by the fluorescent probe DCFH-DA (2′,7′-dichlorodihydrofluorescein diacetate) in an experimental model of reperfusion brain injury in rats [74]. In rats subjected to 45 min of the celiac trunk and superior mesenteric artery ischemia, followed by 60 min of reperfusion, treatment with melatonin (applied 5 min before to reperfusion) significantly reduced ischemia-reperfusion injury (neutrophil-mediated oxidative stress) as indicated by the inhibition of pathways related to $ONOO^-$ measured by the molecular probe DHR123 [75]. Yan et al. used the DCFH-DA fluorescent probe to confirm the attenuation of oxidative stress induced by temporary ischemia of the superior mesenteric artery in mice treated with HO-1-expressing bone mesenchymal stromal cells (BMSC); based on the analysis using the fluorescent probe, it was concluded that BMSC that express HO-1 are more effective than treatment with BMSC alone in limiting intestinal damage and inflammation following ischemia and reperfusion injury [76]. Nagira et al. used DHR123 in the monolayers of human intestinal epithelial cell line to indicate that tight junctions and dysfunction of P-glycoprotein are induced through generation of reactive oxygen metabolites by ischemia and reperfusion in vitro model, and demonstrate the use of lutein as an antioxidant [77].

Recently was performed a study with Wistar rats in a model of small bowel ischemia (established by clamping of branches of the superior mesenteric artery) followed by reperfusion. Using fluorescent molecular probes, we measured the synthesis of ROS and RNS, 80 min after starting the experiment and 45 min after reperfusion. The CM-H_2DCFDA probe was used for general analysis of intracellular ROS, whereas DAF-FM and APF allowed specific evaluation of NO and $ONOO^-$, respectively. Analysis of the results using these fluorescent probes revealed that treatment with the antioxidants sulforaphane and albumin significantly reduced levels of total ROS, NO, and $ONOO^-$ in rats subjected to intestinal ischemia and reperfusion. Furthermore, reduced formation of free radicals and their by-products was shown to protect the intestinal mucosa. Antioxidant treatment decreased the concentration of macrophage-positive cells (ED-1), activation of intracellular NFκB signaling, and increased the amount of iNOS, LDH, and caspase 3 expression. They also observed relevant intestinal mucosal lesions and reduced concentration of goblet cells, a significant increase in apoptosis, greater macrophage infiltration, detachment, and structural disarrangement of the small intestine epithelium [78] (Figure 2).

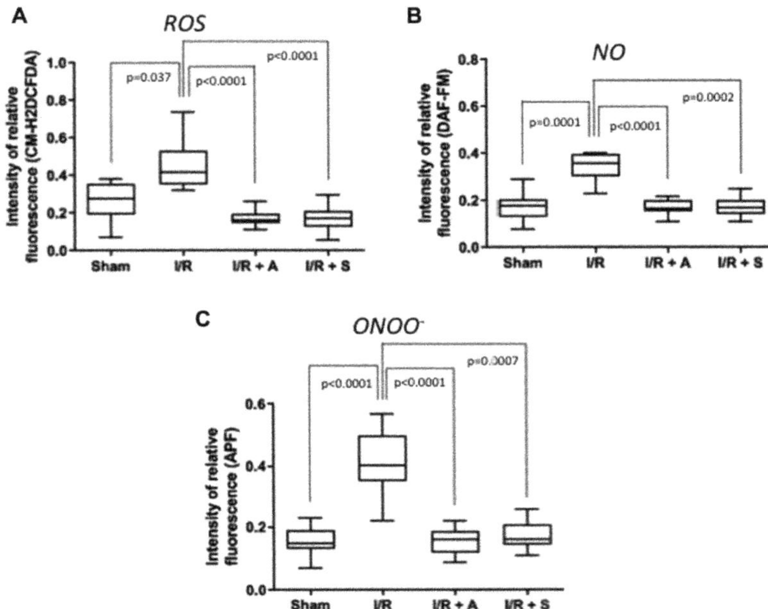

Figure 2. Sulforaphane (S) and albumin (A) administration attenuates the production of reactive oxygen and nitrogen species in intestinal ischemia/ reperfusion (I/R) injury. The administration of S and A before reperfusion prevented increases in reactive oxygen species (ROS) (**A**), nitric oxide (NO) (**B**), and peroxynitrite (ONOOL) (**C**) in the peripheral blood. The horizontal bars represent the medians, the boxes represent the 25th and 75th percentiles, and the vertical lines below and above the boxes represent the minimum and maximum values, respectively. The data are representative of two independent experiments (8 animals per group) [78]. The value of each "*p*" is showing in the figure its value related to the groups shown in blox plot graph.

The development of new versatile fluorescent probes with the possibility of high yield, high photostability, fast response time, low detection limit, high sensitivity and selectivity, low cytotoxicity, is what has been pursued by research aimed at diagnosing and interpreting evolution [79]. It is imperative for redox researchers to understand the detection mechanism and limitations of fluorescent probes in order to draw appropriate conclusions.

A clinically useful probe to identify biomarker(s) of mesenteric ischemia should have diagnostic specificity, exhibit prognostic value, be reasonably stable in various biological samples, and correlate with disease severity. Application and measurements would also need to be cost-effective with high reproducibility. Despite dozens of recognized markers and methods, results using fluorescent probes for the detection of oxidative stress are inconsistent among authors and thus weaken the overall translational value for clinical-surgical practice [80]. Therefore, additional and uniform research with consistent sampling will be necessary to avoid biases and identify the limited values of molecular probes as well as disease-specific diagnostic standards. Concerning AMI, future investigations using selective fluorescent probes, in parallel with proteomic and metabolomic approaches, will considerably improve our understanding of the signaling mechanisms that underpin the disease and facilitate the identification of clinically relevant biomarkers.

Table 2 lists some studies that support the use of fluorescent probes, especially in the pathophysiology of ischemia reperfusion, based on various clinical conditions, showing benefits from their use. What we need is the translational extrapolation to clinical practice, with clinical works that support the proper use in some specific conditions, mainly because there is a technological effort to improve the quality of fluorescent probes. Acting on the pathophysiological basis of some diseases seems to be better supported.

Table 2. Fluorescent probes for reactive oxygen species use in translational studies. Several studies have been carried out to justify the use of fluorescent probes in experimental models with the possibility of use in clinical practice.

Author	Fluorescent Probe	Results	Graphics
Childs EW, et al. [73]	Dihydrorhodamine 123 i.v. and observes in vivo mesenteric endothelium	Reactive oxygen species production in the mesenteric microvascular endothelium, attributed to hemorrhagic shock and reperfusion injury, after resuscitation, and mediated by the administration of a platelet activating factor antagonist	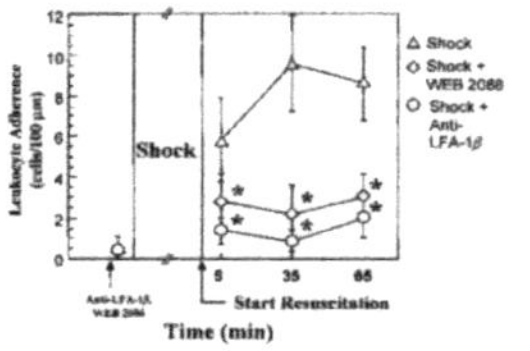 Effect of platelet activating factor (anti-LFA-1_ and WEB 2086) on leukocyte adherence ROS given 10 min prior to the shock period versus the hemorrhagic shock alone group. * $p < 0.05$ compared with the hemmorhagic shock alone group.
Tang Y, et al. [74]	DCFH-DA (2′, 7′-dichlorodihydrofluorescein diacetate) used in fresh tissue homogenates	Human albumin intravenous administration, in ROS attenuation, in a global cerebral ischemia reperfusion model by Wnt/β-Catenin pathway signaling	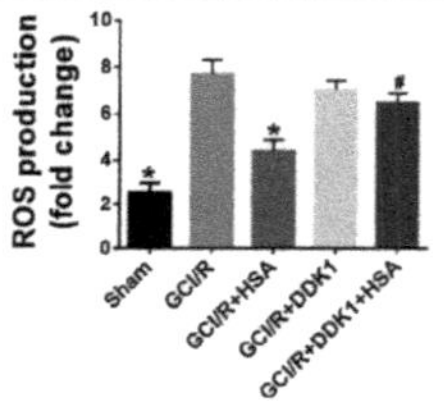 Effect of human albumin treatment on oxidative stress following global cerebral ischemia/reperfusion, ($p < 0.05$) in contrast to the Global Cerebral Ischemia/Reperfusion (GCI/R) group, * $p < 0.05$ in contrast to GCI/R group, # $p < 0.05$, in contrast to the GCI/R+Human Serum Albumin group.
Cuzzocrea S, et al. [75]	Dihydrorodamine 123 i.v. plasma analysed	Melatonin infusion attenuated the reperfusion injury produced by splanchnic artery occlusion	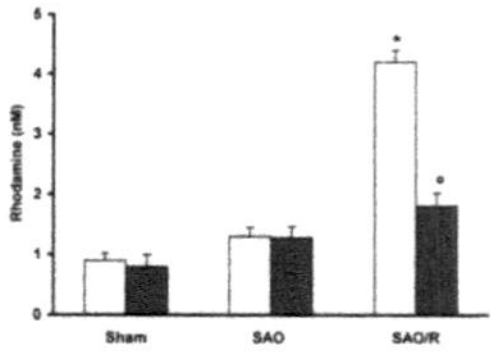 Plasma peroxynitrite production assessed by oxidation of dihydrorhodamine 123 to rhodamine. Peroxynitrite production in the Splancnic Arterial Oclusion (SAO)-shocked rats was significantly increased versus sham group. Melatonin-treated rats show a significant reduction of the SAO-induced elevation of the plasma peroxynitrite production. * $p < 0.01$ versus vehicle. ° $p < 0.01$ versus SAO.

Table 2. *Cont.*

<table>
<tr><th>Author</th><th>Fluorescent Probe</th><th>Results</th><th>Graphics</th></tr>
<tr><td>Yan XT, et al. [76]</td><td>DCFH-DA used in homogenized intestinal tissue</td><td>Heme Oxygenase-1-expressing Bone Marrow Steam Cell after intestinal I/R performed by temporary occlusion of the superior mesenteric artery</td><td>DCFH-DA Fluorescence
200
150
100
50
0
**
+
Sham I/R BMSC BMSC/HO-1
Bone Marrow Steam Cell/HemeOxygenase-1 (BMSC/HO-1) attenuated production of ROS in intestine and serum. Levels of ROS in intestine at 24 h of reperfusion were significantly higher than those in sham group and decreased after treatment of BMSC/HO-1. ** $p < 0.01$ vs. Sham; # $p < 0.05$ vs. I/R; + $p < 0.05$ vs. BMSC.</td></tr>
<tr><td>Nagira M, et al. [77]</td><td>Rhodamine 123</td><td>Lutein effects In vitro ischemia reperfusion injury, using monolayers of human colon cancer intestinal epithelial cell line</td><td><table><tr><th></th><th>A to B (× 10^{-5} cm/sec)</th><th>B to A (× 10^{-5} cm/sec)</th></tr><tr><td>control</td><td>0.044 ± 0.006</td><td>1.550 ± 0.086</td></tr><tr><td>t-BuOOH (10 mM) only</td><td>0.787 ± 0.241</td><td>0.881 ± 0.052</td></tr><tr><td>+ lutein 1 μM</td><td>0.559 ± 0.013</td><td>0.806 ± 0.114</td></tr><tr><td>5 μM</td><td>0.513 ± 0.054*</td><td>1.164 ± 0.017**</td></tr><tr><td>+ biliverdin 10 μM</td><td>0.484 ± 0.036*</td><td>0.896 ± 0.063</td></tr></table>The effects of lutein and biliverdin on rhodamine 123 permeability in the apical to basal direction in cell monolayers. * $p < 0.05$ and ** $p < 0.01$ vs. lipid peroxidation inducer.</td></tr>
</table>

Author Contributions: Conceptualization, G.S.d.H., S.d.S.V., J.A.d.M., A.S.; bibliography search and editing, A.M.C., R.C.L.L., B.d.C., O.B.F.; supervision, S.d.S.V., J.A.d.M.; validation, S.d.S.V., A.S.; writing—original draft, G.S.d.H.; writing—review and editing, A.S. All authors have read and agreed to the published version of the manuscript.

Funding: Supported by grants from Fundação Carlos Chagas Filho de Amparo à Pesquisa do Estado do Rio de Janeiro (FAPERJ): E-26/202.921/2019; E-26/211.176/2019; E-26/202.516/2019; and Conselho Nacional de Desenvolvimento Científico e Tecnológico (CNPq): 304265/2018-7.

Acknowledgments: The authors are grateful to the Center of Experimental Surgery and Redox Biology Laboratory staff.

Conflicts of Interest: The authors declare no conflict of interest.

References

1. Kuhn, F.; Schiergens, T.S.; Klar, E. Acute mesenteric ischemia. *Visc. Med.* **2020**, *36*, 256–262. [CrossRef] [PubMed]
2. Ehlert, B.A. Acute gut ischemia. *Surg. Clin. North Am.* **2018**, *98*, 995–1004. [CrossRef]
3. Karkkainen, J.M. Acute mesenteric ischemia: A challenge for the acute care surgeon. *Scand. J. Surg.* **2021**, 14574969211007590. [CrossRef]
4. Gnanapandithan, K.; Feuerstadt, P. Review article: Mesenteric ischemia. *Curr. Gastroenterol. Rep.* **2020**, *22*, 17. [CrossRef] [PubMed]
5. Memet, O.; Zhang, L.; Shen, J. Serological biomarkers for acute mesenteric ischemia. *Ann. Transl. Med.* **2019**, *7*, 394. [CrossRef]
6. Abdal Dayem, A.; Hossain, M.K.; Lee, S.B.; Kim, K.; Saha, S.K.; Yang, G.M.; Choi, H.Y.; Cho, S.-G. The role of reactive oxygen species (ROS) in the biological activities of metallic nanoparticles. *Int. J. Mol. Sci.* **2017**, *18*, 120. [CrossRef] [PubMed]
7. Dhatt, H.S.; Behr, S.C.; Miracle, A.; Wang, Z.J.; Yeh, B.M. Radiological evaluation of bowel ischemia. *Radiol. Clin. N. Am.* **2015**, *53*, 1241–1254. [CrossRef]
8. Peoc'h, K.; Corcos, O. Biomarkers for acute mesenteric ischemia diagnosis: State of the art and perspectives. *Ann. Biol. Clin.* **2019**, *77*, 415–421. [CrossRef]
9. Haber, F.; Willstätter, R. Unpaarigkeit und Radikalketten im Reaktionsmechanismus organischer und enzymatischer Vorgänge. *Ber. Der Dtsch. Chem. Ges. A B Ser.* **1931**, *64*, 2844–2856. [CrossRef]

10. Haber, F.; Weiss, J.; Pope, W.J. The catalytic decomposition of hydrogen peroxide by iron salts. *Proc. R. Soc. Lond. Ser. A Math. Phys. Sci.* **1934**, *147*, 332–351.
11. McCord, J.M.; Fridovich, I. Superoxide dismutase. An enzymic function for erythrocuprein (hemocuprein). *J. Biol. Chem.* **1969**, *244*, 6049–6055. [CrossRef]
12. De Flora, S.; Grassi, C.; Carati, L. Attenuation of influenza-like symptomatology and improvement of cell-mediated immunity with long-term *N*-acetylcysteine treatment. *Eur. Respir. J.* **1997**, *10*, 1535–1541. [CrossRef] [PubMed]
13. Acosta, S. Mesenteric ischemia. *Curr. Opin. Crit. Care.* **2015**, *21*, 171–178. [CrossRef] [PubMed]
14. Karkkainen, J.M.; Acosta, S. Acute mesenteric ischemia (part I)—Incidence, etiologies, and how to improve early diagnosis. *Best Pract. Res. Clin. Gastroenterol.* **2017**, *31*, 15–25. [CrossRef]
15. Karkkainen, J.M.; Acosta, S. Acute mesenteric ischemia (Part II)—Vascular and endovascular surgical approaches. *Best Pract. Res. Clin. Gastroenterol.* **2017**, *31*, 27–38. [CrossRef]
16. Prakash, V.S.; Marin, M.; Faries, P.L. Acute and chronic ischemic disorders of the small bowel. *Curr. Gastroenterol. Rep.* **2019**, *21*, 27. [CrossRef]
17. Granger, D.N. Intestinal microcirculation and transmucosal fluid transport. *Am. J. Physiol.* **1981**, *240*, G343–G349. [CrossRef]
18. Granger, D.N.; Hollwarth, M.E.; Parks, D.A. Ischemia-reperfusion injury: Role of oxygen-derived free radicals. *Acta Physiol. Scand. Suppl.* **1986**, *548*, 47–63.
19. Kajino-Sakamoto, R.; Omori, E.; Nighot, P.K.; Blikslager, A.T.; Matsumoto, K.; Ninomiya-Tsuji, J. TGF-beta-activated kinase 1 signaling maintains intestinal integrity by preventing accumulation of reactive oxygen species in the intestinal epithelium. *J. Immunol.* **2010**, *185*, 4729–4737. [CrossRef]
20. VanDeinse, W.H.; Zawacki, J.K.; Phillips, D. Treatment of acute mesenteric ischemia by percutaneous transluminal angioplasty. *Gastroenterology* **1986**, *91*, 475–478. [CrossRef]
21. Beaulieu, R.J.; Arnaoutakis, K.D.; Abularrage, C.J.; Efron, D.T.; Schneider, E.; Black, J.H., 3rd. Comparison of open and endovascular treatment of acute mesenteric ischemia. *J. Vasc. Surg.* **2014**, *59*, 159–164. [CrossRef] [PubMed]
22. Yan, L.; Xie, Y.; Li, J. A colorimetric and fluorescent probe based on rhodamine B for detection of Fe(3+) and Cu(2+) ions. *J. Fluoresc.* **2019**, *29*, 1221–1226. [CrossRef] [PubMed]
23. Nadatani, Y.; Watanabe, T.; Shimada, S.; Otani, K.; Tanigawa, T.; Fujiwara, Y. Microbiome and intestinal ischemia/reperfusion injury. *J. Clin. Biochem. Nutr.* **2018**, *63*, 26–32. [CrossRef]
24. Bertoni, S.; Ballabeni, V.; Barocelli, E.; Tognolini, M. Mesenteric ischemia-reperfusion: An overview of preclinical drug strategies. *Drug Discov. Today* **2018**, *23*, 1416–1425. [CrossRef] [PubMed]
25. Mester, A.; Magyar, Z.; Sogor, V.; Tanczos, B.; Stark, Y.; Cherniavsky, K.; Laszlo, B.; Katalin, P.; Norbert, N. Intestinal ischemia-reperfusion leads to early systemic micro-rheological and multiorgan microcirculatory alterations in the rat. *Clin. Hemorheol. Microcirc.* **2018**, *68*, 35–44. [CrossRef]
26. Papezikova, I.; Lojek, A.; Cizova, H.; Ciz, M. Alterations in plasma antioxidants during reperfusion of the ischemic small intestine in rats. *Res. Vet. Sci.* **2006**, *81*, 140–147. [CrossRef]
27. Kalogeris, T.; Baines, C.P.; Krenz, M.; Korthuis, R.J. Ischemia/reperfusion. *Compr. Physiol.* **2016**, *7*, 113–170. [PubMed]
28. Granger, D.N.; Kvietys, P.R. Reperfusion injury and reactive oxygen species: The evolution of a concept. *Redox. Biol.* **2015**, *6*, 524–551. [CrossRef]
29. Wu, Z.; Wang, H.; Fang, S.; Xu, C. Roles of endoplasmic reticulum stress and autophagy on H_2O_2 induced oxidative stress injury in HepG2 cells. *Mol. Med. Rep.* **2018**, *18*, 4163–4174. [CrossRef]
30. Valko, M.; Morris, H.; Cronin, M.T. Metals, toxicity and oxidative stress. *Curr. Med. Chem.* **2005**, *12*, 1161–1208. [CrossRef]
31. Liskova, A.; Samec, M.; Koklesova, L.; Kudela, E.; Kubatka, P.; Golubnitschaja, O. Mitochondriopathies as a clue to systemic disorders-analytical tools and mitigating measures in context of predictive, preventive, and personalized (3P) medicine. *Int. J. Mol. Sci.* **2021**, *22*, 2007. [CrossRef]
32. Battelli, M.G.; Polito, L.; Bolognesi, A. Xanthine oxidoreductase in atherosclerosis pathogenesis: Not only oxidative stress. *Atherosclerosis* **2014**, *237*, 562–567. [CrossRef]
33. Kleinert, H.; Schwarz, P.M.; Forstermann, U. Regulation of the expression of inducible nitric oxide synthase. *Biol. Chem.* **2003**, *384*, 1343–1364. [CrossRef]
34. Valenca, S.S.; Pimenta, W.A.; Rueff-Barroso, C.R.; Ferreira, T.S.; Resende, A.C.; Moura, R.S.; Porto, L.C. Involvement of nitric oxide in acute lung inflammation induced by cigarette smoke in the mouse. *Nitric Oxide* **2009**, *20*, 175–181. [CrossRef] [PubMed]
35. Pires, K.M.; Lanzetti, M.; Rueff-Barroso, C.R.; Castro, P.; Abrahao, A.; Koatz, V.L.; Valença, S.S.; Porto, L.C. Oxidative damage in alveolar macrophages exposed to cigarette smoke extract and participation of nitric oxide in redox balance. *Toxicol. In Vitro* **2012**, *26*, 791–798. [CrossRef] [PubMed]
36. Nesi, R.T.; Barroso, M.V.; Souza Muniz, V.; de Arantes, A.C.; Martins, M.A.; Brito Gitirana, L.; Neves, J.S.; Benjamim, C.F.; Lanzetti, M.; Valenca, S.S. Pharmacological modulation of reactive oxygen species (ROS) improves the airway hyperresponsiveness by shifting the Th1 response in allergic inflammation induced by ovalbumin. *Free Radic Res.* **2017**, *51*, 708–722. [CrossRef]
37. Valenca, S.S.; Rueff-Barroso, C.R.; Pimenta, W.A.; Melo, A.C.; Nesi, R.T.; Silva, M.A.; Porto, L.C. L-NAME and L-arginine differentially ameliorate cigarette smoke-induced emphysema in mice. *Pulm. Pharmacol. Ther.* **2011**, *24*, 587–594. [CrossRef] [PubMed]

38. Lanzetti, M.; da Costa, C.A.; Nesi, R.T.; Barroso, M.V.; Martins, V.; Victoni, T.; Lagente, V.; Pires, K.M.P.; Silva, P.M.R.e.; Resende, A.C.; et al. Oxidative stress and nitrosative stress are involved in different stages of proteolytic pulmonary emphysema. *Free Radic. Biol. Med.* **2012**, *53*, 1993–2001. [CrossRef] [PubMed]
39. Li, H.; Forstermann, U. Nitric oxide in the pathogenesis of vascular disease. *J. Pathol.* **2000**, *190*, 244–254. [CrossRef]
40. Guzik, T.J.; Korbut, R.; Adamek-Guzik, T. Nitric oxide and superoxide in inflammation and immune regulation. *J. Physiol. Pharmacol.* **2003**, *54*, 469–487.
41. Moller, M.N.; Rios, N.; Trujillo, M.; Radi, R.; Denicola, A.; Alvarez, B. Detection and quantification of nitric oxide-derived oxidants in biological systems. *J. Biol. Chem.* **2019**, *294*, 14776–14802. [CrossRef]
42. Barzilai, A.; Yamamoto, K. DNA damage responses to oxidative stress. *DNA Repair.* **2004**, *3*, 1109–1115. [CrossRef] [PubMed]
43. Luo, C.C.; Huang, C.S.; Ming, Y.C.; Chu, S.M.; Chao, H.C. Calcitonin gene-related peptide downregulates expression of inducible nitride oxide synthase and caspase-3 after intestinal ischemia-reperfusion injury in rats. *Pediatr. Neonatol.* **2016**, *57*, 474–479. [CrossRef]
44. Eiserich, J.P.; Patel, R.P.; O'Donnell, V.B. Pathophysiology of nitric oxide and related species: Free radical reactions and modification of biomolecules. *Mol. Aspects Med.* **1998**, *19*, 221–357. [CrossRef]
45. Adams, L.; Franco, M.C.; Estevez, A.G. Reactive nitrogen species in cellular signaling. *Exp. Biol. Med.* **2015**, *240*, 711–717. [CrossRef]
46. Harris, C.M.; Sanders, S.A.; Massey, V. Role of the flavin midpoint potential and NAD binding in determining NAD versus oxygen reactivity of xanthine oxidoreductase. *J. Biol. Chem.* **1999**, *274*, 4561–4569. [CrossRef]
47. Laursen, J.B.; Somers, M.; Kurz, S.; McCann, L.; Warnholtz, A.; Freeman, B.A.; Tarpey, M.; Fukai, T.; Harrison, D.G. Endothelial regulation of vasomotion in apoE-deficient mice: Implications for interactions between peroxynitrite and tetrahydrobiopterin. *Circulation* **2001**, *103*, 1282–1288. [CrossRef] [PubMed]
48. Stocker, R.; Perrella, M.A. Heme oxygenase-1: A novel drug target for atherosclerotic diseases? *Circulation* **2006**, *114*, 2178–2189. [CrossRef] [PubMed]
49. Huang, Y.; Wu, Z.; Riwanto, M.; Gao, S.; Levison, B.S.; Gu, X.; Fu, X.; Wagner, M.A.; Besler, C.; Gerstenecker, G.; et al. Myeloperoxidase, paraoxonase-1, and HDL form a functional ternary complex. *J. Clin. Investig.* **2013**, *123*, 3815–3828. [CrossRef]
50. Schiffrin, E.L. Remodeling of resistance arteries in essential hypertension and effects of antihypertensive treatment. *Am. J. Hypertens.* **2004**, *17*, 1192–1200. [CrossRef] [PubMed]
51. Nauseef, W.M. Assembly of the phagocyte NADPH oxidase. *Histochem. Cell Biol.* **2004**, *122*, 277–291. [CrossRef]
52. Lee, M.Y.; Griendling, K.K. Redox signaling, vascular function, and hypertension. *Antioxid. Redox Signal.* **2008**, *10*, 1045–1059. [CrossRef]
53. Liao, G.; Chen, S.; Cao, H.; Wang, W.; Gao, Q. Review: Acute superior mesenteric artery embolism: A vascular emergency cannot be ignored by physicians. *Medicine* **2019**, *98*, e14446. [CrossRef]
54. Singh, M.; Long, B.; Koyfman, A. Mesenteric ischemia: A deadly miss. *Emerg. Med. Clin. North Am.* **2017**, *35*, 879–888. [CrossRef]
55. Li, R.; Jia, Z.; Zhu, H. Regulation of Nrf2 signaling. *React. Oxyg. Species* **2019**, *8*, 312–322. [CrossRef]
56. Kehrer, J.P. The Haber-Weiss reaction and mechanisms of toxicity. *Toxicology* **2000**, *149*, 43–50. [CrossRef]
57. Duanghathaipornsuk, S.; Farrell, E.J.; Alba-Rubio, A.C.; Zelenay, P.; Kim, D.S. Detection technologies for reactive oxygen species: Fluorescence and electrochemical methods and their applications. *Biosensors* **2021**, *11*, 30. [CrossRef]
58. Katerji, M.; Filippova, M.; Duerksen-Hughes, P. Approaches and methods to measure oxidative stress in clinical samples: Research applications in the cancer field. *Oxid. Med. Cell Longev.* **2019**, *2019*, 1279250. [CrossRef] [PubMed]
59. Jiang, X.; Wang, L.; Carroll, S.L.; Chen, J.; Wang, M.C.; Wang, J. Challenges and opportunities for small-molecule fluorescent probes in redox biology applications. *Antioxid. Redox Signal.* **2018**, *29*, 518–540. [CrossRef] [PubMed]
60. Hempel, S.L.; Buettner, G.R.; O'Malley, Y.Q.; Wessels, D.A.; Flaherty, D.M. Dihydrofluorescein diacetate is superior for detecting intracellular oxidants: Comparison with 2′,7′-dichlorodihydrofluorescein diacetate, 5(and 6)-carboxy-2′,7′-dichlorodihydrofluorescein diacetate, and dihydrorhodamine 123. *Free Radic. Biol. Med.* **1999**, *27*, 146–159. [CrossRef]
61. Kalyanaraman, B.; Darley-Usmar, V.; Davies, K.J.; Dennery, P.A.; Forman, H.J.; Grisham, M.B.; Mann, G.E.; Moore, K.; Roberts, J., II; Ischiropoulss, H. Measuring reactive oxygen and nitrogen species with fluorescent probes: Challenges and limitations. *Free Radic. Biol. Med.* **2012**, *52*, 1–6. [CrossRef] [PubMed]
62. Li, M.; Han, H.; Zhang, H.; Song, S.; Shuang, S.; Dong, C. Boronate based sensitive fluorescent probe for the detection of endogenous peroxynitrite in living cells. *Spectrochim. Acta A Mol. Biomol. Spectrosc.* **2020**, *243*, 118683. [CrossRef]
63. Deshwal, S.; Antonucci, S.; Kaludercic, N.; di Lisa, F. Measurement of mitochondrial rOS Formation. *Methods Mol. Biol.* **2018**, *1782*, 403–418. [PubMed]
64. Fluorescence imaging of oxidative stress in live cells. *BioProbes J. Cell Biol. Appl.* **2011**, *65*, 10.
65. Schenk, B.; Fulda, S. Reactive oxygen species regulate Smac mimetic/TNFalpha-induced necroptotic signaling and cell death. *Oncogene* **2015**, *34*, 5796–5806. [CrossRef]
66. Kageyama, S.; Hirao, H.; Nakamura, K.; Ke, B.; Zhang, M.; Ito, T.; Aziz, A.; Oncel, D.; Kaldas, F.M.; Bussutil, R.W.; et al. Recipient HO-1 inducibility is essential for posttransplant hepatic HO-1 expression and graft protection: From bench-to-bedside. *Am. J. Transplant.* **2019**, *19*, 356–367. [CrossRef] [PubMed]
67. Peshavariya, H.M.; Dusting, G.J.; Selemidis, S. Analysis of dihydroethidium fluorescence for the detection of intracellular and extracellular superoxide produced by NADPH oxidase. *Free Radic. Res.* **2007**, *41*, 699–712. [CrossRef]

68. Zhao, H.; Joseph, J.; Fales, H.M.; Sokoloski, E.A.; Levine, R.L.; Vasquez-Vivar, J.; Kalyanaraman, B. Detection and characterization of the product of hydroethidine and intracellular superoxide by HPLC and limitations of fluorescence. *Proc. Natl. Acad. Sci. USA* **2005**, *102*, 5727–5732. [CrossRef] [PubMed]
69. Dikalov, S.I.; Harrison, D.G. Methods for detection of mitochondrial and cellular reactive oxygen species. *Antioxid. Redox Signal.* **2014**, *20*, 372–382. [CrossRef] [PubMed]
70. Nagano, T. Bioimaging probes for reactive oxygen species and reactive nitrogen species. *J. Clin. Biochem. Nutr.* **2009**, *45*, 111–124. [CrossRef]
71. Cohn, C.A.; Simon, S.R.; Schoonen, M.A. Comparison of fluorescence-based techniques for the quantification of particle-induced hydroxyl radicals. *Part Fibre Toxicol.* **2008**, *5*, 2. [CrossRef]
72. Wiederschain, G.Y. The molecular probes handbook. A guide to fluorescent probes and labeling technologies. *Biochemistry* **2011**, *76*, 1276. [CrossRef]
73. Childs, E.W.; Udobi, K.F.; Wood, J.G.; Hunter, F.A.; Smalley, D.M.; Cheung, L.Y. In vivo visualization of reactive oxidants and leukocyte-endothelial adherence following hemorrhagic shock. *Shock* **2002**, *18*, 423–427. [CrossRef]
74. Tang, Y.; Shen, J.; Zhang, F.; Yang, F.Y.; Liu, M. Human serum albumin attenuates global cerebral ischemia/reperfusion-induced brain injury in a Wnt/beta-Catenin/ROS signaling-dependent manner in rats. *Biomed. Pharmacother.* **2019**, *115*, 108871. [CrossRef]
75. Cuzzocrea, S.; Costantino, G.; Mazzon, E.; Micali, A.; de Sarro, A.; Caputi, A.P. Beneficial effects of melatonin in a rat model of splanchnic artery occlusion and reperfusion. *J. Pineal Res.* **2000**, *28*, 52–63. [CrossRef]
76. Yan, X.T.; Cheng, X.L.; He, X.H.; Zheng, W.Z.; Xiao-Fang, Y.; Hu, C. The HO-1-expressing bone mesenchymal stem cells protects intestine from ischemia and reperfusion injury. *BMC Gastroenterol.* **2019**, *19*, 124. [CrossRef]
77. Nagira, M.; Tomita, M.; Mizuno, S.; Kumata, M.; Ayabe, T.; Hayashi, M. Ischemia/reperfusion injury in the monolayers of human intestinal epithelial cell line caco-2 and its recovery by antioxidants. *Drug Metab. Pharmacokinet.* **2006**, *21*, 230–237. [CrossRef]
78. Sampaio de Holanda, G.; dos Santos Valenca, S.; Maran Carra, A.; Lopes Lichtenberger, R.C.; Franco, O.B.; Ribeiro, B.E.; Rosas, S.L.P.; Santana, P.T.; Castelo-Branco, M.T.L.; de Souza, H.S.F.; et al. Sulforaphane and albumin attenuate experimental intestinal ischemia-reperfusion injury. *J. Surg. Res.* **2021**, *262*, 212–223. [CrossRef]
79. Fang, Y.; Dehaen, W. Fluorescent probes for selective recognition of hypobromous acid: Achievements and future perspectives. *Molecules* **2021**, *26*, 363. [CrossRef]
80. Frijhoff, J.; Winyard, P.G.; Zarkovic, N.; Davies, S.S.; Stocker, R.; Cheng, D.; Knight, A.R.; Taylor, E.L.; Oettrich, J.; Ruskovska, T.; et al. Clinical relevance of biomarkers of oxidative stress. *Antioxid. Redox Signal* **2015**, *23*, 1144–1170. [CrossRef]

 *metabolites*

Article

The Proteomic Signature of Intestinal Acute Rejection in the Mouse

Mihai Oltean [1,2,*], Jasmine Bagge [2], George Dindelegan [3,4], Diarmuid Kenny [5], Antonio Molinaro [6], Mats Hellström [2], Ola Nilsson [7], Carina Sihlbom [5], Anna Casselbrant [8], Marcela Davila [9] and Michael Olausson [1]

1 The Transplant Institute, Sahlgrenska University Hospital, 413 45 Gothenburg, Sweden; michael.olausson@transplant.gu.se
2 Laboratory for Transplantation and Regenerative Medicine, Institute of Clinical Sciences, Sahlgrenska Academy at the University of Gothenburg, Sahlgrenska Science Park Medicinaregatan 8, 413 90 Gothenburg, Sweden; jasmine.bagge@gu.se (J.B.); mats.hellstrom@gu.se (M.H.)
3 First Surgical Clinic, Str. Clinicilor 3-5, 400006 Cluj-Napoca, Romania; george.dindelegan@umfcluj.ro
4 Faculty of Medicine, University of Medicine and Pharmacy Cluj-Napoca, 400000 Cluj-Napoca, Romania
5 Proteomics Core Facility, Sahlgrenska Academy, University of Gothenburg, Medicinaregatan 5, 413 90 Gothenburg, Sweden; diarmuidkenny@gmail.com (D.K.); carina.sihlbom@gu.se (C.S.)
6 Wallenberg Laboratory, Department of Molecular and Clinical Medicine, Sahlgrenska Academy, University of Gothenburg, 413 45 Gothenburg, Sweden; antonio.molinaro@wlab.gu.se
7 Sahlgrenska Center for Cancer Research, Department of Laboratory Medicine, Institute of Biomedicine, Sahlgrenska Academy, University of Gothenburg, 405 30 Gothenburg, Sweden; ola.nilsson@llcr.med.gu.se
8 Department of Surgery, Institute of Clinical Sciences, Sahlgrenska Academy, University of Gothenburg, 413 45 Gothenburg, Sweden; anna.casselbrant@gastro.gu.se
9 Bioinformatics Core Facility, University of Gothenburg, Medicinaregatan 5, 413 90 Gothenburg, Sweden; marcela.davila@gu.se
* Correspondence: mihai.oltean@surgery.gu.se

Citation: Oltean, M.; Bagge, J.; Dindelegan, G.; Kenny, D.; Molinaro, A.; Hellström, M.; Nilsson, O.; Sihlbom, C.; Casselbrant, A.; Davila, M.; et al. The Proteomic Signature of Intestinal Acute Rejection in the Mouse. *Metabolites* **2022**, *12*, 23. https://doi.org/10.3390/metabo12010023

Academic Editor: Amedeo Lonardo

Received: 30 October 2021
Accepted: 23 December 2021
Published: 27 December 2021

Abstract: Intestinal acute rejection (AR) lacks a reliable non-invasive biomarker and AR surveillance is conducted through frequent endoscopic biopsies. Although citrulline and calprotectin have been suggested as AR biomarkers, these have limited clinical value. Using a mouse model of intestinal transplantation (ITx), we performed a proteome-wide analysis and investigated rejection-related proteome changes that may eventually be used as biomarkers. ITx was performed in allogenic (Balb/C to C57Bl) and syngeneic (C57Bl) combinations. Graft samples were obtained three and six days after transplantation (n = 4/time point) and quantitative proteomic analysis with iTRAQ-labeling and mass spectrometry of whole tissue homogenates was performed. Histology showed moderate AR in all allografts post-transplantation at day six. Nine hundred and thirty-eight proteins with at least three unique peptides were identified in the intestinal grafts. Eighty-six proteins varying by >20% between time points and/or groups had an alteration pattern unique to the rejecting allografts: thirty-seven proteins and enzymes (including S100-A8 and IDO-1) were significantly upregulated whereas forty-nine (among other chromogranin, ornithine aminotransferase, and arginase) were downregulated. Numerous proteins showed altered expression during intestinal AR, several of which were previously identified to be involved in acute rejection, although our results also identified previously unreported proteome changes. The metabolites and downstream metabolic pathways of some of these proteins and enzymes may become potential biomarkers for intestinal AR.

Keywords: intestinal transplantation; rejection; biomarkers; enzymes; chromogranin A

1. Introduction

The enterocytes abundantly express class II major histocompatibility complex molecules, making the intestinal lining highly immunogenic and susceptible to acute rejection (AR). During AR, recipient T-cells initially leave the intravascular compartment and infiltrate the lamina propria of the graft and ultimately attack graft enterocytes, eventually leading to mucosal loss. Unlike kidney and liver grafts, the acute rejection (AR) of the intestinal allograft lacks reliable non-invasive markers. The need for a biomarker of intestinal AR is pressing as AR may lead to graft and patient loss [1,2]. Given this lack of biomarkers, the current strategy for rejection surveillance still relies heavily on frequent protocol endoscopies and mucosal biopsies, which incur logistic issues, risks, and costs [1,3]. Depending on its stage, the histology of intestinal AR reveals various degrees of inflammation in the lamina propria, increased crypt apoptosis, crypt damage, crypt loss, villous blunting with edema, and congestion, culminating in mucosal sloughing. Alterations in the metabolism of several biomolecules such as decrease in plasma citrulline or increased fecal calprotectin have been reported during intestinal AR, and these two parameters were suggested as non-invasive rejection biomarkers. Unfortunately, both are influenced by numerous factors including the renal function, body surface area, infectious enteritis, or reperfusion injury. Hence, the significant inter-individual variation and their low specificity (50–75%) make them unreliable [4,5] and a search for other reliable biomarkers, suitable for safe, routine measurements is warranted.

Changes in the cellular expression of various molecules have been reported during intestinal AR in both the experimental and clinical setting. These changes appear secondary to different biological processes such as inflammation [6–8], tissue injury and repair [9,10], cell metabolism or apoptosis [11]. Two studies using proteomics and metabolomics to analyze the stomal effluent in intestinal transplant patients revealed complex patterns in the metabolism of various molecules released into the intestinal lumen, many of which seemed related to the AR. One study showed a clear interclass separation of metabolites detected during episodes of rejection with several metabolites related to leukotriene E4 and water soluble vitamins [12]. Another study identified 17 distinct protein expression profiles altered during rejection including human neutrophil peptide (HNP) 1, HNP 2, and human α-defensin 5 [13]. Despite the interesting and comprehensive data, and the undeniable clinical relevance, clinical studies have several limitations such as concurrent medication, the contribution of other digestive organs, or patient and sampling heterogeneicity [14,15].

Experimental transplantation in mice offers unique advantages such as highly standardized experimental conditions, a similar surgical procedure, and the advantage of a fully mapped proteome. Moreover, the investigator may easily influence the timing of rejection by the use of various strain combinations [16], select endpoints, and analyze rejection-related changes without the involvement of immunosuppressants. In the current study, we performed a proteome-wide analysis of the protein changes that occurred within the intestinal graft prior to, and during the acute rejection using isobaric tags for relative and absolute quantitation (iTRAQ). We then analyzed and compared protein expression changes in rejecting grafts (allogeneic combination) with non-rejecting (syngeneic) grafts.

2. Results

2.1. Intestinal Graft Histology

Syngeneic grafts did not show any histological alterations at any of the two time points and revealed long, slender villi, continuous epithelium, and no apoptosis in the crypts. Three days after transplantation, the histology of the intestinal allografts was unremarkable, and did not show any signs of acute rejection or other abnormalities. However, at post-transplantation day six, all allografts showed swollen villus tips, significant lymphocytic infiltrate in the lamina propria, widespread crypt apoptosis, and focal crypt destruction, findings consistent with moderate rejection (Figure 1).

2.2. The Proteomic Analysis

The proteomic analysis identified 3172 proteins, 1522 of which were identified based on three or more unique peptides. Of these, 1087 proteins were detected in all samples. We identified 109 proteins that demonstrated a significant change (either increase or decrease) in the allogeneic grafts between day 3 and day 6. Ninety-four proteins were found differentially expressed between the rejecting (allogenic) grafts and the non-rejecting (syngeneic) grafts at day 6 after transplantation (Figure 2). Of these, the following eight proteins had the same alteration pattern as that found in syngeneic grafts (and were likely unrelated to AR) and were therefore excluded from the analysis: mitochondrial 3-ketoacyl-CoA thiolase (THIM), non-specific lipid-transfer protein (NLTP), Arginase 2 (ARGI2), cytoplasmic isocitrate dehydrogenase (IDHC), hydroxyacyl-coenzyme A dehydrogenase (HCDH), alcohol dehydrogenase 1 (ADH1), carbonyl reductase 1 (CBR1), and UDP-glucose 6-dehydrogenase (UGDH). The remaining 86 proteins had a changed pattern only found in the rejecting allografts. Eighteen proteins (21%) had a >2-fold change (either increase or decrease) while the other forty-six proteins (53%) revealed a 1.5–2-fold change compared with the levels before rejection. Interestingly, the expression of four proteins, serine protease inhibitor A3N (SPA3N), fibrinogen gamma (FIBG), annexin 13 (ANX13,) and NADPH-cytochrome P450 reductase (NCPR), increased in rejecting grafts at day 6 whereas they decreased in non-rejecting syngeneic grafts at the same time point compared with day 3.

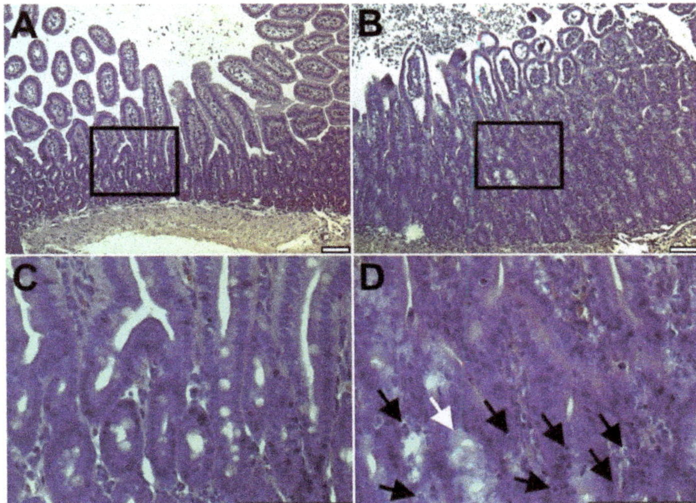

Figure 1. Histology of the intestinal allografts three days (**A**,**C**) and six days after transplantation (**B**,**D**) at low (100×) and high (400×) magnification showing essentially normal histology after three days and moderate acute rejection six days after transplantation with swollen villi, lymphocytic infiltration of lamina propria, crypt apoptosis (black arrows), and focal crypt destruction (white arrow). Scale bar = 100 µm.

In brief, the number of proteins that showed more than a 1.2-fold increased expression included enzymes (11), regulatory proteins, or transcription factors (16) and structural proteins (11). The identified number of proteins that decreased their expression by more than 20% during the rejection event included enzymes (28), regulatory proteins (9), and structural or functional proteins (10). Further details are presented in Table 1.

IPA interaction analysis found two hundred and ten canonical pathways differentially expressed between the syngenic (no rejection) and allogeneic grafts (moderate rejection) at day 6. In addition, two hundred and fourteen canonical pathways were found differentially expressed between the allogeneic grafts at day 3 (no rejection) and allogeneic grafts (moderate rejection) at day 6. The first 20 canonical pathways (ordered according to the magnitude of changes between the first and the second time point) in both allogenic and syngenic grafts are shown in Figure 3. Pathways involved in energy metabolism, TCA, substrate metabolism (glucose, fatty acid, amines and amino acid metabolism), and mitochondrial oxidative phosphorylation, all linked to oxidative stress response were among the most altered (mostly downregulated proteins).

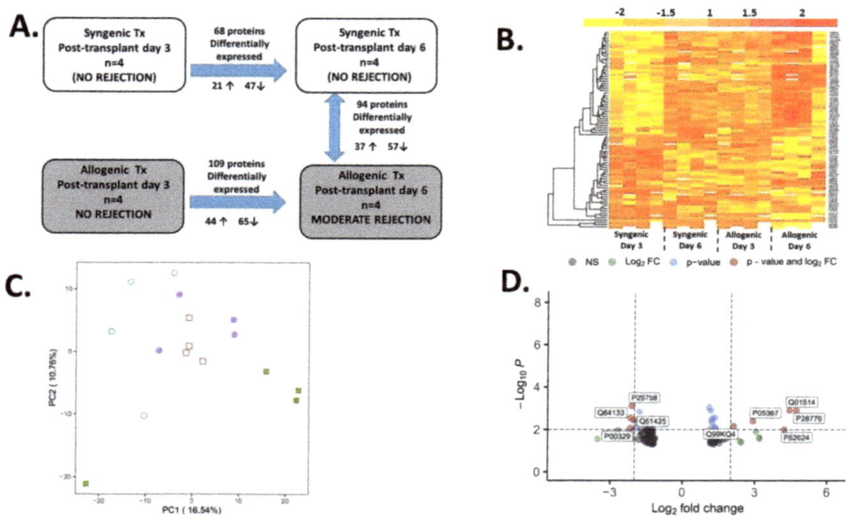

Figure 2. An outline of the proteomics analysis. (**A**) Summary of proteins differing between the two time points and groups; (**B**) Heat map showing the relative abundance and clustering of the 86 proteins identified across all four groups; (**C**) Principal component analysis (PCA) of the samples in the syngeneic group post-transplant day 3 (open circle) and day 6 (closed circle), and in the allogenic group at post-transplant day 3 (open square) and day 6 (closed square); (**D**) Volcano plot illustrating the fold change (log base 2) in protein expression in relation to the *p*-value (−log base 10) between non-rejecting (syngeneic) vs. rejecting (allogenic) grafts at day 6. Each dot represents a protein. Proteins at a significance level greater than 0.01 are in blue, those with a log2 fold change less than −2 and greater than 2 are in green, while proteins fulfilling both thresholds are in red, and their names are displayed.

Table 1. List of proteins in the rejecting allografts with an altered protein expression relative to that found in syngeneic, non-rejecting grafts at the same time point (post-transplant day 6) as identified by iTRAQ-based quantitation.

Accession	Symbol	Description	Fold Change	Molecular Function	Biological Process
Upregulated tissue expression					
P28776	Ido1	Indoleamine 2,3-dioxygenase 1	4.72	Dioxygenase, Oxidoreductase	Inflammatory response
Q01514	Gbp1	Interferon-induced guanylate-binding protein 1	4.44	Hydrolase	Inflammatory response
P52624	Upp1	Uridine phosphorylase 1	4.21	Glycosyltransferase, Transferase	Inflammatory response
Q91WP6	Ser3na	Serine protease inhibitor	3.19	Protease inhibitor	Inflammatory response
P27005	S100a8	Protein S100-A8	3.17	Antimicrobial	Cell death and survival
P42225	Stat1	Signal transducer and activator of transcription 1	3.04	Activator, DNA-binding	Cell death and survival
P05367	Saa2	Serum amyloid A-2 protein	2.91	Cytokine	Inflammatory response
Q8VCM7	Fgg	Fibrinogen gamma chain	2.44	Binding protein	Hemostasis
O35744	Chi3l1	Chitinase-3-like protein 3	2.38	Antimicrobial	Inflammatory response
Q99KQ4	Nampt	Nicotinamide phosphoribosyltransferase	2.12	Cytokine, Glycosyltransferase	Cell death and survival
E9Q555	Rnf213	E3 ubiquitin-protein ligase RNF213	1.82	Hydrolase, Transferase	Angiogenesis

Table 1. *Cont.*

Accession	Symbol	Description	Fold Change	Molecular Function	Biological Process
P01899	H2-d1	H-2 class I histocompatibility antigen, D-B alpha chain	1.69	Binding protein	Immunology
Q9R233	Tapbp	Tapasin	1.59	Binding protein	Immunology
Q9JIK5	Ddx21	Nucleolar RNA helicase 2	1.56	Binding protein	Immunology
P17918	Pcna	Proliferating cell nuclear antigen	1.55	DNA-Binding	Cell death and survival
P31001	Des	Desmin	1.52	Muscle protein	Cell structure
P26041	Msn	Moesin	1.49	Signal protein	Inflammatory response
Q60590	Orm1	Alpha-1-acid glycoprotein 1	1.46	Transport protein	Inflammatory response
P25206	Mcm3	DNA replication licensing factor MCM3	1.4	DNA-binding, Helicase, Hydrolase	Cell death and survival
P09405	Ncl	Nucleolin	1.36	Binding protein	Angiogenesis
P16858	Gapdh	Glyceraldehyde-3-phosphate dehydrogenase	1.35	Oxidoreductase, Transferase	Inflammatory response
P68033	Actc1	Actin, alpha cardiac muscle 1	1.35	Muscle protein	Cell movement
Q6NZJ6	Eif4g1	Eukaryotic translation initiation factor 4 gamma 1	1.35	Initiation factor, RNA-binding, Translational shunt	Cell death and survival
P52480	Pkm	Pyruvate kinase isozymes M1/M2	1.35	Allosteric enzyme, Kinase, Transferase	Cancer
Q9CPY7	Lap3	Cytosol aminopeptidase	1.32	Aminopeptidase, Hydrolase, Protease	Cell death and survival
Q9Z1Q5	Clic1	Chloride intracellular channel protein 1	1.32	Ion channel	Cellular growth and proliferation
Q99JG3	Anxa13	Annexin A13	1.31	Binding protein	Cell death and survival
P97372	Psme2	Proteasome activator complex subunit 2	1.29	Immunoproteasome assembly	Cell death and survival
Q61029	Tmpo	Lamina-associated polypeptide 2	1.27	DNA-Binding	Cell structure
P11499	Hsp90ab1	Heat shock protein HSP 90-beta	1.25	Chaperon	Cell death and survival
P05784	Krt18	Keratin, type I cytoskeletal 18	1.24	Structural protein	Cell structure
Q80 × 90	Flnb	Filamin-B	1.24	Actin-binding,	Cell movement
P60710	Actb	Actin, cytoplasmic 1	1.21	Muscle protein	Cell movement
P97371	Psme1	Proteasome activator complex subunit 1	1.21	Immunoproteasome assembly	Inflammatory response
P62137	Ppp1ca	Serine/threonine-protein phosphatase PP1-alpha catalytic subunit	1.2	Hydrolase, phosphatase	Cellular growth and proliferation
P37040	Por	NADPH–cytochrome P450 reductase	1.2	Oxidoreductase	Cellular function and maintenance
O08808	Diaph1	Protein diaphanous homolog 1	1.2	Actin-binding	Cell structure
Downregulated tissue expression					
Q9CZ13	Uqcrc1	Cytochrome b-c1 complex subunit 1, mitochondrial	0.79	Electron transport, Respiratory chain, Transport	Cellular function and maintenance
Q9ERG0	Lima1	LIM domain and actin-binding protein 1	0.79	Binding protein	Lipid metabolism
Q9D0F3	Aldh1b1	Protein ERGIC-53	0.79	Oxidoreductase	Cancer
Q921H8	Acaa1	3-ketoacyl-CoA thiolase A, peroxisomal	0.79	Acyltransferase, Transferase	Lipid metabolism
Q02819	Nucb1	Nucleobindin-1	0.78	DNA-binding, Guanine-nucleotide releasing factor	Cellular growth and proliferation
P09103	Pdia1	Protein disulfide-isomerase	0.78	Isomerases	Cell death and survival
Q9CY27	Tecr	Trans-2,3-enoyl-CoA reductase	0.78	Oxidoreductase	Lipid metabolism
Q9JII6	Ak1a1	Alcohol dehydrogenase	0.77	Dehydrogenase/reductase	Small molecule biochemistry
P28271	Aco1	Cytoplasmic aconitate hydratase	0.77	Lyase, RNA-binding	Cellular growth and proliferation
Q9JLQ0	Cd2ap	CD2-associated protein	0.76	Adapter protein	Cell cycle
Q99KI0	Acon	Aconitate hydratase, mitochondrial	0.76	Lyase	Cellular growth and proliferation
P47738	Aldh2	Aldehyde dehydrogenase, mitochondrial	0.75	Oxidoreductase	Small molecule biochemistry
P19783	Cox41	Cytochrome c oxidase subunit 4 isoform 1	0.75	Oxidoreductase	Cellular function and maintenance
P35700	Prdx1	Peroxiredoxin-1	0.74	Peroxidase	Cellular function and maintenance
P45952	Acadm	Medium-chain specific acyl-CoA dehydrogenase, mitochondrial	0.74	Oxidoreductase	Lipid metabolism

Table 1. *Cont.*

Accession	Symbol	Description	Fold Change	Molecular Function	Biological Process
P24270	Cat	Catalase	0.74	Catalase	Cellular function and maintenance
Q60598	Cttn	Src substrate cortactin	0.72	Unknown	Cell structure
Q80XN0	Bdh1	D-beta-hydroxybutyrate dehydrogenase, mitochondrial	0.71	Dehydrogenase/reductase	Lipid metabolism
Q9EPB4	Pycard	Apoptosis-associated speck-like protein containing a CARD	0.71	Unknown	Inflammatory response
Q9DBS5	Klc4	Kinesin light chain 4	0.71	Motor protein	Cell movement
P99028	Uqcrh	Cytochrome b-c1 complex subunit 6, mitochondrial	0.71	Oxidoreductase	Cellular function and maintenance
Q9D855	Uqcrb	Cytochrome b-c1 complex subunit 7	0.71	Electron transport, Respiratory chain, Transport	Cellular function and maintenance
Q5SYD0	Myo1d	Myosin-Id	0.71	Motor protein	Cell structure
Q8K2B3	Sdha	Succinate dehydrogenase [ubiquinone] flavoprotein subunit, mitochondrial	0.7	Oxidoreductase	Cellular function and maintenance
Q9CQW5	Lgals2	Galectin-2	0.7	Binding protein	Unknown
Q99K01	Pdxdc1	Pyridoxal-dependent decarboxylase domain-containing protein 1	0.69	Decarboxylase, Lyase	Cell cycle
P10852	Slc3a2	4F2 cell-surface antigen heavy chain	0.68	transport protein	Cellular function and maintenance
Q3UMR5	Mcu	Coiled-coil domain-containing protein 109A	0.68	Calcium channel, Ion channel	Cellular function and maintenance
Q9Z2I8	Suclg2	Succinyl-CoA ligase subunit beta, mitochondrial	0.68	Ligase	Cellular function and maintenance
Q8VC30	Tkfc	Bifunctional ATP-dependent dihydroxyacetone kinase	0.66	Multifunctional enzyme	Cellular function and maintenance
Q9R100	Cdh17	Cadherin-17	0.65	Adhesion protein	Cell structure
P56391	Cox6b1	Cytochrome c oxidase subunit 6B1	0.65	Oxidoreductase	Cellular function and maintenance
P14152	Mdh1–2	Malate dehydrogenase, cytoplasmic	0.64	Oxidoreductase	Cellular function and maintenance
Q8C196	Cps1	Carbamoyl-phosphate synthase, mitochondrial	0.64	Ligase	Cellular function and maintenance
P31786	Acbp	Acyl-CoA-binding protein	0.63	Binding protein	Unknown
Q9DCN2	Nb5r3	NADH-cytochrome b5 reductase 3	0.61	Oxidoreductase	Lipid metabolism
P57016	Lad1	Ladinin-1	0.59	Anchoring filament	Cell structure
O09131	Gsto1	Glutathione S-transferase omega-1	0.58	Oxidoreductase, Transferase	Oxidative stress
Q8K0C9	Gmds	GDP-mannose 4,6 dehydratase	0.57	Lyase	Cellular function and maintenance
Q9CZS1	Al1b1	Aldehyde dehydrogenase X, mitochondrial	0.57	Oxidoreductase	Small molecule biochemistry
Q9CQ62	Decr	2,4-dienoyl-CoA reductase, mitochondrial	0.54	Oxidoreductase	Lipid metabolism
Q8R0Y6	Fthfd	10-formyltetrahydrofolate dehydrogenase	0.54	Oxidoreductase	Cellular function and maintenance
Q9D8W7	Ocad2	OCIA domain-containing protein 2	0.5	Unknown	Cancer
Q9QWG7	St1b1	Sulfotransferase family cytosolic 1B member 1	0.49	Sulfotransferase	Cellular function and maintenance
P29758	Oat	Ornithine aminotransferase, mitochondrial	0.47	Aminotransferase, Transferase	Cellular function and maintenance
Q64133	Aofa	Amine oxidase A	0.46	Oxidoreductase	Cellular function and maintenance
O88310	Itl1a	Intelectin-1a	0.37	Antimicrobial	Inflammatory response
P26339	Cmga	Chromogranin-A	0.31	Inhibitor protein	Immunology
P35230	Reg3b	Regenerating islet-derived protein 3-beta	0.28	Antibacterial protein	Immunology

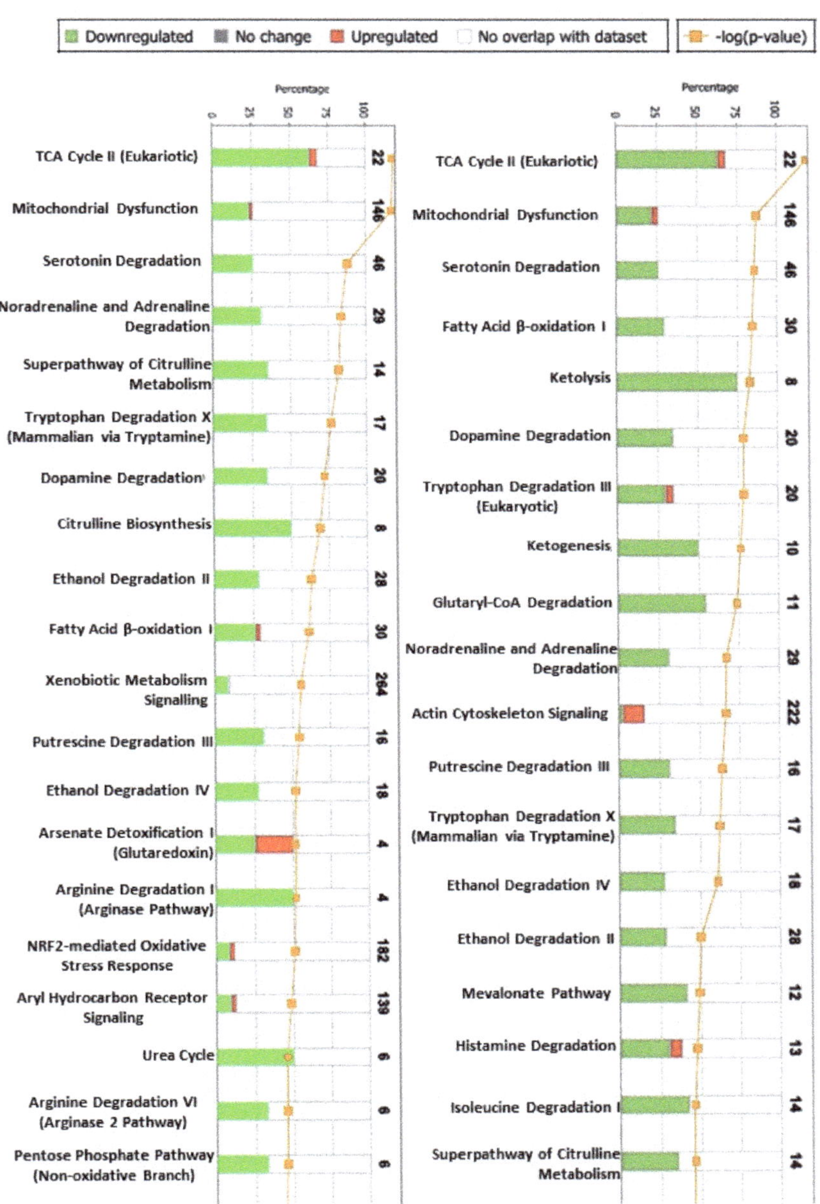

Figure 3. The alterations in the first 20 canonical pathways in allogenic grafts between post-transplant day 3 and day 6 (**left**) and between rejecting (allogeneic grafts) and non-rejecting (syngeneic grafts) at post-transplant day 6 (**right**) as revealed by the interactive pathway analysis. Downregulated pathways shown in green, upregulated pathways shown in red.

The four biological networks that differed the most between allografts with or without rejection, as indicated by the IPA, were (i) cell-to-cell signaling and interaction, tissue development and cell cycle (35 proteins differing); (ii) energy production, lipid metabolism, small molecule biochemistry (33 proteins differing); (iii) cell death and survival, carbohydrate metabolism, lipid metabolism (21 proteins); and (iv) lipid metabolism, small molecule biochemistry, organismal functions (21 proteins).

2.3. Confirmatory Analysis: Western Blot Analysis and Immunohistochemistry

To confirm the iTRAQ data from the pooled samples, we assessed the expression level of chromogranin A in individual samples using western blot analysis and immunofluorescence. This protein was selected as it was the second most downregulated protein as indicated by the proteomics analysis. In addition, chromogranin A is an analyte for which reliable, routine laboratory tests are already available. Immunoblotting for chromogranin A revealed downregulation in allogeneic, rejecting transplants on day 6 post-transplant when compared with the normal intestines (9.4-fold change, $p = 0.05$, Mann–Whitney U test) or syngeneic controls at the same time point (3.32-fold change, $p = 0.1$, Mann–Whitney U test; Figure 4).

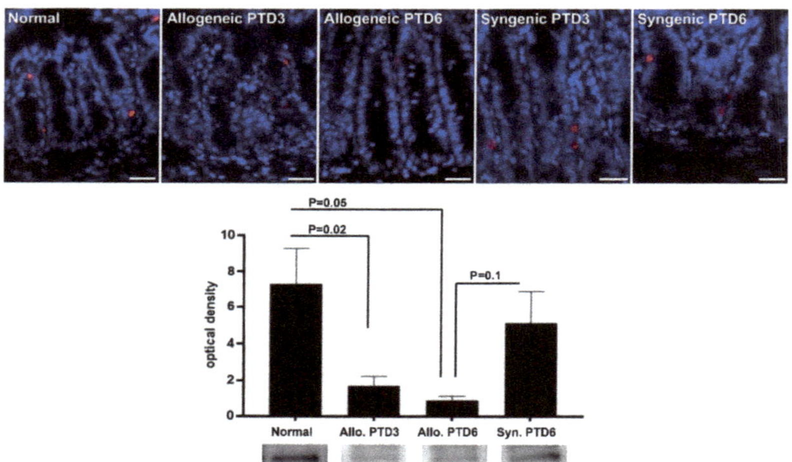

Figure 4. (**Upper panel**) Immunofluorescence microphotographs showing crypt cells positive for chromogranin A (red). Nuclei were stained blue using 40,6-diamidino-2-phenylindole. Original magnification 400×, scale bar 25 µm. (**Lower panel**) Western blot analysis for chromogranin A and representative immunoblot bands. The results from four separate experiments in each group are shown. PTD–post-transplant day.

Normal intestines had 1–2 cells positive for chromogranin A in each crypt (Figure 4). Syngeneic grafts had similar density and distribution of chromogranin-positive cells with that of normal intestines at both time points. At day 3 post-transplant, allografts had lower density and size of chromogranin A-positive cells whereas at during moderate rejection (post-transplant 6), most crypts were devoid of positive cells

3. Discussion

This study deciphers novel molecular mechanisms of intestinal allograft rejection in mice and represents the most comprehensive proteomic investigation on this topic to date. The analysis offers a simultaneous snapshot of several major processes and events during the acute rejection and maps its key players (e.g., showing the upregulation of specific proteins responsible for apoptosis and cell death, inflammation, cell migration, and antigen presentation). A few of these alterations such as impaired citrulline or tryptophan metabolism or increased calprotectin have been identified earlier in experimental and clinical studies [4,5], while several of our findings are novel for the intestinal transplantation setting.

The current results also revealed significant alterations in the tissue expression of several structural proteins. Due to the ongoing cell and tissue injury during rejection, this was an anticipated finding that we considered in the experimental design and explains why grafts with only a moderate rejection were assessed. Advanced AR would have involved significant tissue injury and mucosal loss, whereas during moderate AR, all mucosal compartments are preserved and the enterocyte mass is nearly intact. Hence, our experimental model allowed us to study a broad array of changes that occurred in a largely retained mucosa.

Besides changes in structural proteins, the analysis found quantitative changes of various enzymes and components of several cellular signaling pathways. Some of these pathways were identified as parts of the ongoing inflammation and tissue injury. As an example, we found a greatly increased expression of signal transducer and activator of transcription (STAT)-1, a transcription factor known to promote intestinal allograft rejection [17] or the upregulation of H-2 class I histocompatibility antigen, the murine correspondent of the human major histocompatibility complex class II. Furthermore, rejecting intestines showed increased expression of serum amyloid A, an acute-phase lipoprotein induced during inflammation or infection and has been shown to correlate with ICAM-1 or VCAM in patients with Crohn's disease [18].

Several of the identified proteins are involved in maintaining the cellular redox status, modulate the oxidative stress, and provide cytoprotection against pro-oxidant stimuli. It is known that inflammation increases the level of reactive oxygen metabolites, resulting in oxidative stress due to an imbalance between antioxidants and reactive oxygen. Many of these classical antioxidant enzymes such as catalase, glutathione peroxidases, and peroxiredoxins directly inactivate reactive oxygen and nitrogen species, and have been suggested to be suitable biomarkers themselves for oxidative stress [19,20]. Other antioxidant enzymes such as 2,4-dienoyl-CoA reductase, 3-ketoacyl-CoA thiolase A, or glutathione-S-transferase recycle thiols or detoxify endogenous compounds such as peroxidized lipids and reactive secondary metabolites (such as aldehydes, peroxides, epoxides) [20]. The notable decrease in the tissue expression of several key antioxidant enzymes shown herein reflects a significant, ongoing oxidative stress and suggests that this is a mechanistic cause for cell injury related to intestinal acute rejection. This mirrors an earlier proteomics analysis that assessed rejection after rat liver transplantation. That study detected decreased tissue levels of catalase and aldehyde dehydrogenase and advocated that the imbalance in the reactive oxygen species scavenging machinery may have contributed to the dysfunction of hepatocytes and liver allografts [21]. Our results are also similar to a proteomics analysis of rejecting mouse hearts that found significantly decreased aconitate hydratase and increased pyruvate kinase isozyme M2 and glyceraldehyde-3-phosphate dehydrogenase (all related to the energy metabolism) [22]. Taken together, these findings suggest an impending energetic failure, in addition to the deteriorating antioxidant defense.

Our proteome analysis also revealed an obvious ongoing stress response in the rejecting grafts as indicated by the upregulation of several stress-related proteins, most notably heat shock protein (HSP) 90. Although acute rejection has been previously shown to induce various heat shock proteins in the intestine [10,23] and other organs [24], the significance of the heat shock response and its pathways remains unclear [25,26]. In addition, we found significant increases in the tissue expression of several central enzymes including uridine phosphorylase 1 (UPP1) and nicotinamide phosphoribosyltransferase (NAMPT), which are involved in key cellular functional mechanisms such as DNA repair and chromatin remodeling, secondary to the progressive, ongoing tissue injury (apoptosis, necrosis). Interestingly, the expression of the four proteins serine protease inhibitor A3N (SPA3N), fibrinogen gamma (FIBG), annexin 13 (ANX13), and NADPH-cytochrome P450 reductase (NCPR) increased in the rejecting grafts at day 6, while they decreased in non-rejecting syngeneic grafts at the same time point compared with day 3. Whereas the PCA analysis revealed a certain overlay of the non-rejecting groups, rejecting grafts did not overlap with any of the other datasets, indicating an alteration pattern rather specific to AR and giving hope to the search for new candidate biomarkers for intestinal rejection.

Chromogranin A, a neuroendocrine secretory protein produced by the enteroendocrine crypt cells, was selected for the confirmatory study as it was the second most downregulated protein in the rejecting intestines, as revealed by the proteomics analysis. Immunoblotting and immunofluorescence confirmed the decreasing trend in the rejecting allografts. This may be followed by lower levels in the blood or feces, suggesting a potential use of chromogranin A in blood or feces as a non-invasive rejection biomarker. Interestingly, the confirmatory analyses also indicated a trend toward lower chromogranin in syngeneic, non-rejecting grafts. The significance of this finding is unclear, although previous studies found a lower density of chromogranin A-positive cells in the mouse intestine following vagotomy [27] or in the prostate after its peripheral denervation [28]. Hence, lower chromogranin A expression may be, at least in part, secondary to graft denervation following transplantation.

The current analysis was restricted to proteins identified on the basis of three or more unique peptides. This arbitrary threshold made protein identification extremely accurate, but this selection may have omitted numerous other relevant proteins. Hence, an extended analysis of the remaining >200 proteins identified based on two unique peptides is mandated. A relative shortcoming of the present study is the use of homogenate from whole tissue samples. Thus, it is difficult to discriminate between the tissue compartments (i.e., mucosa, submucosa, muscular layer, vasculature, immune cells) where these changes have occurred. The results were obtained using intestinal transplantation in mice and without the use of immunosuppressive medication, and the intestine was transplanted heterotopically, without exposure to luminal alimentary stimuli or proximal trophic factors. However, despite the experimental setting, the clinical relevance of the current findings is apparent when considering the current findings of impaired citrulline metabolism during intestinal acute rejection [29] and the increased expression of S100-A8 (a constitutive part of calprotectin) [4]. As discussed earlier, citrulline has been shown to decrease following significant intestinal mucosal damage during severe rejection supposedly through the loss of enterocyte mass [5]. Our analysis identified altered expression of several enzymes central for citrulline biosynthesis. Hence, the current data suggest that, besides the loss of enterocytes, citrulline decrease may also be functional due to lower expression of upstream enzymes.

This study confirmed the proteomics finding for only one protein out of the 86 proteins, revealing an altered expression during intestinal AR. This investigation needs to be expanded to other proteins in the list, and analyzed with respect to corresponding alterations in the intestinal luminal content and in the blood, using both animal and human samples. In conclusion, this proteome-wide analysis indicated a significant ongoing oxidative stress during the acute rejection of the murine intestinal allograft and the exhaustion of several key redox mechanisms. Almost 100 proteins involved in numerous metabolic pathways altered their expression during allograft rejection. These pathways may include metabolites with potential new, non-invasive biomarkers for intestinal acute rejection.

4. Materials and Methods

4.1. Animals and Study Design

Male BalbC (donors) and C57BL6 (donors and recipients) mice weighing 25–30 g were used. Donor mice were fasted overnight before the explantation, while recipients had unrestricted access to food and water. The study closely followed the ethical regulations outlined by the NIH and the European Union and were reviewed and approved by the local committee of the Swedish Animal Welfare Agency (Dnr 287/99).

Heterotopic intestinal transplantation was performed in either allogeneic (BalbC donor to C57BL6 recipient) or syngeneic (C57BL6 to C57BL6) combinations. Previous experiments revealed that this fully allogeneic, high responder strain combination resulted in advanced rejection at post-transplant day six and severe rejection and animal death due to graft perforation and peritonitis by post-transplant day eight [16]. Mice (n = 4 per group and time point) were sacrificed at either day 3 and day 6 and graft segments were stored in formalin (histology) or snap-frozen (proteomics and western blot).

4.2. Surgery

Surgery was performed under 2% isoflurane anesthesia using a technique previously described [30]. In brief, the proximal half of the donor intestine was isolated by removing the duodenum, ileum, and colon and freeing the portal vein and the aorta above and below the emergence of the superior mesenteric artery. The intestinal graft was perfused in situ with cold, heparinized saline via the infrarenal aorta and stored in saline at 4 °C until recipient preparation (around 1 h). The graft was transplanted into the recipients using microvascular end-to-side anastomoses between the aortic patch containing the emergence of the superior mesenteric artery of the graft and the infrarenal recipient aorta. Venous drainage was achieved by anastomosing the portal vein of the graft and the recipient infrarenal vena cava, respectively, using 11/0 nylon sutures. The extrinsic, splanchnic nerves were not reconstructed. The proximal graft end was brought out as a stoma, while the distal end was anastomosed to the terminal ileum of the recipient. The recipient mice received a single intraperitoneal dose of cefuroxim (40 mg/kg) (Zinacef®, Glaxo Wellcome, UK) at the end of surgery. No immunosuppression was used.

4.3. Histology

Formalin-fixed intestinal graft segments were embedded in paraffin and cut into 5-micron sections. Hematoxylin-eosin slides were examined blindly by an experienced transplant pathologist and scored using a previously described scheme [30].

4.4. Immunofluorescence

Paraffin sections were deparaffinized and rehydrated, then antigen retrieval was performed by pressure cooking the slides in citrate buffer (pH 6) for 20 min. After species-specific blocking, slides were incubated overnight at 4 °C with antibodies against chromogranin A (PA5-77917, 1:250, Invitrogen AB, Lidingö, Sweden). Thereafter, slides were incubated with secondary antibody conjugated with Alexa 488 (1:500; Invitrogen). The sections were counterstained with 4′6′-diamidino-2-phenylindole, mounted with aqueous mounting medium (Vector Laboratories, Burlingame, CA, USA) and examined by fluorescence microscopy (Leica). Image acquisition and processing were performed using Leica LAS software.

4.5. Proteomic Analysis

4.5.1. Sample Preparation

Samples were homogenized using a FastPrep-24 System (MP Biomedicals, Santa Ana, CA, USA). Samples were transferred to Lysis Matrix B tubes (MP Biomedicals, Santa Ana, CA, USA) containing 0.1 mm silica spheres. A total of 300 µL of lysis buffer (50 mM Triethylammoinium bicarbonate (TEAB) (Fluka, Sigma Alrdich, St Louis, MO), 8 M urea, 4% 3-[(3-cholamidopropyl)dimethylammonio]-1-propanesulfonate (CHAPS), 0.2% sodium dodecyl sulfate (SDS), 5 mM ethylenediaminetetraacetic acid (EDTA), adjusted to pH 8.5, was added to the tubes.

The protein concentration was determined using the Pierce 660 nm Protein Kit (Thermo Scientific, Basel, Switzerland) according to the manufacturer's guidelines. For Tandem Mass Tag (TMT) labeling, 100 µg of the total protein of each sample and 100 µg of a pool containing equal amounts of all samples were diluted with two volumes of 0.5 M TEAB and 1 volume of 18 mΩH_2O. One µL of 2% SDS was also added to each tube. The samples were reduced by the addition of 2 µL of 50 mM tris-(2-carboxyethyl) phosphine (TCEP) (Thermo Scientific, Basel, Switzerland), which was incubated at 37 °C for 1 h. The samples were subsequently alkylated with the addition of 1 µL of 200 mM methyl methanethiosulfonate (MMTS) (Fluka, Sigma Aldrich, St Louis, MO, USA) in gradient grade acetonitrile (Merck KGaA, Darmstadt, Germany) and incubated at room temperature for 10 min. The samples were digested with 4 µg of sequencing grade modified porcine trypsin (Promega, Madison, WI, USA) in 18 mΩH_2O water to (Promega, Madison, WI, USA) overnight at 37 °C.

The digested samples were dried in a SpeedVac to ~25–30 µL and 70 µL of 0.5 M was then added to each tube as a preparation for labeling with the TMT® (Thermo Scientific, Basel, Switzerland). TMT reagents were allowed to equilibrate to room temperature, and thereafter, 42 µL of gradient grade acetonitrile was added to each tube. The TMT reagent was transferred to the appropriate sample tube (see Table 1 for labeling details) and incubated at room temperature for 1 h. The reaction was quenched by the addition of 8 µL of 5% hydroxylamine (Thermo Scientific, Basel, Switzerland), which was incubated for 15 min at room temperature. After TMT labeling, the labeled samples were pooled and concentrated to ~50 µL in a SpeedVac in preparation for strong cation exchange (SCX) fractionation.

4.5.2. Strong Cation Exchange Chromatography (SCX) of TMT Labeled Peptides

The concentrated peptides were acidified to below pH 3 by the addition of 25 µL 10% formic acid. The acidified labeled peptides were diluted with 25% gradient grade acetonitrile (Merck KGaA, Darmstadt, Germany) and injected onto a 2.1 mm i.d. × 10 cm length, 5 µm particle size, 300 Å pore size PolySULFOETHYL ATM strong cation-exchange (SCX) column (PolyLc Inc., Columbia, MD, USA) at a flowrate of 0.25 mL/min. SCX chromatography and fractionation was performed using an ÄKTA purifier system (GE Healthcare life science, Uppsala, Sweden) at 0.25 mL/min flow rate using the following gradient: 100% A (25 mM ammonium formate, pH 2.8 in 25% acetonitrile) for 10 min; 0–20% B (500 mM ammonium formate, pH 2.8 in 25% acetonitrile) for 20 min; 20–40% B for 10 min, and 40–100% B for 10 min and 100% B held for 10 min. UV absorbance at 280 nm was monitored while fractions were collected in tubes at 0.5 mL intervals and dried down in a Speedvac to approximately 50 µL. The 16 peptide containing fractions were desalted on PepClean C18 spin columns (Thermo Fisher Scientific, Inc., Waltham, MA, USA) according to the manufacturer's guidelines.

4.5.3. LC-MS/MS Analysis on LTQ-Orbitrap Velos

The desalted and dried fractions were reconstituted with 15 µL of 0.1% formic acid (Sigma Aldrich, St. Louis, MO, USA) in 3% acetonitrile and analyzed on a LTQ-Orbitrap Velos mass spectrometer (Thermo Fisher Scientific, Inc., Waltham, MA) interfaced with an in-house constructed nano-LC column. For each sample, a two micro-liter sample injection was made with an Easy-nLC autosampler (Thermo Fisher Scientific, Inc., Waltham, MA, USA), running at 200 nL/min. The peptides were trapped on a precolumn (45 × 0.075 mm i.d.) and separated on a reversed phase column, 190 × 0.075 mm, both packed in-house with 3 µm Reprosil-Pur C18-AQ particles (Dr. Maisch, Ammerbuch, Germany). The gradient was set to 0–70 min 5–35% acetonitrile in 0.2% formic acid, 70–80 min 35–80% acetonitrile in 0.2% formic acid, and the last 10 min at 80% acetonitrile in 0.2% formic acid.

Ions were injected into the LTQ-Orbitrap Velos mass spectrometer by electron spray ionization (ESI) under a spray voltage of 1.6 kV in positive ion mode with a capillary temperature of 250 °C. For MS scans, one microscan was performed at a 30,000 resolution (at m/z 400), and ions were detected within the mass range of m/z 400–1800. MS analysis was performed in a data-dependent mode. The top 10 most abundant doubly or multiply charged precursor ions, with a threshold count greater than 2000 and an isolation width (m/z) of 2.0 in each MS scan was selected for fragmentation (MSn) by stepped high energy collision dissociation (HCD). For MSn scans, one microscan was performed at a 7500 resolution (at m/z 400) with a mass range between m/z 120–2000 with stepped collision energies of 25%, 35%, and 45%, and a maximum injection time of 100 ms, and one repeat count was performed with a 30 s dynamic exclusion.

4.5.4. Database Search and TMT Quantification

MS raw data files from all 16 SCX fractions for each TMT set were merged for relative quantification and identification using Proteome Discoverer^{TM} version 1.3 (Thermo Fisher Scientific, Inc., Waltham, MA, USA). A database search for each set was performed with the Mascot search engine (Matrix Science LTD., London, UK) using the Swissprot Database version 2.3 (Swiss Institute of Bioinformatics, Switzerland) with MS peptide tolerance of 10 ppm and MS/MS tolerance of 100 molecular mass units. Tryptic peptides with a maximum of one missed cleavage were accepted and variable modifications of methionine oxidation, cysteine methylthiolation, and fixed modifications of N-terminal TMT6plex and lysine TMT6plex were selected. Only spectra with a precursor mass between 400 and 8000 Da and a minimum peak count of 10 were chosen for identification.

The detected peptide threshold in the software was set to 1% false discovery rate by searching against a reversed database. Criteria used for positive protein identification were ≥3 peptides match, and an averaged ratio-fold change ≥1.2. For TMT quantification, the ratios of the TMT reporter ion intensities in MS/MS spectra ([M + H]+ m/z 126–131) from raw datasets were used to calculate fold changes between samples. Ratio was derived by Proteome Discoverer^{TM} using the following criteria: fragment ion tolerance as 80 ppm for the most confident centroid peak, TMT reagent purity correction factors were used, and missing quantification values were replaced with minimum intensity. To correct for experimental bias, all peptide rations were normalized by the median protein ratio, assuming a minimum count of 20 proteins had been observed. The co-isolation exclusion threshold of 30% was accepted for co-isolation interference. Only peptides unique for a given protein were considered for relative quantitation excluding those common to other isoforms or proteins of the same family. The quantification was normalized using the protein median. The results were then exported into MS Excel 2016 (Microsoft, Redmond, WA, USA) for manual data interpretation and statistical analysis.

4.5.5. Bioinformatics Analysis of the Differentially Expressed Proteins

Pathway analysis (Ingenuity Pathway Analysis, IPA, Qiagen) was used to obtain further insight into potential cellular pathways that might be modified as a result of protein changes identified in present experiments. IPA automatically generated networks of gene, protein, small molecule, drug, and disease associations on the basis of "hand-curated" data held in a proprietary database. The identifiers (GI mouse identification number) of DEPs were uploaded as an Excel spreadsheet file onto the Ingenuity software (Ingenuity Systems, Redwood City, CA, USA). Each GI mouse identification number was mapped to its corresponding molecule in the Ingenuity Pathway Knowledge Base. The biological functions assigned to each network were ranked according to the significance of that biological function to the network. Networks of these proteins were algorithmically generated based on their connectivity and assigned a score. The score was used to rank networks according to how relevant they were to the proteins in the input dataset. The network identified was then presented as a graph indicating the molecular relationship between proteins. Finally, we compared the proteins differentially expressed between rejecting and non-rejecting grafts varying more than 20% between groups and time points

4.6. Western Blot Analyses of Intestinal Mucosa

Western blot protein analysis was performed using whole tissue frozen specimens as described earlier [31]. In brief, after electrophoresis and protein transfer on poly-vinyl-difluoride membranes, the membranes were blocked, then incubated overnight at 4 °C with primary antibody against chromogranin A (PA5-77917, 1:500, Invitrogen AB, Lidingö, Sweden). After repeated washings, a secondary antibody was applied for 1 h at room temperature and visualization was carried out using the chemiluminescent enzyme substrate CDP-Star (Tropix, Bedford, MA, USA). After repeated washings, a secondary antibody was applied for 1 h at room temperature and visualization was carried out using the chemiluminescent Clarity Western ECL (1705062, Bio-rad). The signal intensities of specific bands were detected and analyzed using a Chemi-doc Imaging system. Data were obtained via ImageLab software using Stain-free technology to perform total protein normalization. The Mann–Whitney U test was used for statistical analysis of western blot data; GraphPad Prism 6 (GraphPad Software, La Jolla, CA, USA) was used, and p values less than 0.05 were considered statistically significant.

Author Contributions: Conceptualization, M.O. (Mihai Oltean), G.D., D.K., M.H. and C.S.; Methodology, M.O. (Mihai Oltean), J.B., G.D., D.K., A.M., O.N. and A.C.; Software, M.D. and C.S.; Formal analysis, M.D., M.O. (Mihai Oltean), J.B. and A.C.; Resources, M.O. (Mihai Oltean) and M.O. (Michael Olausson); Writing—original draft preparation, M.O. (Mihai Oltean), J.B., D.K., M.H. and A.M.; Writing—review and editing, G.D., M.D. and M.O. (Michael Olausson); Visualization, M.H., M.O. (Mihai Oltean) and J.B.; Supervision, M.O. (Mihai Oltean), M.H. and A.C. All authors have read and agreed to the published version of the manuscript.

Funding: Funding for this study was provided by grants from the Swedish state under the agreement between the Swedish government and the country councils (grants ALFGBG-518371 and ALFGBG-812881) and Gothenburg Society of Medicine (GLS-170411) (all to M.O.).

Institutional Review Board Statement: The study closely followed the ethical regulations outlined by the NIH and the European Union and was reviewed and approved by the local committee of the Swedish Animal Welfare Agency and approve code is 287/99.

Informed Consent Statement: Not applicable.

Data Availability Statement: The data presented in this study are available on request from the corresponding author. The data are not publicly available due to further analysis ongoing.

Conflicts of Interest: The authors declare no conflict of interest. The funders had no role in the design of the study; in the collection, analyses, or interpretation of data; in the writing of the manuscript, or in the decision to publish the results.

References

1. Venick, R.S. Grant monitoring after intestinal transplantation. *Curr. Opin. Organ Transplant.* **2021**, *26*, 234–239. [CrossRef] [PubMed]
2. Raghu, V.K.; Beaumont, J.L.; Everly, M.J.; Venick, R.S.; Lacaille, F.; Mazariegos, G.V. Pediatric intestinal transplantation: Analysis of the intestinal transplant registry. *Pediatr. Transplant.* **2019**, *23*, e13580. [CrossRef] [PubMed]
3. Varkey, J. Graft assessment for acute rejection after intestinal transplantation: Current status and future perspective. *Scand. J. Gastroenterol.* **2021**, *56*, 13–19. [CrossRef] [PubMed]
4. Mercer, D.F. Hot topics in postsmall bowel transplantation: Noninvasive graft monitoring including stool calprotectin and plasma citrulline. *Curr. Opin. Organ Transplant.* **2011**, *16*, 316–322. [CrossRef] [PubMed]
5. Gondolesi, G.; Ghirardo, S.; Raymond, K.; Hoppenhauer, L.; Surillo, D.; Rumbo, C.; Fishbein, T.; Sansaricq, C.; Sauter, B. The value of plasma citrulline to predict mucosal injury in intestinal allografts. *Am. J. Transplant.* **2006**, *6*, 2786–2790. [CrossRef]
6. Fujisaki, S.; Park, Y.J.; Kimizuka, Y.; Inoue, M.; Tomita, R.; Fukuzawa, M.; Matsumoto, K. Expression of mucosal addressin cell adhe-sion molecule-1 (MAdCAM-1) during small-bowel graft rejection in rats. *Scand. J. Gastroenterol.* **2003**, *38*, 437–442. [CrossRef]
7. Asaoka, T.; Island, E.R.; Tryphonopoulos, P.; Selvaggi, G.; Moon, J.; Tekin, A.; Amador, A.; Levi, D.M.; Garcia, J.; Smith, L.; et al. Characteristic immune, apoptosis and inflammatory gene profiles associated with intestinal acute cellular rejection in formalin-fixed paraffin-embedded mucosal biopsies. *Transpl. Int.* **2011**, *24*, 697–707. [CrossRef]
8. Joshi, M.; Dindelegan, G.; Olausson, M.; Oltean, M. Natural killer group 2 member D cell recruitment driven by major histocompatibility complex class I chain-related antigens A and B: A possible mechanism during acute intestinal allograft rejection in the mouse. *Transplant. Proc.* **2010**, *42*, 4467–4469. [CrossRef]
9. Mueller, A.; Platz, K.-P.; Heckert, C.; Häusler, M.; Guckelberger, O.; Schuppan, D.; Lobeck, H.; Neuhaus, P. Extracellular matrix: An early target of preservation/reperfusion injury and acute rejection after small bowel transplantation. *Transplant. Proc.* **1998**, *30*, 2569–2571. [CrossRef]
10. Oltean, M.; Dindelegan, G.; Kurlberg, G.; Nilsson, O.; Karlsson-Parra, A.; Olausson, M. Intragraft heat shock protein-60 expression after small bowel transplantation in the mouse. *Transplant. Proc.* **2004**, *36*, 350–352. [CrossRef]
11. D'Errico, A.; Corti, B.; Pinna, A.; Altimari, A.; Gruppioni, E.; Gabusi, E.; Fiorentino, M.; Bagni, A.; Grigioni, W. Granzyme B and perforin as predictive markers for acute rejection in human intestinal transplantation. *Transplant. Proc.* **2003**, *35*, 3061–3065. [CrossRef]
12. Girlanda, R.; Cheema, A.K.; Kaur, P.; Kwon, Y.; Li, A.; Guerra, J.; Matsumoto, C.S.; Zasloff, M.; Fishbein, T.M. Metabolomics of human intestinal transplant rejection. *Am. J. Transplant.* **2012**, *12*, S18–S26. [CrossRef]

13. Kumar, A.R.; Li, X.; Leblanc, J.F.; Farmer, D.G.; Elashoff, D.; Braun, J.; Ziring, D. Proteomic Analysis Reveals Innate Immune Activity in Intestinal Transplant Dysfunction. *Transplantation* **2011**, *92*, 112–119. [CrossRef]
14. Thomson, A.W.; Bonham, C.A.; Zeevi, A. Mode of Action of Tacrolimus (FK506): Molecular and Cellular Mechanisms. *Ther. Drug Monit.* **1995**, *17*, 584–591. [CrossRef]
15. Yandza, T.; Gerhardt, M.F.; Saint-Paul, M.-C.; Braud, V.; Gugenheim, J.; Hébuterne, X. Significance of Serum Bile Acids in Small Bowel Allograft Rejection in Pigs. *Transplantation* **2009**, *87*, 24–28. [CrossRef]
16. Zhang, Z.; Zhu, L.; Quan, D.; Garcia, B.; Ozcay, N.; Duff, J.; Stiller, C.; Lazarovits, A.; Grant, D.; Zhong, R. Pattern of liver, kidney, heart, and intestine allograft rejection in different mouse strain combinations. *Transplantation* **1996**, *62*, 1267–1272. [CrossRef]
17. Stojanovic, T.; Scheele, L.; Wagner, A.H.; Middel, P.; Bedke, J.; Lautenschläger, I.; Leister, I.; Panzner, S.; Hecker, M.; Lautenschl, I. STAT-1 decoy oligonucleotide improves microcirculation and reduces acute rejection in allogeneic rat small bowel transplants. *Gene Ther.* **2007**, *14*, 883–890. [CrossRef]
18. Yarur, A.J.; Quintero, M.A.; Jain, A.; Czul, F.; Barkin, J.S.; Abreu, M.T. Serum Amyloid A as a Surrogate Marker for Mucosal and Histologic Inflammation in Patients with Crohn's Disease. *Inflamm. Bowel Dis.* **2017**, *23*, 158–164. [CrossRef]
19. Poynton, R.A.; Hampton, M.B. Peroxiredoxins as biomarkers of oxidative stress. *Biochim. Biophys. Acta (BBA) Gen. Subj.* **2014**, *1840*, 906–912. [CrossRef]
20. Mittal, M.; Siddiqui, M.R.; Tran, K.; Reddy, S.P.; Malik, A.B. Reactive Oxygen Species in Inflammation and Tissue Injury. *Antioxid. Redox Signal.* **2014**, *20*, 1126–1167. [CrossRef]
21. Cheng, J.; Zhou, L.; Jiang, J.; Qin, Y.; Xie, H.; Feng, X.; Gao, F.; Zheng, S. Proteomic Analysis of Differentially Expressed Proteins in Rat Liver Allografts Developed Acute Rejection. *Eur. Surg. Res.* **2010**, *44*, 43–51. [CrossRef]
22. Kienzl, K.; Sarg, B.; Golderer, G.; Obrist, P.; Werner, E.R.; Werner-Felmayer, G.; Lindner, H.; Maglione, M.; Schneeberger, S.; Margreiter, R.; et al. Proteomic Profiling of Acute Cardiac Allograft Rejection. *Transplantation* **2009**, *88*, 553–560. [CrossRef]
23. Ogita, K.; Hopkinson, K.; Nakao, M.; Wood, R.F.M.; Pockley, A.G. Stress responses in graft and native intestine after rat heterotopic small bowel transplantation. *Transplantation* **2000**, *69*, 2273–2277. [CrossRef]
24. Mehta, N.; Carroll, M.; Sykes, D.; Tan, Z.; Bergsland, J.; Canty, J.; Bhayana, J.; Hoover, E.; Salerno, T. Heat Shock Protein 70 Expression in Native and Heterotopically Transplanted Rat Hearts. *J. Surg. Res.* **1997**, *70*, 151–155. [CrossRef]
25. Tamaki, K.; Otaka, M.; Takada, M.; Yamamoto, S.; Odashima, M.; Itoh, H.; Watanabe, S. Evidence for Enhanced Cytoprotective Function of HSP90-Overexpressing Small Intestinal Epithelial Cells. *Dig. Dis. Sci.* **2011**, *56*, 1954–1961. [CrossRef]
26. Duquesnoy, R.J.; Liu, K.; Fu, X.-F.; Murase, N.; Ye, Q.; Demetris, A.J. Evidence for heat shock protein immunity in a rat cardiac allograft model of chronic rejection. *Transplantation* **1999**, *67*, 156–164. [CrossRef]
27. Qian, B.F.; El-Salhy, M.; Danielsson, A.; Shalaby, A.; Axelsson, H. Changes in intestinal endocrine cells in the mouse after unilateral cervical vagotomy. *Histol. Histopathol.* **1999**, *14*, 453–460.
28. Acosta, S.; Dizeyi, N.; Pierzynowski, S.; Alm, P.; Abrahamsson, P.A. Neuroendocrine cells and nerves in the prostate of the guinea pig: Effects of peripheral denervation and castration. *Prostate* **2001**, *46*, 191–199. [CrossRef]
29. David, A.I.; Selvaggi, G.; Ruiz, P.; Gaynor, J.J.; Tryphonopoulos, P.; Kleiner, G.I.; Moon, J.I.; Nishida, S.; Pappas, P.A.; Conanan, L.; et al. Blood Citrulline Level Is an Exclusionary Marker for Significant Acute Rejection After Intestinal Transplantation. *Transplantation* **2007**, *84*, 1077–1081. [CrossRef] [PubMed]
30. Dindelegan, G.; Liden, H.; Kurlberg, G.; Oltean, M.; Nilsson, O.; Åneman, A.; Lycke, N.; Olausson, M. Laser-Doppler flowmetry is reliable for early diagnosis of small-bowel acute rejection in the mouse. *Microsurgery* **2003**, *23*, 233–238. [CrossRef] [PubMed]
31. Casselbrant, A.; Söfteland, J.M.; Hellström, M.; Malinauskas, M.; Oltean, M. Luminal polyethylene glycol alleviates intestinal preservation injury irrespective of molecular size. *J. Pharmacol. Exp. Ther.* **2018**, *366*, 29–36. [CrossRef]

metabolites

MDPI

Article

Characterizing Autophagy in the Cold Ischemic Injury of Small Bowel Grafts: Evidence from Rat Jejunum

Ibitamuno Caleb [1,*], Luca Erlitz [1], Vivien Telek [1], Mónika Vecsernyés [2], György Sétáló, Jr. [2], Péter Hardi [1], Ildikó Takács [1], Gábor Jancsó [1] and Tibor Nagy [1]

[1] Institute of Surgical Research and Techniques, University of Pécs Medical School, 7624 Pécs, Hungary; lucaerlitz@gmail.com (L.E.); telek.vivien@pte.hu (V.T.); hardipet@gmail.com (P.H.); dr.takacsildi@gmail.com (I.T.); jancsogabor@hotmail.com (G.J.); ntibor85@gmail.com (T.N.)

[2] Central Electron Microscope Laboratory, Institute of Medical Biology, University of Pécs Medical School, 7624 Pécs, Hungary; kali29@gmail.com (M.V.); gyorgy.setalo.jr@aok.pte.hu (G.S.J.)

* Correspondence: ibical@yahoo.com

Abstract: Cold ischemic injury to the intestine during preservation remains an unresolved issue in transplantation medicine. Autophagy, a cytoplasmic protein degradation pathway, is essential for metabolic adaptation to starvation, hypoxia, and ischemia. It has been implicated in the cold ischemia (CI) of other transplantable organs. This study determines the changes in intestinal autophagy evoked by cold storage and explores the effects of autophagy on ischemic grafts. Cold preservation was simulated by placing the small intestines of Wistar rats in an IGL-1 (Institute George Lopez) solution at 4 °C for varying periods (3, 6, 9, and 12 h). The extent of graft preservation injury (mucosal and cellular injury) and changes in autophagy were measured after each CI time. Subsequently, we determined the differences in apoptosis and preservation injury after activating autophagy with rapamycin or inhibiting it with 3-methyladenine. The results revealed that ischemic injury and autophagy were induced by cold storage. Autophagy peaked at 3 h and subsequently declined. After 12 h of storage, autophagic expression was reduced significantly. Additionally, enhanced intestinal autophagy by rapamycin was associated with less tissue, cellular, and apoptotic damage during and after the 12-h long preservation. After reperfusion, grafts with enhanced autophagy still presented with less injury. Inhibiting autophagy exhibited the opposite trend. These findings demonstrate intestinal autophagy changes in cold preservation. Furthermore, enhanced autophagy was protective against cold ischemia–reperfusion damage of the small bowels.

Keywords: autophagy; apoptosis; ischemia–reperfusion; cold preservation; mucosal injury; small bowel grafts

Citation: Caleb, I.; Erlitz, L.; Telek, V.; Vecsernyés, M.; Sétáló, G., Jr.; Hardi, P.; Takács, I.; Jancsó, G.; Nagy, T. Characterizing Autophagy in the Cold Ischemic Injury of Small Bowel Grafts: Evidence from Rat Jejunum. *Metabolites* **2021**, *11*, 396. https://doi.org/10.3390/metabo11060396

Academic Editor: Norbert Nemeth

Received: 1 April 2021
Accepted: 14 June 2021
Published: 17 June 2021

1. Introduction

Intestinal transplant (Itx) is indicated for patients presenting life-threatening complications due to long-term total parenteral nutrition, short bowel syndrome, or intestinal failure [1,2]. Recent advancements in immunosuppressive pharmacology, surgical techniques, and postoperative care have contributed greatly to making this procedure a valid therapeutic option [2]. However, complications still exist with Itx, some of which are essentially caused by the unavoidable ischemia and the ensuing reperfusion of the organ [3]. Minimizing ischemic damage to the intestinal graft is, therefore, crucial. Thus, effective measures should be undertaken to ensure its viability [1]. Currently, the widespread strategy for preserving the intestine involves standard retrograde vascular perfusion followed by static cold storage (SCS) [4]. Although hypothermic storage is deployed to avert adverse intestinal insult, with time, grafts still deteriorate [3,5]. This is primarily due to the total absence of tissue perfusion and microcirculation during preservation. The developing tissue hypoxia will result in reduced adenosine triphosphate (ATP) formation, decreased glycolysis (shift to anaerobic type), increased acidosis, and dysfunction of energy-dependent enzymes [6]. The alterations present at the end of the storage period are worsened by reperfusion, initiating a complex injury cascade referred to as ischemia–reperfusion injury (IRI) [7]. Several parameters are used in addition to the mucosal changes to evaluate this ischemic damage of the intestine [8,9]. Ischemic injuries may result in graft dysfunction or non-function and chronic organ failure in the recipient. Advanced IRI of the intestinal graft is also followed by bacterial translocation, post-reperfusion syndrome, and massive fluid and electrolyte shifts, secondary to the impairment of its mucosal barrier [3]. Hence, the current clinical consensus on intestinal preservation allows only for a storage time of less than 10 h [3,10]. Within this limit, evidence suggests that the intestines will only exhibit moderate pathological changes [10].

Ischemic events deprive the affected cell populations of their oxygen and nutrient requirements, resulting in states of starvation and stress. To sustain core cellular function, these cells undergo a starvation-induced "self-eating" process known as autophagy [11]. The autophagy pathway involves the formation of a membrane-bound vesicle, called the autophagosome, that encircles cytoplasmic proteins and organelles. Autophagosome formation is generally under the control of autophagy-related genes (Atg) and associated proteins [12]. Some of the well-studied Atg proteins include Beclin-1 and the microtubule-associated protein 1 light chain 3 (LC3) proteins. By interacting with the autophagy-specific class III PI3k complex, Beclin-1 can initiate autophagy. In contrast, modification of the LC3 protein to LC3II is considered an essential step for the maturation of autophagosomes [13–15]. Both Beclin-1 and LC3II proteins are frequently used to monitor the autophagy process [15,16]. The mature autophagosomes will fuse and empty their contents into lysosomes for degradation. The degradation products are then released back into the cytoplasm and re-utilized to aid metabolism [12,13]. Studies into autophagy using various models show that this pathway is rapidly upregulated under conditions of starvation, hypoxia, or ischemia [14,17–21].

Extended periods of ischemia can also elicit an apoptotic response in cells. Apoptosis is often described as the major form of cell death during warm and cold ischemia [22]. Increased caspase-3 activation and enterocyte apoptosis after varying periods of cold preservation have also been depicted in intestinal models [6,23]. Apoptosis and autophagy have a complex relationship [24,25]. Typically, an early autophagic response is associated with cell survival. In contrast, the upregulation of apoptosis will hasten cell demise, in some cases, by even inactivating autophagy [24,25]. The interaction between both pathways has been described as important for organ survival during ischemia [19,26–28].

The role of autophagy during cold preservation tends to show variation amongst the different organs. In the kidney, for example, a protective effect of autophagy during cold preservation has been described in the literature and that inhibition of this pathway accelerates ischemic injury [26]. Meanwhile, in the lungs, the opposite has been suggested [17,20]. Therefore, understanding the influence of autophagy on intestinal preservation may prove significant. Previous studies on the intestine have shown that disruption of the epithelial barrier by amino acid deprivation can induce protective autophagy in intestinal epithelial cells [29]. More so, the benefits of regulating autophagy have been demonstrated in the warm IRI of the intestine [14]. Hence, this study is designed to explore the changes in autophagy during the cold preservation of small bowel grafts. Furthermore, we determine whether modulating this pathway can play a protective role in the intestinal ischemia induced by cold storage.

2. Results

2.1. Preservation Injury of Small Bowel Grafts

2.1.1. Histology

We evaluated the extent of intestinal histological injury amongst the groups according to Park/Chiu's classification system [8]. Intestinal mucosal injury was significantly higher in all the preservation groups (3, 6, 9, and 12 h) compared to the operated control group (0 h) ($p \leq 0.0001$). High injury grades (Park/Chiu ≥ 4), characterized by frequently denuded villi, loss of villous tissue, and injured crypts, occurred more frequently after 12 h of preservation (4.27 ± 0.10) (Table 1).

Table 1. Mucosal injury scores after cold preservation.

Groups	0 h	3 h	6 h	9 h	12 h
Post-preservation	0.16 ± 0.04	1.13 ± 0.09 *	2.21 ± 0.12 *	2.86 ± 0.10 *	4.27 ± 0.10 *

Data are mean ± SEM, n = 6. * $p < 0.0001$ versus the 0 h group (operated control).

By quantitative analysis, we also measured changes in mucosal thickness and crypt depth as indicators of mucosal atrophy. A significant decrease in mucosal thickness in the 3, 6, 9, and 12 h preservation groups was observed when compared to the operated control ($p < 0.005$) (Figure 1a). Similarly, crypt depth decreased with increasing time. This decrease was significant after 9 and 12 h of preservation ($p = 0.0103$ and $p = 0.0005$, respectively) (Figure 1b).

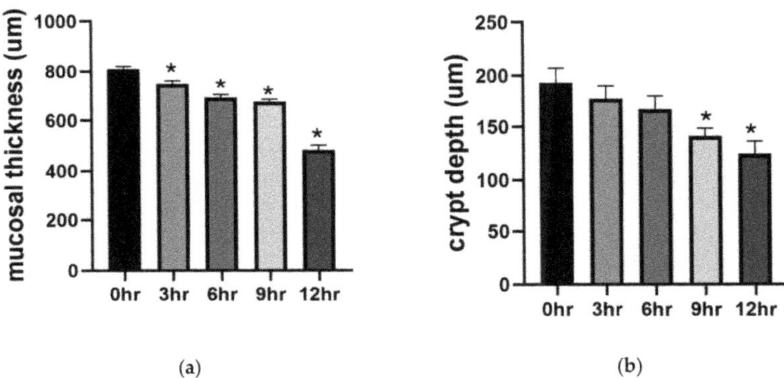

(a) (b)

Figure 1. Morphometric analysis of the mucosa. Panel (**a**) shows changes in mucosal thickness. Panel (**b**) shows changes in crypt depth. Data are mean ± SEM, n = 6. * $p < 0.05$ versus the 0 h group (operated control).

2.1.2. Goblet Cell Abundance

Goblet cells are present in the entire intestinal mucosa and secrete mucin, which is protective against luminal insults and bacteria gaining access to the epithelium. They are rapidly depleted in response to ischemia, whereas they contribute significantly to mucosal restitution after ischemia–reperfusion [30]. Using periodic acidic Schiff (PAS) staining, we were able to estimate the presence of mucin-containing goblet cells in the mucosa of our experimental groups. From the results, the number of goblet cells decreased slightly but significantly in the 3 h group compared to the operated control ($p = 0.0254$). The decrease was more pronounced in the 6, 9, and 12 h groups ($p < 0.0001$) (Figure 2).

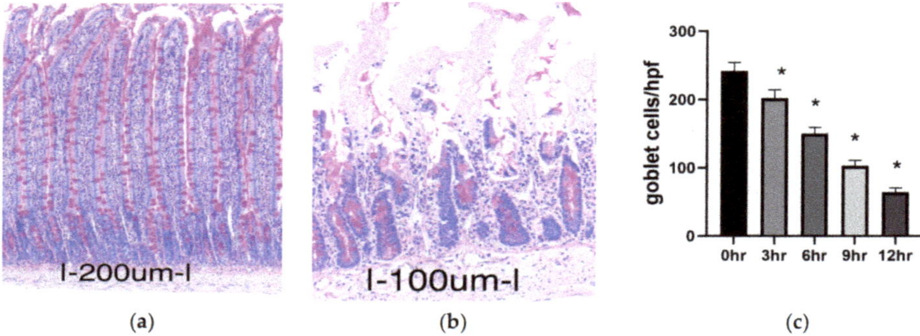

Figure 2. Goblet cell abundance in intestinal mucosa. Panel (**a**) is a representative photograph taken from the operated control (0 h) group. Panel (**b**) is a representative photograph taken from the 12 h group. The goblet cells containing mucin are stained purple (PAS staining positive). Panel (**c**) represents the changes in goblet cell numbers among the groups. Data are mean ± SEM, n = 6. * $p < 0.05$ increase versus the 0 h group (operated control).

2.1.3. Biochemical Analysis

By biochemical analysis of the preservation fluid, we quantified the released lactate dehydrogenase enzyme (LDH) and lactate from the preserved grafts. LDH, as a general marker of cell injury, has been associated with ischemic events in the intestine. Similarly, changes in lactate are frequently used as a marker of cellular hypoxia [9,31]. The results show a significantly increased concentration of both markers in the 6 h group and the 12 h group in contrast to the 0 h group (LDH: $p < 0.0001$; lactate: $p < 0.0001$) (Table 2).

Table 2. Biochemical analysis of preservation fluid.

Groups	0 h	6 h	12 h
LDH (IU/L)	27.44 ± 2.17	349.40 ± 15.86 *	716.30 ± 34.02 *
Lactate (mmol/L)	0.07 ± 0.01	0.68 ± 0.06 *	1.55 ± 0.06 *

Data are mean ± SEM, n = 6. * $p < 0.0001$ versus 0 h.

2.2. *Autophagy Changes during Cold Preservation of Small Bowel Grafts*

To quantify autophagic changes in intestinal tissues following cold preservation, the protein expression of two common markers of autophagy, LC3II and Beclin-1, were examined using immunoblot and immunostaining techniques. Our Western blot results for LC3II illustrated a significant increase in the 3, 6, and 9 h groups compared to the control ($p < 0.0001$) (Figure 3a,d). In the 12 h group, the LC3II protein expression was significantly decreased ($p < 0.0001$). Results for the Western blot analysis of Beclin-1 also showed a significant increase in the 3 and 6 h groups than in the control ($p < 0.0001$ and $p = 0.0052$, respectively). After 9 h, the values were higher than the control but not to the level of significance ($p = 0.4007$) (Figure 3a,e). In the 12 h group, a significant decrease was also observed compared to the control ($p < 0.0001$). A similar trend was noted in the immunohistochemical staining for the Beclin-1 protein. This method stains the cytosolic presence of Beclin-1. A darker stain signifies more prominence of the protein, and this can be quantified by measuring optical density (OD). Compared to the control, average OD scores were significantly increased in the 3 h group and significantly decreased in the 12 h group ($p = 0.0095$ and $p < 0.0001$, respectively) (Figure 3b,c,f).

We also observed the presence of autophagosomes in the cytoplasm of intestinal epithelial cells (those that have microvilli) using electron microscopy. When compared, more autophagosomes could be observed in the 6 h group than the 0 h or 12 h group. Representative electron microscopy photographs are available as Supplementary Material (S1).

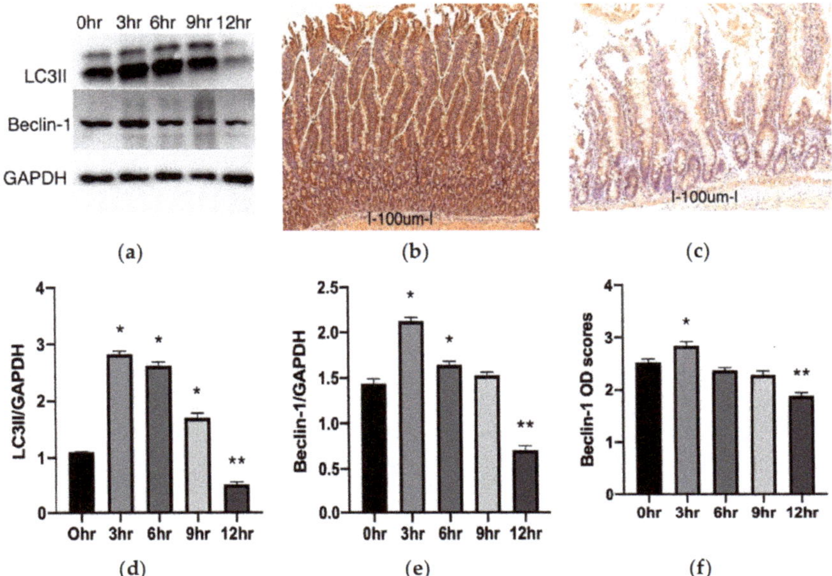

Figure 3. Changes of autophagy in intestinal mucosa. Panel (**a**) shows representative electrophoresis patterns for LC3II and Beclin-1. Panels (**b**,**c**) are representative of immunostaining for Beclin-1 at 0 h (**b**) and 12 h (**c**). Scale bar: 100 µm; Mg: 14.0x. Panels (**d**,**e**) show changes in LC3II and Beclin-1, respectively, with time. Each protein is represented as a ratio of GAPDH. Panel (**f**) illustrates changes in optical density (OD) scores for Beclin-1 immunostaining. Data are mean ± SEM, n = 6. * $p < 0.05$ increase versus the 0 h group (operated control). ** $p < 0.05$ decrease versus the 0 h group.

2.3. Effects of Regulating Autophagy during Cold Preservation of Small Bowel Grafts

2.3.1. Changes in Autophagic Activity

Significant mucosal disintegration and autophagy decline were observed after 12 h of storage in our first study. Therefore, we adopted this extended preservation duration for our second experiment series. To access the influence of autophagy, we treated separate groups with rapamycin (Rapa) or 3-methyladenine (3-MA) shortly before organ retrieval. Rapamycin is frequently used for promoting autophagy, while 3-MA is a known inhibitor of starvation-induced autophagy [14,15]. An untreated group served as a cold preservation control (PC). We verified the influence of these pharmacological agents on autophagic activity by performing immunostaining for Beclin-1 and p62/sqstmi proteins at the beginning (0 h), during (6 h), and at the end of cold preservation (12 h). p62/sqstmi (simply, p62) is another widely used autophagy marker that is selectively incorporated into the autophagosome and degraded by autophagy. The level of p62 protein inversely correlates with autophagic activity [14]. Additionally, at the end of preservation, we performed immunoblot staining for LC3II and Beclin-1 proteins.

Immunostaining for Beclin-1 and p62 showed no significant difference amongst the groups after organ retrieval. By the 6 h mark, Beclin-1 OD scores for the Rapa group were significantly higher ($p = 0.0072$), while those for 3-MA group were significantly less ($p = 0.0337$) than for the PC group. At the end of the preservation period, immunostaining for Beclin-1 in the Rapa-group was still significantly more intense than the control ($p < 0.0001$). The 3-MA group scores were lower but not to a level of statistical significance ($p = 0.48400$). Immunostaining for p62 in the Rapa group had significantly lower OD scores than the PC group after both 6 and 12 h of preservation ($p = 0.0318$ and $p = 0.0007$, respectively). In contrast, the OD scores for p62 immunostaining in the 3-MA group were significantly higher at both 6 and 12 h time points ($p = 0.0018$ and $p = 0.0194$, respectively). Representative pictures for immunostaining for Beclin-1 and p62 are available as Supplementary Material (S2.1). Western blot analysis indicated that autophagy was enhanced in the group treated with rapamycin, as evidenced by significantly increased LC3II expression ($p < 0.0001$) and significantly increased Beclin-1 expression ($p < 0.0001$) when compared to the preservation control. By contrast, autophagy was inhibited in the group of animals treated with 3-methyladenine. We noted significant LC3II ($p < 0.0001$) and Beclin-1 ($p = 0.0073$) decreases compared to the preservation control. (Figure 4). Taken together, these results confirm that rapamycin enhances autophagy activity in the intestinal grafts, while 3-MA inhibits autophagy.

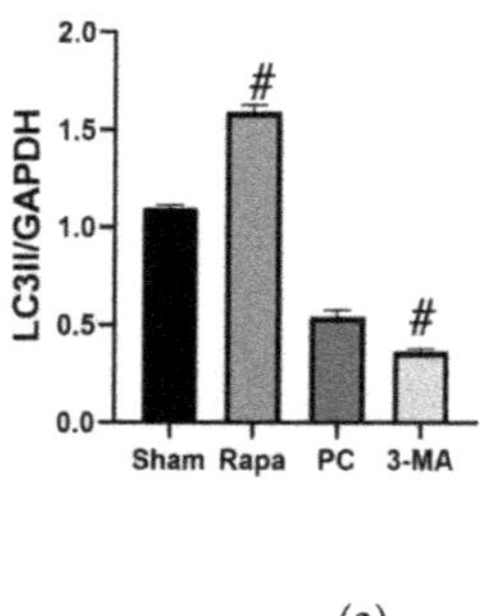

(a)

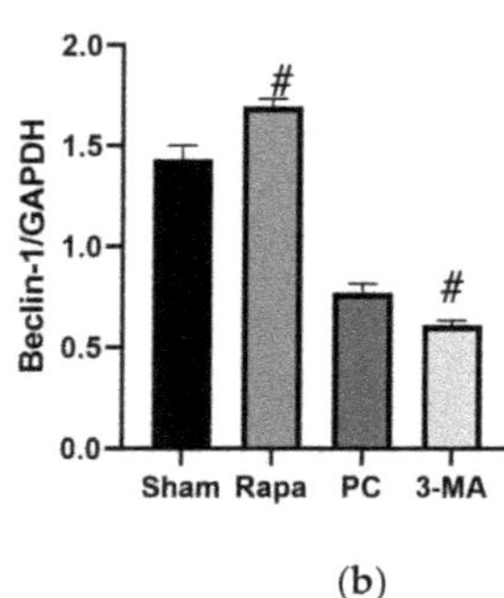

(b)

Figure 4. Changes in autophagy after 12 h long preservation. Panels (**a**,**b**) show changes in immunoblot for LC3II and Beclin-1, respectively. Each protein is represented as a ratio of GAPDH. Data are mean ± SEM, n = 6. # $p \leq 0.05$ versus PC. Electrophoresis patterns for LC3II and Beclin-1 are available as supplementary materials (S2.2).

2.3.2. Effects on Apoptosis

The consequence of promoting or inhibiting autophagy on apoptosis was evaluated at different time points during the preservation period (0, 6, 12 h). Immunostaining for active caspase 3 showed that at the onset of preservation (0 h), there was no significant difference between the Rapa and 3-MA groups versus the PC group ($p = 0.6501$ and $p = 0.8278$, respectively). At 6 h of preservation, there were significantly more positive-stained caspase cells in the 3-MA group than the PC group ($p = 0.0060$), while in the Rapa group, there were significantly fewer positive cells present ($p = 0.0012$) (Figure 5a). After 12 h, the 3-MA group still had a significantly higher amount of positively stained apoptotic cells ($p = 0.0092$) compared to the preservation control, while in the Rapa group, we observed fewer apoptotic cells ($p = 0.0138$) (Figure 5b). Representative pictures for immunostaining for cleaved caspase-3 are available as Supplementary Material (S3). Consistently, at the end of the 12 h long preservation period, the expression of cleaved caspase-3 in Western blot analysis also significantly increased in the 3-MA group ($p = 0.0183$) and decreased in the Rapa group ($p < 0.0001$) when juxtaposed with the PC group (Figure 5c). Altogether, these results suggest that enhancing autophagy attenuates apoptotic damage while inhibiting this autophagy process accelerates apoptotic injury.

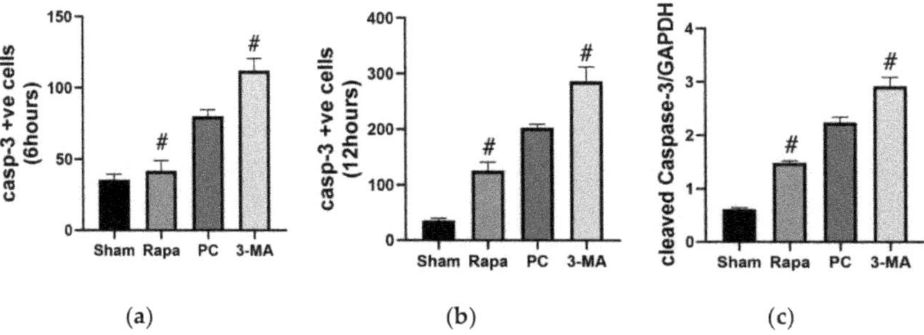

(a) (b) (c)

Figure 5. Changes in apoptosis after autophagy regulation. Panels (**a**,**b**) represent quantitative apoptotic cell counts for positively stained caspase-3 cells at 6 and 12 h, respectively. Panel (**c**) shows changes in expression of cleaved caspase-3 after 12 h of preservation. Data are mean ± SEM, n = 6. # $p \leq 0.05$ versus PC. Representative electrophoresis pattern for cleaved caspase-3 is available as Supplementary Material (S3).

2.3.3. Effects on Intestinal Mucosa

To evaluate the role of autophagy on preservation-induced intestinal mucosa damage, we quantified Park/Chiu scores, mucosal thickness, crypt depth, and goblet cell density after 0, 6, and 12 h of preservation. Mucosal examination was unremarkable between the treated groups (Rapa and 3-MA) and the control group (PC) at the onset of preservation. Injury scores, mucosal thickness, crypt depth, and goblet cell density showed no significant difference between the groups.

After 6 h of preservation, however, the 3-MA group showed higher injury scores, characterized by areas of epithelial breakdown and some denuded villi compared to the PC group ($p = 0.0002$), while the Rapa group displayed only minimal injury in comparison, with some regions containing subepithelial blebbing at the tip of the villus ($p = 0.0307$; Figure 6). Similarly, mucosal thickness and crypt depth decreased significantly in the 3-MA group ($p < 0.0001$ and $p = 0.00339$, respectively). In the Rapa group, the values of mucosal thickness were significantly superior to the PC group ($p = 0.0049$), while the values for crypt depth were higher but not significant enough in comparison ($p = 0.7183$). Goblet cell count was significantly higher in the Rapa group compared to the PC group ($p = 0.0104$). In the 3-MA group, the number of goblet cells reduced significantly ($p = 0.0418$).

At the end of the preservation period (12 h), the Rapa group still displayed significantly lesser tissue injury ($p < 0.0001$), superior mucosal thickness ($p < 0.0001$), and crypt depth ($p = 0.0084$) compared to the control. Similarly, the goblet cell count was significantly higher ($p < 0.0001$)). In contrast, the 3-MA group had worse injury scores, characterized by more regions of crypt damage compared to the control ($p = 0.0050$). Crypt depth was also significantly reduced ($p = 0.0481$), while the mucosal thickness was reduced but not significantly ($p = 0.1577$). Furthermore, the number of goblet cells was less in the 3-MA group ($p = 0.2273$). Picture representation after 12 h of preservation is available as Supplementary Material (S4.1). These results from the intestinal mucosa collectively suggest that enhancing autophagy attenuates preservation-induced mucosal changes while inhibiting this process proves detrimental for the graft mucosa.

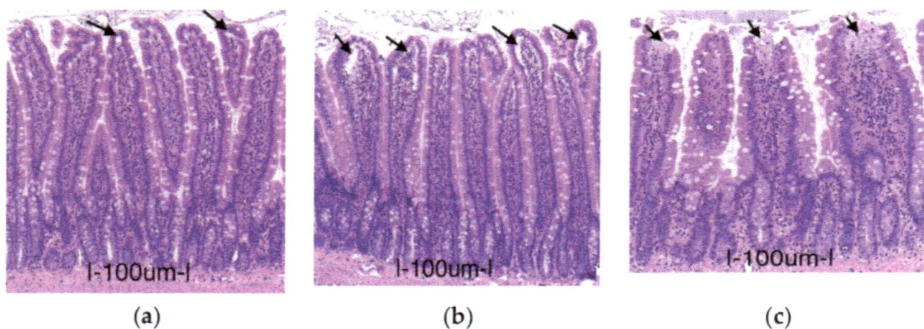

(a) (b) (c)

Figure 6. Representative pictures for mucosa changes after 6 h of preservation. Panel (**a**) represents the Rapa group. The arrows here point to subepithelial bleb at the tip of the villus, which corresponds to a Park/Chiu score of 1. Panel (**b**) represents the PC group. The arrows here point to more extended subepithelial spaces in the upper half of the villus (Park/Chiu score = 2). Panel (**c**) represents the 3-MA group. The arrows here show regions of epithelial breakdown and the denudation of villi (Park/Chiu score = 3).

2.3.4. Effects on LDH and Lactate

To quantify the effect of autophagy on the cellular markers of preservation injury, LDH and lactate release were evaluated after 6 and 12 h of storage. The findings show that the extent of preservation-induced release of LDH and lactate was attenuated in the Rapa group and worsened in the 3-MA group. After 6 h of preservation, there was less release of LDH and lactate into the preservation solution in the Rapa group compared to the PC group ($p = 0.0015$ and $p = 0.0194$, respectively). In contrast, in the 3-MA group, there was significantly more LDH and lactate in the preservation solution ($p < 0.0001$ and $p = 0.0002$, respectively). Similarly, after 12 h of preservation, increased concentrations of LDH and lactate were measured from the preservation fluid in the 3-MA group when juxtaposed with the PC group ($p = 0.0221$ and $p = 0.0090$, respectively). In the Rapa group, both LDH and lactate were significantly less concentrated ($p < 0.0001$). These results indicate that rapamycin-enhanced autophagy attenuates while inhibiting autophagy exacerbates cellular injury during intestinal preservation. Data for changes in LDH and lactate are available as Supplementary Material (S4.2).

2.3.5. Reperfusion Injury

To further confirm the role of autophagy on the preservation-induced ischemia–reperfusion cascade of the intestine, at the end of the 12 h storage period, we re-perfused the grafts briefly. We compared reperfusion injury between the treated and non-treated group by quantifying the histology (Park/Chiu) scores and the release of LDH upon perfusion. The amount of LDH was measured at the end of the first minute and after the 60-min-long perfusion. Compared to the PC group, the grafts with enhanced autophagy (Rapa group) had significantly less mucosal damage ($p < 0.0001$). In contrast, the 3-MA group exhibited worse average injury scores but not significantly ($p = 0.1119$). Similarly, the amount of LDH released in the first minute and 60th minute was significantly less in the Rapa group ($p = 0.0125$ and $p = 0.0005$, respectively) compared to the PC group, while in the 3-MA group, LDH was significantly increased at both times ($p = 0.0233$ and $p = 0.0009$, respectively). These results illustrate that enhancing autophagy limits the extent of reperfusion injury while inhibiting autophagy aggravates this injury. Data for changes after reperfusion are available as Supplementary Material (S5).

2.3.6. The Protective Effects of Rapa-Enhanced Autophagy Are Attenuated by 3-MA

Although rapamycin is well known as a pro-autophagic drug, some other effects, independent of autophagy, have been described in the literature [32]. Hence, we determined whether the effects observed, using rapamycin as an inducer of autophagy, were truly dependent on the autophagic pathway. We compared tissue changes, biochemical changes, and apoptotic changes at the end of the preservation period (12 h) in the Rapa group with a drug control (DC) group (animals here received both 3-methyladenine and rapamycin at the same time). The finding showed that the beneficial effects of administering rapamycin on preservation injuries were indeed due to its effect on autophagy, as evidenced by the Park/Chiu injury scores, which were significantly higher in the DC group compared to the Rapa group ($p < 0.0001$). Similarly, there were also significantly higher concentrations of LDH ($p < 0.0001$) and lactate ($p = 0.0120$) in the DC group than in the Rapa group. Picture representation for mucosal changes and changes in LDH and lactate for the DC group can be found in the Supplementary Material (S4). With regards to apoptosis, the protective role of rapamycin was still due to its pro-autophagic effect. Immunostaining for cleaved caspase-3 revealed more positive cells in the DC group than in the Rapa group ($p = 0.0051$). Similarly, the blot analysis for cleaved caspase 3 showed a significant increase in the DC group in comparison to the Rapa group ($p < 0.0001$). Representative immunostaining and immunoblots for cleaved caspase-3 for the DC group are available in the supplementary material (S5).

3. Discussion

Mucosal damage is the hallmark of tissue injury associated with cold preservation of the intestine. Whatever mucosal injury is present at the end of the preservation period will get worse during reperfusion. Epithelial ulceration, barrier disruption, and increased permeability of the intestinal mucosa may occur and affect clinical outcomes negatively [1,7]. Evaluating the extent of mucosal damage is, therefore, commonly employed to ascertain the integrity of intestinal grafts [1,7]. Our first study examined mucosal changes during cold preservation. We accessed mucosal damage by quantifying changes using the Park/Chiu ischemic injury score system [8], total mucosal thickness, villus depth, and goblet cell counts at different time points. The Park/Chiu scoring system can microscopically describe the progression of mucosal injury during ischemia, while the decrease in mucosal thickness and villus depth indicates atrophic changes. More so, the abundance of mucin-containing goblet cells is an important indicator of the reparative ability of the intestinal mucosa [30]. Our results show that with increasing storage time, the average Park/Chiu scores increased while mucosal thickness, crypt depth, and goblet cell count decreased. At the end of our longest storage time (12 h), we recorded the most apparent mucosal destruction. These results agree with similar studies that have correlated the extent of mucosal damage with the length of the cold ischemic time [5,33].

In addition to mucosal changes, we further characterized the preservation injury of the intestinal grafts by quantifying the release of two general markers of cell hypoxia and injury: lactate and lactate dehydrogenase (LDH) enzymes. Biochemical analysis of the amount of lactate and LDH released into the preservation solution demonstrated that cellular injury worsens with longer cold storage periods. This is consistent with previous studies that have used similar markers [9,31]. Taken together, our results from mucosal and biochemical analyses confirm that lengthening the cold ischemic time worsens preservation injury in small bowel grafts.

The background of ischemic injury is dominated by several metabolic events, including oxidative and inflammatory stress, apoptosis, necrosis, and autophagic responses [22]. Over the past few years, interest in determining the role of autophagy in both warm and cold ischemic events has increased [14,18,28,34]. Autophagy, which has a basal expression level in a variety of cells, is necessary for functional homeostasis. Under conditions of starvation, hypoxemia, and energy depletion, this pathway may be upregulated. The autophagic pathway can then target damaged or redundant cytoplasmic organelles for degradation. The breakdown products of these organelles may serve as important basic metabolites for cellular reuse [11,13,21,35].

In our first experiment, we also focused on whether rat intestinal autophagy can be activated after cold preservation. Our results revealed that autophagic markers LC3II and Beclin-1 proteins showed significant upregulation after 3 and 6 h of preservation compared to the control. Accordingly, using electron microscopy, we could visualize more autophagosomes after 6 h of preservation compared to the control. This seems to confirm the upregulation of the autophagic process during the cold preservation of intestinal grafts. These results agree with other studies that show an increase in autophagy activity during the cold storage of different organs [17,26]. The autophagic pathway is, however, not without constraint. Extended states of starvation will eventually lead to a decline due to enzyme-limiting processes [35]. Accordingly, in our results, after the peak at 3 h of preservation, we noted a steady decrease in LC3II and Beclin-1 proteins. Both LC3II and Beclin-1 proteins show a significant decrease after 12 h of preservation compared to the operated control. We offer two possible explanations for this: First, it is likely that the baseline used for our study (0 h, operated control group) is not the true basal expression of autophagy in intestinal cells. Since our experimental animals were starved for 24 h as part of the protocol and starvation is a strong moderator of autophagy, we suspect that the autophagic process was already upregulated in the intestine before we retrieved the organ. Accordingly, the baseline we adopted is a good one for monitoring changes in autophagy within a limit but should be interpreted cautiously. Secondly, in some instances, prolonged periods of ischemia (starvation) impair autophagy. The consumption and depletion of essential components for autophagy after sustained starvation and the inhibition of key regulators such as Beclin-1 by caspases are said to be behind this phenomenon [25,36]. The exact reason for this occurrence in our results was not investigated further. These time-based differences in the expression of autophagic proteins do, however, suggest a role for autophagy in cold intestinal preservation.

The role of autophagy during warm or cold ischemia varies amongst the different organs [14,19,20,26]. Therefore, in our second experiment series, we investigated the influence of autophagy on the cold preservation ischemia of small bowel grafts. From our results, inhibiting the upregulation of autophagy with 3-methyladenine resulted in a more pronounced preservation injury. The effects of this inhibition were already prominent at 6 h of preservation, lasting until after 12 h of preservation. This suggests that inhibition of the cold-ischemia-induced autophagy accelerates the damage to small bowel grafts. In contrast, by enhancing intestinal autophagy using rapamycin, preservation injury of the intestine was attenuated after 6 h and at the end of the 12-h-long preservation period. Additionally, when the autophagy inhibitor, 3-methyladenine, was added to animals that also received rapamycin, the protective effects of enhancing autophagy were diminished. Collectively, our results indicate that autophagy is protective and that enhancing this pathway will reduce the preservation injury of small bowel grafts.

During severe cold ischemia–reperfusion injury of the intestine, the breakdown of the mucosal barrier function enables bacterial translocation and subsequent septic events. A correlation has previously been established between the rate of this occurrence and the length of the cold ischemic time [23,37]. In view of this, intestinal grafts preserved for extended periods (over 9–10 h) are generally not utilized [10,23]. Ways to prolong the preservation time for intestinal grafts are continually being proposed [3]. In our second study, at the end of the extended preservation period (12 h), we carried out short-term reperfusion. Our results revealed that reperfusion injury was significantly less in the grafts where autophagy was enhanced. Those grafts displayed better mucosal morphology and less release of LDH into the perfusate.

In terms of mucosal damage during ischemia–reperfusion injury, our results contrast those obtained by Li Baochuan et al. They monitored autophagy in the warm ischemia–reperfusion injury of the intestine and concluded that enhanced autophagy was detrimental for the mucosa layer [14]. We measured changes in autophagy after cold storage, as well as mucosa damage after preservation and reperfusion, and we propose a beneficial role. Although the eventual outcome of warm and cold ischemia–reperfusions are similar, their mechanisms develop differently [19]. During cold storage, the preserved organ is entirely cut off from the 'milieu' of tissue perfusion and depends solely on its intrinsic adaptation. In an ongoing warm ischemia–reperfusion injury, however, the possible influence of collateral blood supply on metabolism cannot be ruled out. Furthermore, differences in autophagic adaptation within similar pathologies have been described in the literature [19,38]. Taken together, the need for continuous studies into autophagy is further emphasized.

The protective effects of autophagy have been ascribed, at least in part, to its influence on the apoptotic death pathway [24]. The relationship between autophagy and apoptosis is rather dynamic and depends on the nature of the stimuli [24,25]. In a cell undergoing stress, both pathways can be activated. Autophagy is generally more beneficial and serves to promote cell survival. It can also inhibit apoptosis propagation. However, in some cases, autophagy may induce cell death as well, known as autophagic cell death [25]. Caspase-3 activation and enterocyte apoptosis have been previously reported as facilitators of intestinal damage during cold storage [6,30]. By manipulating autophagy, we determined its effect on cold preservation-induced apoptosis. From the results, enhancing autophagy reduced Caspase-3 activity whilst inhibiting autophagy had the opposite effect. Furthermore, the antiapoptotic role of rapamycin-enhanced autophagy was significantly reduced when 3-methyladenine was also administered in the same animals. Accordingly, we can infer that enhancing autophagy has a positive effect on the cold preservation-induced apoptosis of intestinal mucosa.

In summary, our results point to the involvement and possible role of autophagy in cold preservation injury of the intestine. We observed that enhanced autophagy was protective against cold ischemia and reperfusion damage of small bowel grafts. However, in this experiment, we did not consider all the different cell populations (such as Paneth and immune–modulatory-type cells) present in the intestine. Further studies will be needed to dissect the influence of autophagy on these cells during cold preservation. Nonetheless, the findings of this study provide new insights for researchers in this field.

4. Materials and Methods

4.1. Experimental Animals

Healthy male Wistar rats (n = 60), weighing between 250–300 g, were used for this study. They were housed under standard conditions and fed rat chow and water ad libitum. Food was withdrawn 24 h prior to the experiment. Animals were anesthetized with an intraperitoneal (i.p) mixture of ketamine hydrochloride (0.075 mg/g of body weight) and diazepam (0.075 mg/g of body weight). All procedures were performed in accordance with ethical guidelines (BA02/2000-02/2021) to minimize the pain and suffering of the animals.

4.2. Intestinal Procurement and Grouping

After median laparotomy, the intestine was retrogradely perfused via the aorta at 6 mL/min with ice-cold IGL-1 solution (Institute George Lopez) for 2 min. The portal vein was cut to facilitate venous venting. At the end of the perfusion, small bowel grafts were resected from the ligament of Treitz and stored in the same solution at 4 °C.

In the first series of the experiment, the rats were randomly divided into an operated control group (0 h; n = 6) and a preservation group. The preservation group was subdivided into 4 groups—3, 6, 9, and 12 h groups (n = 6 for each group)—based on the duration of cold storage. Intestinal samples were collected after the resection (0 h; control), and at the end of the preservation periods for Western blot (wb), histology, immunohistochemistry (IHC), and biochemical and transmission electron microscopy (TEM) analyses.

In the second series of the experiment, rats were randomly divided into 5 groups (n = 6 for each group): the sham-operated group (Sham), the preservation control group (PC), the preservation group treated with rapamycin (rapamycin group), the preservation group treated with 3-methyladenine (3-MA group), and the drug control group (DC). A promoter of autophagy, rapamycin (Hb2779 Hellobio; 2 mg/kg dissolved in 1 mL dimethyl sulfoxide solution (DMSO), i.p), was injected into rats belonging to the rapamycin group, 30 min before organ retrieval; an inhibitor of autophagy, 3-MA (Hb2267 Hellobio; 2 mg/kg dissolved in 1 mL ddH20, i.p), was injected into the 3-MA group, 30 min before organ retrieval [14]. The drug control group received the same dose of both rapamycin and 3-methyladenine at the same time. The sham-operated group and the preservation control groups received the same volume of the solvents (ddH20 and DMSO) used. The total time of preservation was 12 h. After cold storage, grafts were perfused using oxygenated Krebs Henslet buffer solution (KHBS) for 60 min according to an ex-vivo method described previously [39]. Intestinal samples and preservation and perfusion fluids were taken at various time points for further analysis.

4.3. *Histology (Hematoxylin–Eosin)*

Tissues were fixed in 10% neutral buffered formalin and embedded in paraffin. They were cut in 3-micrometer-thick sections and stained with hematoxylin and eosin. The slides were digitized with a Mirax scanner, and photographs were taken with CaseViewer 2.4 software (3DHISTECH Ltd. Budapest, Hungary). Intestinal mucosa damage was evaluated blindly by two individuals. The degree of injury was determined using the Park/Chiu system described by Park et al. [8]. A minimum of three fields randomly selected from four quadrants of each intestinal sample was evaluated.

Morphometric analysis of total mucosa thickness and villous depth was analyzed using CaseViewer 2.4 software (3DHISTECH Ltd. Budapest, Hungary). Total mucosa thickness was assessed by measuring the distance between the villus tip to the lamina-muscularis mucosae in at least four axially oriented villi in four quadrants. Crypt depth was determined in at least a total of five axially oriented, open, non-destroyed crypts from three quadrants.

4.4. *Goblet Cell Count*

To evaluate the amount of mucus containing goblet cells in the mucosal layer, tissues were fixed in 10% neutral buffered formalin, embedded in paraffin, cut into 3-micrometer-thick sections with a rotational microtome and mounted on coated glass microscope slides. After deparaffinization and rehydration, the samples were incubated in 1% periodic acid for 20 min, followed by a 0.5 min rinse in distilled water. The samples were stained with Schiff reagent for 20 min, differentiated in Schiff-rinsing solution for 2 min, and immersed in tap water for 5 min to further evolve their color. Slides were then incubated in Meyer's hematoxylin for 10 min, and bluing was performed with tap water for 5 min. The samples were dehydrated in alcohol, cleared in xylene, and mounted with permanent mounting medium. The amount of blue/purple-stained goblet cells was evaluated by manually counting the number of visible cells in at least three high-power fields (hpfs) selected randomly from four different quadrants using Caseviewer 2.4 software (3DHISTECH Ltd., Budapest, Hungary).

4.5. *Immunohistochemistry (IHC)*

Intestinal tissues, fixed in 10% neutral buffered formalin and embedded in paraffin, were cut in serial 3-micrometer-thick sections. After deparaffinization and rehydration, samples were pretreated with the heat-induced epitope retrieval method in 1 mM (pH = 6.0) citrate buffer (Histopathology Ltd.Pécs, Hungary) in a microwave oven for 15 min at 750 W. After cooling at room temperature, the tissues were washed in TRIS buffered saline solution (TBS) (pH = 7.6). For immunohistochemistry, samples were incubated in Beclin-1 antibody (Cat. Nr. bs-1353R, Bioss Antibodies Inc., 1:2000, 1 h at room temperature), p62/sqstmi (Cat. Nr. p0067. Sigma-Aldrich Ltd., USA 1:2000, 1 h at room temperature), and cleaved caspase-3 (Asp175, Cat. Nr. 9661, Cell Signaling Technology, Inc., USA 1:100, 1 h at room temperature). The sections were washed in TBS and incubated with a HISTOLS-R anti-rabbit HRP-labeled detection system (Cat. Nr. 30011.R500, Histopathology Ltd., 30 min at room temperature). After washing in TBS, the reaction was developed with a HISTOLS Resistant AEC Chromogen/Substrate System (Cat. Nr. 30015, Histopathology Ltd.) while controlling the intensity of the staining under a microscope. Sections were counterstained with hematoxylin solution, and bluing was performed with tap water. Samples were dehydrated in alcohol, cleared in xylene, and mounted with permanent mounting medium. Slides were digitized with a Mirax scanner, and photographs were taken with CaseViewer 2.4 software (3DHISTECH Ltd.Budapest, Hungary).

Analysis of Beclin-1 and p62/sqstmi stains was performed with the help of the IHC profiler plug-in of Image J software, and the optical density (OD) was scored according to the method described previously [40].

Analysis of the cleaved-caspase-3-stained cells was achieved by manually counting the visibly stained apoptotic cells in at least four high-power fields (hpfs) randomly selected from four different quadrants using CaseViewer 2.4 software.

4.6. Western Blot (wb)

Intestinal tissue samples were frozen in liquid nitrogen and then manually pulverized in a mortar and dissolved in ice-cold lysis buffer (containing 50 mM Tris, pH 7.4, 150 mM NaCl, 1 mm EGTA, 1 mM Na_3VO_4, 100 mM NaF, 5 μM $ZnCl_2$, 10% glycerol, and 1% Triton X-100 plus 10 μg/mL of the protease inhibitor aprotinin). Lysates were subjected to centrifugation at 40,000× g at 4 °C for 30 min, then the protein concentration of the supernatants was determined using Protein Assay Dye Reagent Concentration (Bio-Rad Gmbh München, Germany) and light absorption measurement at 595 nm. Samples containing 30 ug of denatured total protein were prepared and loaded onto 10% polyacrylamide gels. Proteins separated based on size were electro-blotted for half an hour onto PVDF membranes using the Trans-Blot Turbo semi-dry system (Bio-Rad), then blocked in 3% BSA dissolved in Tris-buffered saline containing 0.2% Tween 20. This was followed by the probing of the membranes with the primary antibodies (Beclin-1, LC3 (Cat. Nr. 2775 and Cat. Nr.4108 Cell Signaling Technology), and cleaved caspase-3), diluted 1:1000 in the blocking solution, at 4 °C overnight. The binding of the antibodies to the membrane was detected by a secondary anti-rabbit IgG conjugated to horseradish peroxidase (Santa Cruz Biotechnology. Texas, USA), diluted 1:10,000. The enhanced chemiluminescent signal was visualized using a Gbox gel documentation system (Syngene, UK). All membranes were then stripped from the antibodies and detected again, as above, for possible loading differences using a primary antibody against GAPDH (Cell Signaling Technology) at a dilution rate of 1:3000. Analysis of band densities was performed using Image J software. Each of the densities was further quantified in relation to GAPDH.

4.7. TEM

The intestines were cut into large blocks of approximately 1 mm^3 and fixed overnight at 4 °C in 4% paraformaldehyde with 2.5% glutaraldehyde in phosphate buffer (PB). These blocks were then fixed in 1% osmium tetroxide in 0.1 M PB for 35 min after dehydration in an ascending ethanol series, with uranyl acetate (1%) included in the 70% ethanol stage to increase contrast. The blocks were transferred to propylene oxide before being placed into aluminum foil boats containing Durcupan resin (Sigma) and then embedded in gelatin capsules containing the same resin. Semithin sections were cut with a Leica ultramicrotome and mounted either on mesh or Collodion-coated (Parlodion, Electron Microscopy Sciences, Fort Washington, PA, USA) single-slot copper grids. Additional contrast was provided to these sections with uranyl acetate and lead citrate, and they were examined in a JEM-1400 flash transmission electron microscope.

4.8. Biochemical Analysis

At different time points during preservation and reperfusion, fluid samples were obtained and analyzed for the presence of lactate and lactate dehydrogenase enzyme. After centrifugation (10 min, room temperature, 1500 rcf), both parameters were quantified using the Cobas Integra 400 Plus Analyzer (Roche Diagnostics, GmbH, Mannheim, Germany), following the manufacturer's instructions.

4.9. Statistical Analysis

For statistical evaluation, a one-way analysis of variance (ANOVA) was used, followed by adequate posthoc tests for multiple comparisons. The Kruskal–Wallis test was used for the analysis of histological (Chiu) scores. Comparing changes within a group was performed using the paired *t*-test. All data are represented as the mean ± SEM. The difference was considered statistically significant when the *p*-value was less than 0.05.

Supplementary Materials: The following are available online at https://www.mdpi.com/article/10.3390/metabo11060396/s1. Figure S1: Changes in intestinal mucosa after varying periods of preservation. Figure S2: Impact of autophagy regulation on the expression of LC3II, Beclin-1, and caspase-3 proteins. Figure S3: Influence of autophagy regulation on cold ischemia and reperfusion mucosa damage.

Author Contributions: Conceptualization, I.C., I.T., G.J. and T.N.; Data curation, L.E. and V.T.; Formal analysis, L.E., V.T., M.V. and I.T.; Funding acquisition, T.N.; Investigation, I.C., L.E., V.T., G.S.J. and P.H.; Methodology, I.C., L.E., G.S.J., P.H., G.J. and T.N.; Project administration, M.V. and T.N.; Supervision, G.S.J., P.H. and G.J.; Validation, M.V., I.T. and G.J.; Writing—original draft, I.C.; Writing—review and editing, I.T. and T.N. All authors have read and agreed to the published version of the manuscript.

Funding: Supported by University of Pécs Medical School (AOK-KA2020-06-03—Dr. Nagy Tibor). JEOL JEM-1400Flash TEM electron microscope was funded by the GINOP-2.3.3-15-2016-0002 (New generation electron microscope: 3D ultrastructure).

Institutional Review Board Statement: The study was conducted according to government guidelines (BA02/2000-02/2021).

Informed Consent Statement: Not applicable.

Data Availability Statement: Data are available in the text and upon request from the corresponding author.

Acknowledgments: The authors thank Hajnalka Ábrahám MD, PhD (Institute of Medical Biology and the Central Electron Microscopy Laboratory) for her technical support with the TEM process. We also thank Tania Nengo Luanza Mpati Ndongala and Amirreza Ghamatitavil (Institute of Surgical Research and Techniques) for their help with data collation and processing.

Conflicts of Interest: The authors declare no conflict of interest.

References

1. Balaz, P.; Matia, I.; Jackanin, S.; Rybarova, E.; Kron, I.; Pomfy, M.; Fronek, J.; Ryska, M. Preservation injury of jejunal grafts and its modulation by custodiol and University of Wisconsin perfusion solutions in Wistar rats. *Eur. Surg. Res.* **2004**, *36*, 192–197. [CrossRef] [PubMed]
2. Loo, L.; Vrakas, G.; Reddy, S.; Allan, P. Intestinal transplantation: A review. *Curr. Opin. Gastroenterol.* **2017**, *33*, 203–211. [CrossRef] [PubMed]
3. Oltean, M. Intestinal preservation for transplantation: Current status and alternatives for the future. *Curr. Opin. Organ Transplant.* **2015**, *20*, 308–313. [CrossRef] [PubMed]
4. Abu-Elmagd, K.; Fung, J.; Bueno, J.; Martin, D.; Madariaga, J.R.; Mazariegos, G.; Bond, G.; Molmenti, E.; Corry, R.J.; Starzl, T.E.; et al. Logistics and technique for procurement of intestinal, pancreatic, and hepatic grafts from the same donor. *Ann. Surg.* **2000**, *232*, 680–687. [CrossRef] [PubMed]
5. Lysyy, T.; Finotti, M.; Maina, R.M.; Morotti, R.; Munoz-Abraham, A.S.; Bertacco, A.; Ibarra, C.; Barahona, M.; Agarwal, R.; D'Amico, F.; et al. Human Small Intestine Transplantation: Segmental Susceptibility to Ischemia Using Different Preservation Solutions and Conditions. *Transplant. Proc.* **2020**, *52*, 2934–2940. [CrossRef] [PubMed]
6. Salehi, P.; Spratlin, J.; Chong, T.F.; Churchill, T.A. Beneficial effects of supplemental buffer and substrate on energy metabolism during small bowel storage. *Cryobiology* **2004**, *48*, 245–253. [CrossRef] [PubMed]
7. Varga, J.; Stasko, P.; Toth, S.; Pristasova, Z.; Bujdos, M.; Pomfy, M. Development of jejunal graft damage during intestinal transplantation. *Ann. Transplant.* **2009**, *14*, 62–69. [PubMed]
8. Park, P.O.; Haglund, U.; Bulkley, G.B.; Falt, K. The sequence of development of intestinal tissue-injury after strangulation ischemia and reperfusion. *Surgery* **1990**, *107*, 574–580.
9. Leuvenink, H.G.D.; van Dijk, A.; Freund, R.L.; Ploeg, R.J.; van Goor, H. Luminal preservation of rat small intestine with University of Wisconsin or Celsior solution. *Transplant. Proc.* **2005**, *37*, 445–447. [CrossRef]

10. Tesi, R.J.; Jaffe, B.M.; McBride, V.; Haque, S. Histopathologic changes in human small intestine during storage in viaspan organ preservation solution. *Arch. Pathol. Lab. Med.* **1997**, *121*, 714–718. [PubMed]
11. Mizushima, N.; Komatsu, M. Autophagy: Renovation of Cells and Tissues. *Cell* **2011**, *147*, 728–741. [CrossRef]
12. Yang, Y.; Karsli-Uzunbas, G.; Poillet-Perez, L.; Sawant, A.; Hu, Z.S.; Zhao, Y.H.; Moore, D.; Hu, W.W.; White, E. Autophagy promotes mammalian survival by suppressing oxidative stress and p53. *Genes Dev.* **2020**, *34*, 688–700. [CrossRef] [PubMed]
13. Levine, B.; Kroemer, G. Autophagy in the pathogenesis of disease. *Cell* **2008**, *132*, 27–42. [CrossRef] [PubMed]
14. Li, B.C.; Yao, X.; Luo, Y.H.; Niu, L.J.; Lin, L.; Li, Y.S. Inhibition of Autophagy Attenuated Intestinal Injury After Intestinal I/R via mTOR Signaling. *J. Surg. Res.* **2019**, *243*, 363–370. [CrossRef]
15. Zhang, D.Y.; Qiu, W.; Jin, P.; Wang, P.; Sun, Y. Role of autophagy and its molecular mechanisms in mice intestinal tract after severe burn. *J. Trauma Acute Care Surg.* **2017**, *83*, 716–724. [CrossRef] [PubMed]
16. Klionsky, D.J.; Abdelmohsen, K.; Abe, A.; Abedin, M.J.; Abeliovich, H.; Arozena, A.A.; Adachi, H.; Adams, C.M.; Adams, P.D.; Adeli, K.; et al. Guidelines for the use and interpretation of assays for monitoring autophagy (3rd edition). *Autophagy* **2016**, *12*, 1–222. [CrossRef] [PubMed]
17. Chen, X.; Wu, J.X.; You, X.J.; Zhu, H.W.; Wei, J.L.; Xu, M.Y. Cold ischemia-induced autophagy in rat lung tissue. *Mol. Med. Rep.* **2015**, *11*, 2513–2519. [CrossRef]
18. Chen-Scarabelli, C.; Agrawal, P.R.; Saravolatz, L.; Abuniat, C.; Scarabelli, G.; Stephanou, A.; Loomba, L.; Narula, J.; Scarabelli, T.M.; Knight, R. The role and modulation of autophagy in experimental models of myocardial ischemia-reperfusion injury. *J. Geriatr. Cardiol.* **2014**, *11*, 338–348. [PubMed]
19. Cursio, R.; Colosetti, P.; Gugenheim, J. Autophagy and Liver Ischemia-Reperfusion Injury. *Biomed Res. Int.* **2015**, *2015*. [CrossRef] [PubMed]
20. Liu, S.; Zhang, J.; Yu, B.T.; Huang, L.; Dai, B.; Liu, J.C.; Tang, J. The role of autophagy in lung ischemia/reperfusion injury after lung transplantation in rats. *Am. J. Transl. Res.* **2016**, *8*, 3593–3602. [PubMed]
21. Mejlvang, J.; Olsvik, H.; Svenning, S.; Bruun, J.A.; Abudu, Y.P.; Larsen, K.B.; Brech, A.; Hansen, T.E.; Brenne, H.; Hansen, T.; et al. Starvation induces rapid degradation of selective autophagy receptors by endosomal microautophagy. *J. Cell Biol.* **2018**, *217*, 3640–3655. [CrossRef] [PubMed]
22. Kalogeris, T.; Baines, C.P.; Krenz, M.; Korthuis, R.J. Cell biology of ischemia/reperfusion injury. *Int. Rev. Cell Mol. Biol.* **2012**, *298*, 229–317. [CrossRef]
23. Softeland, J.M.; Bagge, J.; Padma, A.M.; Casselbrant, A.; Zhu, C.L.; Wang, Y.F.; Hellstrom, M.; Olausson, M.; Oltean, M. Luminal polyethylene glycol solution delays the onset of preservation injury in the human intestine. *Am. J. Transplant.* **2021**, *21*, 2220–2230. [CrossRef]
24. Maiuri, M.C.; Zalckvar, E.; Kimchi, A.; Kroemer, G. Self-eating and self-killing: Crosstalk between autophagy and apoptosis. *Nat. Rev. Mol. Cell Biol.* **2007**, *8*, 741–752. [CrossRef]
25. Marino, G.; Niso-Santano, M.; Baehrecke, E.H.; Kroemer, G. Self-consumption: The interplay of autophagy and apoptosis. *Nat. Rev. Mol. Cell Biol.* **2014**, *15*, 81–94. [CrossRef]
26. Turkmen, K.; Martin, J.; Akcay, A.; Nguyen, Q.; Ravichandran, K.; Faubel, S.; Pacic, A.; Ljubanovic, D.; Edelstein, C.L.; Jani, A. Apoptosis and Autophagy in Cold Preservation Ischemia. *Transplantation* **2011**, *91*, 1192–1197. [CrossRef] [PubMed]
27. Wang, K. Autophagy and apoptosis in liver injury. *Cell Cycle* **2015**, *14*, 1631–1642. [CrossRef]
28. Zhu, J.J.; Lu, T.F.; Yue, S.; Shen, X.D.; Gao, F.; Busuttil, R.W.; Kupiec-Weglinski, J.W.; Xia, Q.; Zhai, Y. Rapamycin Protection of Livers From Ischemia and Reperfusion Injury Is Dependent on Both Autophagy Induction and Mammalian Target of Rapamycin Complex 2-Akt Activation. *Transplantation* **2015**, *99*, 48–55. [CrossRef] [PubMed]
29. Yang, Y.; Li, W.; Sun, Y.L.; Han, F.; Hu, C.A.A.; Wu, Z.L. Amino acid deprivation disrupts barrier function and induces protective autophagy in intestinal porcine epithelial cells. *Amino Acids* **2015**, *47*, 2177–2184. [CrossRef] [PubMed]
30. Softeland, J.M.; Casselbrant, A.; Biglarnia, A.R.; Linders, J.; Hellstrom, M.; Pesce, A.; Padma, A.M.; Jiga, L.P.; Hoinoiu, B.; Ionac, M.; et al. Intestinal Preservation Injury: A Comparison Between Rat, Porcine and Human Intestines. *Int. J. Mol. Sci.* **2019**, *20*, 3135. [CrossRef]
31. Sola, A.; De Oca, J.; Gonzalez, R.; Prats, N.; Rosello-Catafau, J.; Gelpi, E.; Jaurrieta, E.; Hotter, G. Protective effect of ischemic preconditioning on cold preservation and reperfusion injury associated with rat intestinal transplantation. *Ann. Surg.* **2001**, *234*, 98–106. [CrossRef] [PubMed]
32. Li, J.; Kim, S.G.; Blenis, J. Rapamycin: One Drug, Many Effects. *Cell Metab.* **2014**, *19*, 373–379. [CrossRef]
33. Lopez-Garcia, P.; Pulido, J.C.; Colina, F.; Jimenez-Romero, C.; de Andres, C.I.; Lopez-Alonso, G.; Loinaz, C.; Gonzalez, M.A.M.; Alonso, I.J.; Molero, F.C.; et al. Histologic Evaluation of Organ Preservation Injury and Correlation With Cold Ischemia Time in 13 Intestinal Grafts. *Transplant. Proc.* **2014**, *46*, 2096–2098. [CrossRef]
34. Zhao, G.; Zhang, W.; Li, L.; Wu, S.; Du, G.H. Pinocembrin Protects the Brain against Ischemia-Reperfusion Injury and Reverses the Autophagy Dysfunction in the Penumbra Area. *Molecules* **2014**, *19*, 15786–15798. [CrossRef] [PubMed]
35. Mizushima, N. Autophagy: Process and function. *Genes Dev.* **2007**, *21*, 2861–2873. [CrossRef]
36. Pallet, N. Response Letter to "Autophagy in Renal Ischemia-Reperfusion Injury: Friend or Foe?". *Am. J. Transplant.* **2014**, *14*, 1466–1467. [CrossRef] [PubMed]
37. Cicalese, I.; Sileri, P.; Green, M.; Abu-Elmagd, K.; Fung, J.J.; Starzl, T.E.; Reyes, J. Bacterial translocation in clinical intestinal transplantation. *Transplant. Proc.* **2000**, *32*, 1210. [CrossRef]

38. Decuypere, J.P.; Ceulemans, L.J.; Agostinis, P.; Monbaliu, D.; Naesens, M.; Pirenne, J.; Jochmans, I. Autophagy and the Kidney: Implications for Ischemia-Reperfusion Injury and Therapy. *Am. J. Kidney Dis.* **2015**, *66*, 699–709. [CrossRef]
39. Brown, M.F.; Ross, A.J.; Dasher, J.; Turley, D.L.; Ziegler, M.M.; Oneill, J.A. The role of leukocytes in mediating mucosal injury of intestinal ischemia reperfusion. *J. Pediatric Surg.* **1990**, *25*, 214–217. [CrossRef]
40. Jafari, S.M.S.; Hunger, R.E. IHC Optical Density Score: A New Practical Method for Quantitative Immunohistochemistry Image Analysis. *Appl. Immunohistochem. Mol. Morphol.* **2017**, *25*, E12–E13. [CrossRef] [PubMed]

Article

Hematological, Micro-Rheological, and Metabolic Changes Modulated by Local Ischemic Pre- and Post-Conditioning in Rat Limb Ischemia-Reperfusion

Csaba Korei [1,2,3], Balazs Szabo [2,3], Adam Varga [2,3], Barbara Barath [2,3], Adam Deak [2], Erzsebet Vanyolos [2], Zoltan Hargitai [4], Ilona Kovacs [4], Norbert Nemeth [2,*] and Katalin Peto [2]

1 Department of Traumatology and Hand Surgery, Faculty of Medicine, University of Debrecen, Bartok Bela ut 2-26, H-4031 Debrecen, Hungary; korei.csaba@med.unideb.hu
2 Department of Operative Techniques and Surgical Research, Faculty of Medicine, University of Debrecen, Moricz Zsigmond u. 22, H-4002 Debrecen, Hungary; balazsszabo929@gmail.com (B.S.); varga.adam@med.unideb.hu (A.V.); barath.barbara@med.unideb.hu (B.B.); deak.adam@med.unideb.hu (A.D.); vanyolos@med.unideb.hu (E.V.); kpeto@med.unideb.hu (K.P.)
3 Doctoral School of Clinical Medicine, University of Debrecen, Nagyerdei krt. 98, H-4032 Debrecen, Hungary
4 Clinical Center, Pathology Unit, Kenezy Campus, University of Debrecen, Bartok Bela ut 2-26, H-4031 Debrecen, Hungary; z.hargy@gmail.hu (Z.H.); dr.kovacs.ilona@kenezy.unideb.hu (I.K.)
* Correspondence: nemeth@med.unideb.hu; Tel./Fax: +36-52-416-915

Abstract: In trauma and orthopedic surgery, limb ischemia-reperfusion (I/R) remains a great challenge. The effect of preventive protocols, including surgical conditioning approaches, is still controversial. We aimed to examine the effects of local ischemic pre-conditioning (PreC) and post-conditioning (PostC) on limb I/R. Anesthetized rats were randomized into sham-operated (control), I/R (120-min limb ischemia with tourniquet), PreC, or PostC groups (3 × 10-min tourniquet ischemia, 10-min reperfusion intervals). Blood samples were taken before and just after the ischemia, and on the first postoperative week for testing hematological, micro-rheological (erythrocyte deformability and aggregation), and metabolic parameters. Histological samples were also taken. Erythrocyte count, hemoglobin, and hematocrit values decreased, while after a temporary decrease, platelet count increased in I/R groups. Erythrocyte deformability impairment and aggregation enhancement were seen after ischemia, more obviously in the PreC group, and less in PostC. Blood pH decreased in all I/R groups. The elevation of creatinine and lactate concentration was the largest in PostC group. Histology did not reveal important differences. In conclusion, limb I/R caused micro-rheological impairment with hematological and metabolic changes. Ischemic pre- and post-conditioning had additive changes in various manners. Post-conditioning showed better micro-rheological effects. However, by these parameters it cannot be decided which protocol is better.

Keywords: limb ischemia-reperfusion; hemorheology; metabolites; ischemic pre-conditioning; ischemic post-conditioning

check for updates

Citation: Korei, C.; Szabo, B.; Varga, A.; Barath, B.; Deak, A.; Vanyolos, E.; Hargitai, Z.; Kovacs, I.; Nemeth, N.; Peto, K. Hematological, Micro-Rheological, and Metabolic Changes Modulated by Local Ischemic Pre- and Post-Conditioning in Rat Limb Ischemia-Reperfusion. *Metabolites* **2021**, *11*, 776. https://doi.org/10.3390/metabo11110776

Academic Editor: Amedeo Lonardo

Received: 26 October 2021
Accepted: 11 November 2021
Published: 13 November 2021

1. Introduction

Acute limb ischemia still remains a great challenge in clinical practice. In connection with traumatic injuries, vascular and orthopedic surgery the blood supply to the limbs is often impaired or temporarily stopped (hypoperfusion, ischemia), resulting in ischemia-reperfusion (I/R) injury associated with significant morbidity and mortality [1–4]. The incidence is estimated at 1.5 cases per 10,000 people [1]. The I/R injury is also a major problem in emergency care. It is very important to reduce the time of ischemia to prevent further organ damage. Indirect tissue damage may be exacerbated by shock or resuscitation. Post-cardiac arrest syndrome can also be considered part of systemic I/R injury, which leads to organ and brain failure, death. Systemic ischemia is associated with hypoxia and a lack of energy at the tissue and cellular level. In case of contradictory but successful resuscitation, recurrent circulation may further aggravate organ failure [4–6]. Repeated periods of ischemia and reperfusion may also cause wound healing disorders [7].

It is essential to restore circulation as soon as possible within the ischemic tolerance time of the given organ or tissue. The critical ischemic time for human muscle tissue is about 2.25 h in warm ischemia, and irreversible muscle damage starts after 3 h of ischemia and is nearly complete at 6 h [8,9]. The restoration of blood supply can further aggravate the ischemic damage and results in endothelial and parenchymal injury, referred to as I/R injury. The etiological factors include Ca^{2+} overload, oxidative stress, leukocyte infiltration, release of inflammatory mediators, endothelial dysfunction, and complement activation [9–11].

To prevent or reduce I/R injury, considerable effort has been made in developing various therapeutic strategies, including pharmacological and surgical approaches [3,4]. A simple strategy is hypothermia and the use of chilled heparin saline [6]. During surgeries, I/R damage can be reduced by graduate reperfusion [12]. Ischemic pre-conditioning has emerged as a powerful experimental method in decreasing ischemic injury by Murry et al. in 1986 in canine myocardium [13]. The method consists of inducing brief ischemic insults to the target organ before the subsequent prolonged ischemia. Its beneficial effect was proven in several organs and tissues [14,15]. However, the clinical application is limited due to the unpredictable onset of an ischemic insult or embolic event. A further development of the method is the remote ischemic pre-conditioning, which was described by Przyklenk in 1993 [16]. It can increase the tolerance against I/R injury in many organs, involving the brain, heart, and kidney [17]. It is based on intermittent short-term interruptions of the blood flow of another organ or extremity before the ischemic period of the target organ. The limitation of the method is that it can be applied only in case of elective interventions. The mechanism of post-conditioning was introduced by Zhao et al. [18]. In contrast to pre-conditioning, the brief periods of ischemia and reperfusion are applied at the onset of the reperfusion, after the target organ ischemia. This technique can be easily applied to the ischemic tissue after the surgery.

The red blood cell deformability and aggregation show significant changes in many pathophysiological conditions. Ischemia-reperfusion injury may cause micro-rheological changes due to metabolic changes, free radical reactions, and acute phase reactions [19–21]. Although a huge number of studies have examined the consequences of I/R injury to skeletal muscle and the protective effect of pre- and post-conditioning [22–24], very few studies have addressed the changes in the micro-rheological parameters [25,26].

In our previous studies, we have found significant alterations in blood rheological parameters due to limb ischemia and reperfusion, however, the magnitude of changes was different. The most obvious alterations were found after 3 h of ischemia (vascular clamp combined with tourniquet) that led to histological changes and rheological deterioration together. The morphological changes could be sustained by local cooling, but not the rheological ones [27]. In rat models of 1- and 2-h ischemia and following reperfusion, microrheological deterioration was found in the early reperfusion period (the first hour of reperfusion) and in a second wave on the first to third postoperative days [28,29]. The time factor is important, depending on the ischemic tolerance of the tissues. The skin and muscles are known to have relatively wide ischemic tolerance. However, it is important to note that the endothelium is very sensitive to hypoxia, and endothelial dysfunction may occur in a short time. Kayar et al. found that even 15 min of ischemia may lead to endothelial dysfunction and rheological alterations in rats [19]. So, the picture is very complex and it is not clear where the boundary of reversible and irreversible changes is. It is not known how large hemorheological change leads to a perfusion problem, and how depressed should the perfusion be to result in further changes in the tissues, not talking about the duration, temperature, and extension of complete ischemia.

The present study was therefore designed to evaluate the alterations in the hematological and micro-rheological parameters due to lower limb ischemia and to compare the expected favorable effect of pre- and post-conditioning in a rat model.

2. Results

2.1. Hematological Parameters

Hematological parameters are shown in Table 1. The white blood cell count increased just after the 120-min ischemia in the I/R ($p = 0.018$ vs. base) and in the PreC groups ($p < 0.001$ vs. base) and was elevated further by the end of the first postoperative (p.o.) week, reaching a significant level in the PreC ($p < 0.001$ vs. base) and the PostC groups ($p = 0.008$ vs. base).

Table 1. Selected hematological parameters in the control, the ischemia-reperfusion (I/R), the ischemic pre-conditioning (PreC), and the ischemic post-conditioning (PostC) groups.

Variable	Group	Base	Reperfusion	1st p.o. Week
WBC [G/L]	Control	9.36 ± 1.85	8.62 ± 2.35	10.4 ± 3.59
	I/R	8.05 ± 1.23	10.77 ± 4.23 *	10.17 ± 4.47
	PreC	7.08 ± 1.11	10.67 ± 2.53 *	11.2 ± 4.15 *
	PostC	7.51 ± 1.57	8.77 ± 3.67	10.65 ± 3.1 *
RBC [T/L]	Control	7.78 ± 0.47	7.49 ± 0.41	7.44 ± 0.37
	I/R	8.38 ± 0.64	7.69 ± 0.71 *	7.09 ± 0.28 *
	PreC	8.35 ± 0.39	7.55 ± 0.56 *	6.93 ± 1.37 *
	PostC	8.18 ± 0.73	7.69 ± 0.77	7.09 ± 0.43 *
Hgb [g/dL]	Control	15.23 ± 0.74	14.89 ± 0.67	14.52 ± 0.83 *
	I/R	15.61 ± 0.82	14.75 ± 0.76 *	13.41 ± 0.51 *
	PreC	15.18 ± 0.41	14.14 ± 0.79 *	12.84 ± 2.56 *
	PostC	15.16 ± 0.68	14.25 ± 0.84 *	13.25 ± 0.78 * #
Hct [%]	Control	46.68 ± 2.45	45.23 ± 2.43	43.6 ± 2.07 *
	I/R	47.4 ± 3.07	44.6 ± 3.49	40.39 ± 1.31 * #
	PreC	46.7 ± 1.41	43.81 ± 2.99	38.96 ± 7.56 *
	PostC	45.93 ± 2.87	44.13 ± 3.16	40.21 ± 2.09 * #
Plt [G/L]	Control	741.2 ± 62.1	626.8 ± 99.5 *	833.6 ± 40.9 *
	I/R	810.1 ± 72.9	747.6 ± 80.5 * #	958.5 ± 100.1 * #
	PreC	726.3 ± 82.7	638.3 ± 63 *	873.1 ± 315.4 *
	PostC	712.6 ± 67.2	625.7 ± 57.2 *	1000 ± 44.5 * #

Means ± S.D.; * $p < 0.05$ vs. base (same group), # $p < 0.05$ vs. control group (same time); RBC: red blood cell count; WBC: white blood cell count; Hgb: hemoglobin concentration; Hct: hematocrit; Plt: platelet count.

Red blood cell count, hemoglobin, and hematocrit values showed a moderate decrease at the beginning of the reperfusion versus base values (RBC in I/R group: $p = 0.021$, in PreC group: $p < 0.001$; Hgb in I/R group: $p = 0.009$, in PreC group: $p < 0.001$, in PostC group: $p = 0.004$). A further decrease was observed by the first *p*.o. week (RBC, Hgb and Hct in I/R, PreC and PostC groups: $p < 0.001$). In the PostC groups, reduction of hemoglobin and hematocrit ($p = 0.001$ for both), as well as hematocrit decrease in I/R group was found to be significant compared to the Control group ($p < 0.001$).

Platelet count showed lower values after the 120-min ischemia compared to base values (Control: $p < 0.001$, I/R: $p = 0.041$, PreC: $p = 0.002$, PostC: $p = 0.001$), and rose by the first p.o. week (Control: $p < 0.001$, I/R: $p < 0.001$, PreC: $p = 0.032$, PostC: $p < 0.001$). The values of the I/R group were higher versus the Control group at the beginning of the reperfusion ($p = 0.002$) and on the first p.o. week ($p = 0.006$). The highest values were found in the PostC group one week after surgery ($p < 0.001$ vs. Control).

2.2. Red Blood Cell Deformability

The cumulative elongation index (EI)—shear stress (SS) curves are shown on Figure 1, the comparative parameterization data are summarized in Table 2.

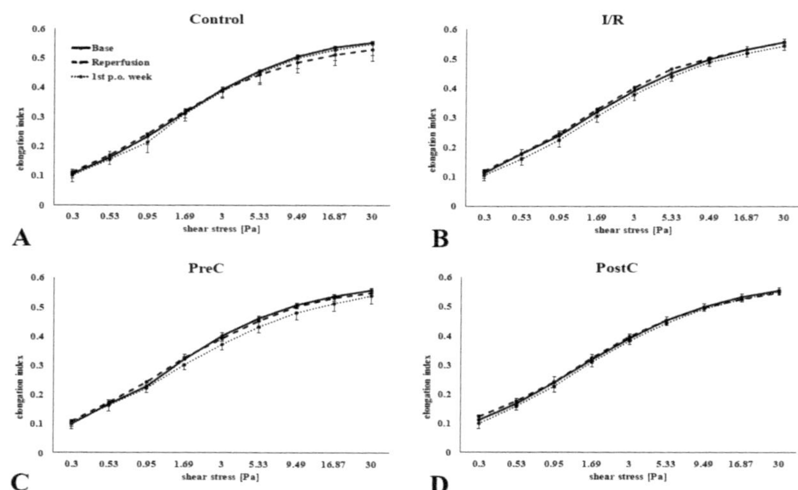

Figure 1. Changes of red blood cell deformability (elongation index in the function of shear stress in the control (**A**), the ischemia-reperfusion (I/R) (**B**), the ischemic pre-conditioning (PreC) (**C**) and the ischemic post-conditioning (PostC) (**D**) groups. Means ± S.D.

Table 2. Erythrocyte deformability values delivered from the elongation index (EI): shear stress (SS) curves in the control, the ischemia-reperfusion (I/R), the ischemic pre-conditioning (PreC), and the ischemic post-conditioning (PostC) groups.

Variable	Group	Base	Reperfusion	1st p.o. Week
EI at 3 Pa	Control	0.393 ± 0.007	0.392 ± 0.024	0.387 ± 0.01
	I/R	0.392 ± 0.013	0.402 ± 0.009	0.378 ± 0.02 *
	PreC	0.400 ± 0.012	0.392 ± 0.012	0.371 ± 0.018
	PostC	0.391 ± 0.013	0.397 ± 0.01	0.382 ± 0.012
EI_{max}	Control	0.582 ± 0.015	0.553 ± 0.049	0.571 ± 0.017
	I/R	0.601 ± 0.02	0.587 ± 0.022	0.581 ± 0.032
	PreC	0.585 ± 0.019	0.589 ± 0.024	0.588 ± 0.046
	PostC	0.595 ± 0.028	0.576 ± 0.023	0.593 ± 0.022
$SS_{1/2}$ [Pa]	Control	1.44 ± 0.25	1.2 ± 0.42	1.53 ± 0.3
	I/R	1.23 ± 0.33	1.26 ± 0.25	1.45 ± 0.27
	PreC	1.36 ± 0.25	1.21 ± 0.23	1.33 ± 0.52
	PostC	1.28 ± 0.46	1.32 ± 0.36	1.34 ± 0.25
$EI_{max}/SS_{1/2}$ [Pa^{-1}]	Control	0.419 ± 0.092	0.52 ± 0.19	0.387 ± 0.087
	I/R	0.524 ± 0.156	0.488 ± 0.136	0.418 ± 0.112
	PreC	0.454 ± 0.143	0.499 ± 0.096	0.553 ± 0.365
	PostC	0.487 ± 0.223	0.473 ± 0.165	0.458 ± 0.103

Means ± S.D.; * $p < 0.05$ vs. base (same group); EI_{max}: maximal EI; $SS_{1/2}$: shear stress at half EI_{max}.

The EI values were lower one week after surgery in the I/R group, and more obviously in the PreC group (Figure 1B,C). However, the differences were small. By the 1st p.o. week EI values at 3 Pa showed significant lowering in the I/R group compared to their base ($p = 0.048$). When analyzing the changes compared to base values, individually, these relative values revealed more differences (Figure 2). The magnitude of lowering in EI at 3 Pa was significant one week after surgery compared to the post-ischemic relative values

(Control: $p = 0.012$, I/R: $p = 0.005$, PreC: $p = 0.025$). The rise in $SS_{1/2}$ was the highest in the I/R group, which was reflected in the $EI_{max}/SS_{1/2}$ ratio ($p = 0.002$) (Figure 2).

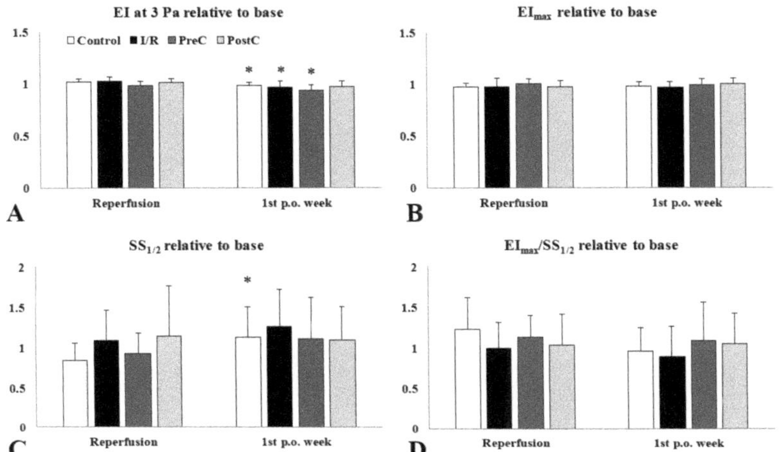

Figure 2. Relative values of elongation index (EI) at 3 Pa (**A**), maximal elongation index (EI_{max}) (**B**), shear stress at half EI_{max} ($SS_{1/2}$) (**C**), and the ratio of EI_{max} and $SS_{1/2}$ (**D**), compared to their base values (as 100%) in the control, the ischemia-reperfusion (I/R), the ischemic pre-conditioning (PreC) and the ischemic post-conditioning (PostC) groups. Means ± S.D., * $p < 0.05$ vs. reperfusion.

2.3. Red Blood Cell Aggregation

Enhanced red blood cell aggregation was found in the ischemic groups (I/R, PreC, PostC) one week after surgery, showing increased M and M1 index values (Figure 3). The I/R and PreC groups expressed the highest values of M 5s ($p < 0.001$ vs. base for both), M1 5s (only in PreC group: $p = 0.038$ vs. base and $p = 0.046$ vs. Control), as well as in M 10s ($p < 0.001$ vs. base for both) and M1 10s index values ($p < 0.001$ vs. base for both) (Figure 3).

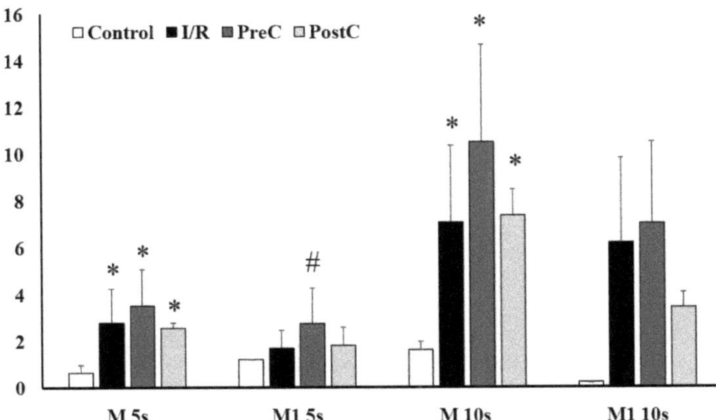

Figure 3. Red blood cell aggregation index values (M 5s, M1 5s, M 10s, M1 10s) in the control, the ischemia-reperfusion (I/R), the ischemic pre-conditioning (PreC), and the ischemic post-conditioning (PostC) groups on the first postoperative week. Means ± S.D., * $p < 0.05$ vs. Control group and # $p < 0.05$ vs. I/R group (same time).

2.4. Blood Gases, Acid-Base Parameters, Electrolytes, and Metabolites

Table 3 summarizes the changes of blood gas, pH, electrolyte, and metabolic parameters.

Table 3. Changes of blood gas (pO_2, pCO_2), pH, electrolytes (Na^+, K^+, Ca^{2+}, Cl^-,) and metabolites (glucose, lactate, creatinine) in the control, the ischemia-reperfusion (I/R), the ischemic pre-conditioning (PreC), and the ischemic post-conditioning (PostC) groups.

Variable	Group	Base	Reperfusion	1st p.o. Week
pO_2 [mmHg]	Control	68.87 ± 13.26	65.73 ± 11.05	55.2 ± 6.93
	I/R	60.14 ± 6.61	61.37 ± 8.49	55.06 ± 9.22
	PreC	65.05 ± 5.75	69.47 ± 15.27	51.78 ± 11.8
	PostC	59.65 ± 9.27	68.08 ± 6.57	57.1 ± 16.21
pCO_2 [mmHg]	Control	41.3 ± 9.04	43.55 ± 5.54	52.62 ± 11.23
	I/R	50.45 ± 6.57	49.94 ± 17.77	49.4 ± 4.72
	PreC	47.64 ± 6.78	48.01 ± 11.17	42.42 ± 4.95
	PostC	47.78 ± 6.62	47.68 ± 12.2	45.12 ± 5.86
pH	Control	7.41 ± 0.03	7.37 ± 0.03	7.36 ± 0.06
	I/R	7.36 ± 0.04	7.33 ± 0.11	7.38 ± 0.04
	PreC	7.38 ± 0.07	7.31 ± 0.07	7.41 ± 0.04
	PostC	7.35 ± 0.03	7.32 ± 0.08	7.39 ± 0.02
Na^+ [mmol/L]	Control	142.12 ± 3.35	141.83 ± 2.13	143.6 ± 2.61
	I/R	141 ± 2.31	140.28 ± 5.4	142 ± 2.34
	PreC	142.28 ± 2.69	140.28 ± 4.46	141.71 ± 2.56
	PostC	142.42 ± 3.2	140.14 ± 2.47	143.85 ± 2.41
K^+ [mmol/L]	Control	4.72 ± 0.32	5.45 ± 0.56 *	4.82 ± 0.29
	I/R	4.32 ± 0.28	5.92 ± 0.57 *	4.04 ± 0.38 #
	PreC	4.25 ± 0.25	5.58 ± 0.86 *	4.61 ± 0.5
	PostC	4.24 ± 0.25	6.05 ± 0.62 *	4.17 ± 0.29 #
Ca^{2+} [mmol/L]	Control	1.36 ± 0.04	1.38 ± 0.11	1.28 ± 0.19
	I/R	1.35 ± 0.04	1.39 ± 0.04	1.23 ± 0.31
	PreC	1.34 ± 0.06	1.39 ± 0.06	1.27 ± 0.17
	PostC	1.39 ± 0.05	1.39 ± 0.03	1.32 ± 0.09
Cl^- [mmol/L]	Control	104.25 ± 1.67	108.16 ± 3.18	104.6 ± 1.34
	I/R	103.14 ± 2.11	105.85 ± 2.19	102.6 ± 1.67
	PreC	103.14 ± 2.03	104.85 ± 2.79	104 ± 1.52
	PostC	104 ± 2.31	106.28 ± 1.6	103.85 ± 2.19
glucose [mmol/L]	Control	19.62 ± 3.54	17.11 ± 3.47	17.94 ± 3.08
	I/R	17.98 ± 2.09	19.6 ± 4.27	12.34 ± 2.21 * #
	PreC	17.22 ± 2.19	22.48 ± 7.61	11.8 ± 2.61 * #
	PostC	16.5 ± 1.09	22.64 ± 4.77 * #	13.12 ± 2.45 * #
lactate [mmol/L]	Control	1.171 ± 0.34	1.52 ± 0.61	1.02 ± 0.32
	I/R	1.61 ± 0.73	1.36 ± 0.34	2.31 ± 0.91 #
	PreC	1.78 ± 0.87	1.72 ± 0.85	3.11 ± 1.89 #
	PostC	1.82 ± 0.99	1.28 ± 0.31	3.39 ± 2.2 #
creatinine [μmol/L]	Control	31 ± 4.37	43.83 ± 10	36.2 ± 3.89
	I/R	35.33 ± 5.68	47.85 ± 8.47	37.2 ± 5.26
	PreC	32.42 ± 3.59	68.42 ± 31.28 *	52.85 ± 25.51
	PostC	43.85 ± 8.45	83 ± 43.77 * # +	35.71 ± 6.15

Means ± S.D.; * $p < 0.05$ vs. base (the same group), # $p < 0.05$ vs. Control group and + $p < 0.05$ vs. I/R group (at the same time).

The values of pO_2, pCO_2 did not change significantly. The pH decreased in the PreC and PostC groups at the start of the reperfusion and normalized by the first p.o. week.

Sodium, calcium, and chloride ion concentrations did not show significant changes. Just after the ischemia, the potassium ion concentration increased significantly in all groups versus the base, with larger magnitude in the groups subjected to ischemia (Control: $p = 0.01$, I/R: $p < 0.001$, PreC: $p = 0.011$, PostC: $p < 0.001$ vs. base). By the first p.o. week, the potassium concentration was lower in I/R and in PostC groups compared to the Control ($p = 0.007$ and $p = 0.004$, respectively).

An increase was observed in the glucose concentration in all groups with ischemia and reperfusion. The rise was significant in the PostC group ($p = 0.002$ vs. base, $p = 0.039$ vs. Control). By the first p.o. week these values significantly decreased in these groups compared to base (I/R: $p = 0.001$, PreC: $p = 0.001$, PostC: $p = 0.006$) and compared to the Control group (I/R: $p = 0.01$, PreC: $p = 0.004$, PostC: $p = 0.013$). Lactate concentration increased significantly by the first p.o. week in the I/R ($p = 0.019$), PreC ($p = 0.037$) and PostC ($p = 0.041$) groups compared to the Control. Creatinine concentration increased significantly only in the PreC ($p = 0.002$ vs. base) and PostC ($p = 0.017$ vs. base, $p = 0.008$ vs. Control, $p = 0.038$ vs. I/R) groups just after the reperfusion.

2.5. Histology

No histological abnormalities attributable to ischemia were seen in any of the groups by light microscopy (Figure 4). Analyzing the postoperative samples preserved striation, regular contour of the muscle fibers was found with the normal distribution of the nuclei without any sign of hypertrophy. No disorganization of myofibrils occurred in the sarcoplasm. Swelling and intense inflammatory infiltration could not be demonstrated. Necrotic muscle fiber sections were not present and no caliber fluctuation was detected. In a pre-conditioned animal, signs of subacute inflammation and fresh bleeding were seen in the perimysium, which may have developed due to the tourniquet compression.

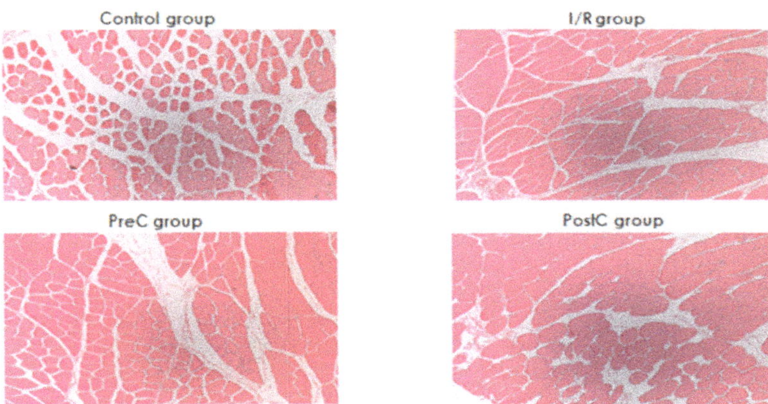

Figure 4. Representative histological slides of the skeletal muscle biopsies in the control, the ischemia-reperfusion (I/R), the ischemic pre-conditioning (PreC), and the ischemic post-conditioning (PostC) groups. Staining: H&E; original magnification: 50×.

3. Discussion

The tourniquet is used in many areas of surgery, such as traumatology, orthopedics, and vascular surgery. In experimental studies, several strategies have been developed to reduce I/R injury, but very few of them have been introduced into clinical practice. Among surgical approaches, pre- and post-conditioning are promising methods [13–18], although there are still many unanswered questions involving the optimal timing of the stimulus and the number and duration of the cycles, etc. It should also be taken into account that the applicability of pre-conditioning in clinical practice is limited to scheduled surgery, while ischemic post-conditioning can be easily applied even in emergency situations [1–4,6].

To answer all questions, it is essential to know the functional and structural changes accompanied by I/R injury. Reviewing the literature, little data is available on changes in the micro-rheological parameters in relation to ischemia-reperfusion injury of extremities and ischemic conditioning surgical maneuvers.

Various models are known for unilateral hind limb ischemia, operating with vascular microvascular clips for clamping the femoral artery, tourniquet or inflated cuff around the thigh [19,27–31]. It is important to note that in rats, remarkable collaterals are existing from the gluteal region [30]. It has been demonstrated that clamping the femoral vessels alone often does not lead to complete ischemia. The tourniquet may compress collaterals as well, but the force applied may cause extended tissue damage. Most of the ischemic pre- and post-conditioning protocols use three or four cycles, the duration of which ranges from 10, 15 and 30 s to 10 min [13,14,18,21,22,24,32]. We have chosen 120-min tourniquet-induced ischemia, preceded or followed by three cycles or 10-min ischemia and reperfusion.

Red blood cell deformability and aggregation show significant changes in many pathophysiological conditions, including ischemia-reperfusion injury. These are mainly due to free radical reactions, metabolic changes and acute phase reactions [20,21,33–37]. Free radicals may damage the red blood cells by lipid peroxidation, methemoglobin formation, as well as by damaging proteins via sulfhydryl cross-linking [19–21]. Metabolic changes may alter the morphological and mechanical properties of red blood cells, which leads to deterioration of red blood cell deformability and disturbed aggregation [21,36]. Acute phase reactions may manifest as a rise in leukocyte count, increase or decrease of platelet count, hemoconcentration, and micro-rheological changes [20,21,34]. Impaired red blood cell deformability and enhanced red blood cell aggregation elevate blood viscosity, increase vascular resistance, and cause perfusion problem in the microcirculatory bed [20,21,38–41].

In this observational study, we found that hemoglobin and hematocrit decreased significantly after reperfusion and one week post-operatively in all ischemic groups. In parallel, the platelet count significantly increased. These changes may be associated with inflammatory processes and acute phase reactions induced by ischemia-reperfusion. The decrease in these values is due not only to I/R damage but also to blood loss caused by surgery and blood sampling. We supposed that the alterations observed post-ischemically are mostly due to redistribution changes, and the later alterations can be originated dominantly from the inflammatory processes.

Similar to our previous findings in other ischemia-reperfusion models [25,26], micro-rheological parameters have deteriorated during and after ischemia in all ischemic groups. Red blood cell aggregation has significantly increased, mostly due to the increased free radical release, acute phase reactions, and inflammatory processes. Interestingly, the most significant increase was found in the PreC group.

Metabolic changes alter the morphological and mechanical properties of blood cells that may result in the deterioration of red blood cells' deformability and disturbed aggregation [20,21,35,36]. Deoxygenated red blood cells have decreased deformability and enhanced aggregation [37], while hypoxia leads to swelling of the cells, altering the cellular surface/volume ratio, and thus the deformability as well [21,36]. In our experiment, the changes in metabolic and micro-rheological parameters were not clearly observed at the same time (post-ischemic vs. first p.o. week values). Therefore, also considering the low sample size, a multivariate regression analysis could not be performed. It is also noted that mathematically significant changes could be detected, but the real in vivo significance of the magnitude in changes of micro-rheological variables are still controversial. It is still not known where the border of reversible and irreversible changes is seen, and it is still obscure whether which magnitude of red blood cell deformability impairment and/or erythrocyte aggregation enhancement causes perfusion problems [21,38–41].

During ischemia, changes in mitochondrial function, enzyme activity, and ion transport may occur. ATP is rapidly dephosphorylated and converted to AMP, which is further degraded. The ion balance of the cells is upset and the intracellular concentration of H^+, Na^+, and Ca^{2+} increases. In the process, free radicals are formed, which also cause damage to enzymes, proteins, carbohydrates, membrane lipids, and DNA. Components released from dead cells initiate inflammatory processes and release cytokines (TNFα, IL-1β, IL-6). Inflammatory processes may be generated and systemic inflammatory response syndrome (SIRS) may develop [9,10]. We harvested tissue biopsies for histological analysis, where no obvious differences were seen between the groups. The compression we used did not cause early complications. However, histological signs of I/R damage were observed in similar animal experiments performed at our institute. Presumably, excellent collateral circulation in the lower limb of the rat [30] reduced the degree of I/R damage we caused to such an extent that, although laboratory abnormalities developed, no histologically detected damage occurred. Ischemia laser Doppler tests confirmed the decreased microcirculatory values on the toe, but not zero values. It is supposed that the applied tourniquet did not cause complete ischemia, only hypoperfusion. The strength of the tourniquet also influences its effect, however, too strong of a compression may cause direct tissue injury, which we wanted to avoid in this model. This is a limitation of the study. We also wished to investigate the possible direct damaging effect of tourniquet applications, therefore, we got biopsies from the thigh muscles. However, in respect of the collaterals [30], examining muscles at the lower region (e.g., plantar flexor complex) would be a better choice for further future studies.

Overviewing the findings, in this study we could see that at the early reperfusion period, significant metabolic and micro-rheological changes occurred. However, their magnitude was not enough to result in visible histological alteration on standard H&E sections. Ischemic pre-conditioning resulted in larger micro-rheological alterations than the post-conditioning protocol. These findings with post-conditioning are comparable to those of other research groups [24,32], however, the issue is still controversial. Some studies have shown its protective effect [23], but enhancement of the damage was also described [42].

Our study used healthy animals without any co-morbidities, with an intact vascular system and normovolemia. Although minimal changes have been observed without significant consequences, they had the potential to have more serious reactions. The effect of the slight changes may add up and be more significant if associated with pathological conditions (arteriosclerosis, bleeding, shock, malnutrition, etc.). There is no perfect model to study all of these aspects. All the animal studies have their own limitations, and numerous factors have to be taken into consideration when planning, conducting the studies and evaluating and extrapolating the results [31]. However, in a human study, we have found comparable results with this recent experimental study, when red blood cell deformability decreased and erythrocyte aggregation enhanced after ischemia-reperfusion by the first and second p.o. days, and non-steroid anti-inflammatory drug administration or ischemic pre-conditioning could moderate the changes in patients with lower extremity operations [43].

4. Materials and Methods

4.1. Experimental Animals

All procedures were approved and registered by the University of Debrecen Committee of Animal Welfare (permission registration Nr.: 25/2016. UDCAW) in accordance with national and EU regulations (Hungarian Animal Protection Act (Law XVIII/1998) and Directive 2010/63/EU). Thirty 8-week male Crl:WI rats were included in the experiment, and were kept in standard cages in alternating day and night light conditions in a 12-h cycle. We provided them with free access to drinking water and conventional rodent chow.

4.2. Operative Techniques and Sampling Protocol

The rats were anesthetized with i.p. injection of ketamine hydrochloride (100 mg/bwkg, CP-Ketamin) and xylazine (10 mg/bwkg, CP-Xylazin), combined with atropine sulfuricum (0.05 mg/bwkg). To maintain anesthesia, one third of the initial dose was administered during the procedure.

The right common carotid artery was cannulated for monitoring blood pressure in all animals. The animals were randomly divided into four groups:

I. Control (C) group (n = 8, 320.4 ± 9 g): besides the common carotid artery cannulation, no other intervention was performed;
II. Ischemia-reperfusion (I/R) group (n = 7, 376.4 ± 42.4 g): unilateral hind limb ischemia was induced by tourniquet application around the thigh, below the right inguinal

region. After 120-min ischemia the tourniquet was completely released to allow full reperfusion;

III. Pre-conditioned (PreC) group (n = 8, 388.6 ± 39.1 g): three cycles of 10-min ischemia and reperfusion (by tightening then releasing the tourniquet, alternately) was applied before the prolonged ischemia, as described in the I/R group;

IV. Post-conditioned (PostC) group (n = 7, 386.7 ± 46 g): the same three cycles of ischemia-reperfusion were introduced at the onset of the reperfusion, after 120-min ischemia as described in the I/R group.

The animals received flunixin meglumine s.c. (10 mg/bwkg) postoperatively.

A laser Doppler device (LD-01 Laser Doppler Tissue Flowmeter, Experimetria Co., Budapest, Hungary; with standard pencil probe MNP100XP, Oxford Optronix Ltd., Abingdon, UK) was used to test microcirculation during the ischemia to confirm ischemia or hypoperfusion. The pencil probe was taken on the skin of the right toe. Artifact-free 20-s recordings were analyzed in each animal. Blood perfusion unit values did not drop to zero, but decreased during tourniquet application versus base values (I/R group: 77.3 ± 19.2%; PreC group: 71.1 ± 9.4%; PostC group: 71 ± 11.3%).

During the experiment, blood samples were taken from the lateral tail vein at the beginning of the procedure (before the 120-min ischemia, as Base), then after the removal of the tourniquet at the beginning of reperfusion, in group IV after the post-conditioning, and one week later in all groups. In the Control group, the timing of the second blood sampling was set 120 min after the preparation and cannulation. Occasionally, 0.3 to 0.5 mL of blood was taken (anticoagulant: 1.8 mg/mL K_3-EDTA). Histological samples were taken from the ischemic muscle one week after the intervention.

4.3. Laboratory Methods

Hematological parameters were determined by a Sysmex K-4500 microcell counter (TOA Medical Electronics Co., Ltd., Kobe, Japan). Red blood cell count (RBC [$\times 10^6$/L]), hematocrit (Hct [%]), hemoglobin (Hgb [g/dL]), white blood cell count (WBC [$\times 10^3$/L]),), and platelet count (Plt [$\times 10^3$/L]) were analyzed in this study.

A LoRRca MaxSis Osmoscan (RR Mechatronics BV, Zwaag, The Netherlands) ektacytometer was used to test red blood cell deformability, determining the elongation index (EI [au]) in the function of shear stress (SS [Pa]) [44,45]. For the tests polyvinylpyrrolidone (PVP), normal phosphate buffered saline (PBS) solution was prepared (PVP: 360 kDa, Sigma-Aldrich Co., St. Luis, MO, USA; PVP-PBS solution viscosity = 29.5 m Pas, osmolality = 300 mOsmol/kg, pH = 7.2). For comparison, the EI values at 3 Pa of shear stress, and by parameterization of EI-SS curves (Lineweaver-Burk equation), maximal elongation index (EI_{max}), the shear stress belonging to the half EI_{max} ($SS_{1/2}$, [Pa]), and their ratio were calculated [46].

To measure red blood cell aggregation in whole blood, we used a Myrenne MA-1 erythrocyte aggregometer (Myrenne GmbH, Roetgen, Germany), based on the light-transmission principle [44,45]. Aggregation was tested in M (shear rate = 0 s^{-1}) and M1 modes (shear rate = 3 s^{-1}) at the 5th and 10th seconds, accordingly, the M 5s, M 10s, M1 5s and M1 10s index parameters were determined.

An epoc® Blood Analysis System (Siemens Healthineers, Erlangen, Germany) was used to test blood partial oxygen and carbon-dioxide tensions (pO_2, pCO_2 [mmHg]), blood pH, electrolytes (Na^+, K^+, Ca^{2+} and Cl^-) and metabolites (glucose [mmol/L], lactate [mmol/L], creatinine [μmol/L]). The test card (per sample) required 0.1 mL of native blood.

4.4. Histology

One week postoperatively, tissue samples were taken from the biceps femoris muscle under general anesthesia. The samples were fixed in 5% formaldehyde, embedded in paraffin, microtomed into 3–5 μm sections, stained with hematoxylin and eosin (H&E), and evaluated under an optical microscope.

4.5. Statistical Analysis

Data are expressed as means ± standard deviation (S.D.). In case of variables when the base values showed high variety, we also analyzed the ratio of changes (relative values vs. its own base in each case). A GraphPad Prism software for Windows, version 8.0 (GraphPad Software Inc., La Jolla, CA, USA) was used for statistical analysis. Differences within and between the groups were analyzed by two-way ANOVA followed by post-hoc Bonferroni-test or Dunn's method, depending on the result of normality test. Probability values (p) less than 0.05 were considered as statistically significantly different.

5. Conclusions

In conclusion, 2-h tourniquet-induced hind limb ischemia and reperfusion caused impairment in red blood cell deformability with enhanced erythrocyte aggregation, accompanied by hematological and metabolic changes promptly after ischemia. Ischemic pre- and post-conditioning (3 × 10-min ischemia with 10-min reperfusion periods prior or after 120-min ischemia) resulted in additive changes of various manners. Post-conditioning showed better micro-rheological effects. However, by these parameters, it cannot be clearly decided which ischemic conditioning protocol is better. Using a longer limb ischemic period (3 or 4 h, with, e.g., vascular clamping plus tourniquet application) would be recommended to use for better studying this issue.

Author Contributions: Conceptualization, C.K., K.P. and N.N.; methodology, investigation, C.K., B.S., A.D. and E.V.; sample preparation, laboratory and histological investigations, A.V., B.B., Z.H., I.K.; data analysis, C.K., A.V., A.D. and N.N.; writing—original draft preparation, C.K., N.N. and K.P.; writing—review and editing, K.P. and N.N.; supervision, N.N. and K.P.; funding acquisition, N.N. All authors have read and agreed to the published version of the manuscript.

Funding: The work was supported by the Bridging Fund of the Faculty of Medicine, University of Debrecen and the National Research, Development and Innovation Office (NKFI-1 "OTKA" K-139184).

Institutional Review Board Statement: The animal experiments were registered and approved by the University of Debrecen Committee of Animal Welfare and by the National Food Chain Safety Office (registration Nr. 25/2016/UDCAW) in accordance with the national (Act XXVIII of 1998 on the protection and sparing of animals) and EU (Directive 2010/63/EU) regulations.

Informed Consent Statement: Not applicable.

Data Availability Statement: The data presented in this study are available on request from the corresponding author.

Acknowledgments: Authors are grateful to the staff of the Department of Operative Techniques and Surgical Research, Faculty of Medicine, University of Debrecen.

Conflicts of Interest: The authors declare no conflict of interest.

References

1. Santistevan, J.R. Acute limb ischemia: An emergency medicine approach. *Emerg. Med. Clin. N. Am.* **2017**, *35*, 889–909. [CrossRef]
2. Olinic, D.M.; Stanek, A.; Tataru, D.A.; Homorodean, C.; Olinic, M. Acute limb ischemia: An update on diagnosis and management. *J. Clin. Med.* **2019**, *8*, 1215. [CrossRef]
3. Naito, H.; Nojima, T.; Fujisaki, N.; Tsukahara, K.; Yamamoto, H.; Yamada, T.; Aokage, T.; Yumoto, T.; Osako, T.; Nakao, A. Therapeutic strategies for ischemia reperfusion injury in emergency medicine. *Acute Med. Surg.* **2020**, *7*, e501. [CrossRef]
4. Natarajan, B.; Patel, P.; Mukherjee, A. Acute lower limb ischemia-etiology, pathology, and management. *Int. J. Angiol.* **2020**, *29*, 168–174. [CrossRef]
5. Walker, A.C.; Johnson, N.J. Critical care of the post-cardiac arrest patient. *Cardiol. Clin.* **2018**, *36*, 419–428. [CrossRef]
6. De Groot, H.; Rauen, U. Ischemia-reperfusion injury: Processes in pathogenetic networks: A review. *Transplant. Proc.* **2007**, *39*, 481–484. [CrossRef]
7. Mustoe, T. Understanding chronic wounds: A unifying hypothesis on their pathogenesis and implications for therapy. *Am. J. Surg.* **2004**, *187*, 65S–70S. [CrossRef]
8. Eckert, P.; Schnackerz, K. Ischemic tolerance of human skeletal muscle. *Ann. Plast. Surg.* **1991**, *26*, 77–84. [CrossRef]

9. Blaisdell, F.W. The pathophysiology of skeletal muscle ischemia and the reperfusion syndrome: A review. *Cardiovasc. Surg.* **2002**, *10*, 620–630. [CrossRef]
10. Eltzschig, H.; Eckle, T. Ischemia and reperfusion—From mechanism to translation. *Nat. Med.* **2011**, *17*, 1391–1401. [CrossRef]
11. Gillani, S.; Cao, J.; Suzuki, T.; Hak, D.J. The effect of ischemia reperfusion injury on skeletal muscle. *Injury* **2012**, *43*, 670–675. [CrossRef] [PubMed]
12. Okamoto, F.; Allen, B.S.; Buckberg, G.D.; Bugyi, H.; Leaf, J. Reperfusion conditions: Importance of ensuring gentle versus sudden reperfusion during relief of coronary occlusion. *J. Thorac. Cardiovasc. Surg.* **1986**, *92*, 613–620. [CrossRef]
13. Murry, C.E.; Jennings, R.B.; Reimer, K.A. Preconditioning with ischemia: A delay of lethal cell injury in ischemic myocardium. *Circulation* **1986**, *74*, 1124–1136. [CrossRef]
14. Pasupathy, S.; Homer-Vanniasinkam, S. Surgical implications of ischemic preconditioning. *Arch. Surg.* **2005**, *140*, 405–409. [CrossRef]
15. Ulus, A.T.; Yavas, S.; Sapmaz, A.; Sakaoğullari, Z.; Simsek, E.; Ersoz, S.; Koksoy, C. Effect of conditioning on visceral organs during indirect ischemia/reperfusion injury. *Ann. Vasc. Surg.* **2014**, *28*, 437–444. [CrossRef]
16. Przyklenk, K.; Bauer, B.; Ovize, M.; Kloner, R.A.; Whittaker, P. Regional ischemic 'preconditioning' protects remote virgin myocardium from subsequent sustained coronary occlusion. *Circulation* **1993**, *87*, 893–899. [CrossRef]
17. Zhang, Y.; Liu, X.R.; Yan, F.; Min, L.Q.; Ji, X.M.; Luo, Y.M. Protective effects of remote ischemic preconditioning in rat hindlimb on ischemia- reperfusion injury. *Neural Regen. Res.* **2012**, *7*, 583–587.
18. Zhao, Z.Q.; Corvera, J.S.; Halkos, M.E.; Kerendi, F.; Wang, N.P.; Guyton, R.A.; Vinten-Johansen, J. Inhibition of myocardial injury by ischemic postconditioning during reperfusion: Comparison with ischemic preconditioning. *Am. J. Physiol. Heart Circ. Physiol.* **2003**, *285*, H579–H588. [CrossRef]
19. Kayar, E.; Mat, F.; Meiselman, H.J.; Baskurt, O.K. Red blood cell rheological alterations in a rat model of ischemia-reperfusion injury. *Biorheology* **2001**, *38*, 405–414.
20. Baskurt, O.K. Mechanisms of blood rheology alterations. In *Handbook of Hemorheology and Hemodynamics*; Baskurt, O.K., Hardeman, M.R., Rampling, M.W., Meiselman, H.J., Eds.; IOS Press: Amsterdam, The Netherlands, 2007; pp. 170–190.
21. Nemeth, N.; Peto, K.; Magyar, Z.; Klarik, Z.; Varga, G.; Oltean, M.; Mantas, A.; Czigany, Z.; Tolba, R.H. Hemorheological and microcirculatory factors in liver ischemia-reperfusion injury—An update on pathophysiology, molecular mechanisms and protective strategies. *Int. J. Mol. Sci.* **2021**, *22*, 1864. [CrossRef]
22. McAllister, S.E.; Ashrafpour, H.; Cahoon, N.; Huang, N.; Moses, M.A.; Neligan, P.C.; Forrest, C.R.; Lipa, J.E.; Pang, C.Y. Postconditioning for salvage of ischemic skeletal muscle from reperfusion injury: Efficacy and mechanism. *Am. J. Physiol. Regul. Integr. Comp. Physiol.* **2008**, *295*, R681–R689. [CrossRef] [PubMed]
23. Park, J.W.; Kang, J.W.; Jeon, W.J.; Na, H.S. Postconditioning protects skeletal muscle from ischemia-reperfusion injury. *Microsurgery* **2010**, *30*, 223–229. [CrossRef]
24. Lintz, J.A.; Dalio, M.B.; Joviliano, E.E.; Piccinato, C.E. Ischemic pre and postconditioning in skeletal muscle injury produced by ischemia and reperfusion in rats. *Acta Cir. Bras.* **2013**, *28*, 441–446. [CrossRef] [PubMed]
25. Magyar, Z.; Mester, A.; Nadubinszky, G.; Varga, G.; Ghanem, S.; Somogyi, V.; Tanczos, B.; Deak, A.; Bidiga, L.; Oltean, M.; et al. Beneficial effects of remote organ ischemic preconditioning on micro-rheological parameters during liver ischemia-reperfusion in the rat. *Clin. Hemorheol. Microcirc.* **2018**, *70*, 181–190. [CrossRef]
26. Varga, G.; Ghanem, S.; Szabo, B.; Nagy, K.; Pal, N.; Tanczos, B.; Somogyi, V.; Barath, B.; Deak, A.; Peto, K.; et al. Renal ischemia-reperfusion-induced metabolic and micro-rheological alterations and their modulation by remote organ ischemic preconditioning protocols in the rat. *Clin. Hemorheol. Microcirc.* **2019**, *71*, 225–236. [CrossRef] [PubMed]
27. Nemeth, N.; Szokoly, M.; Acs, G.; Brath, E.; Lesznyak, T.; Furk, I.; Miko, I. Systemic and regional hemorheological consequences of warm and cold hind limb ischemia-reperfusion in a canine model. *Clin. Hemorheol. Microcirc.* **2004**, *30*, 133–145.
28. Nemeth, N.; Lesznyak, T.; Szokoly, M.; Furka, I.; Miko, I. Allopurinol prevents erythrocyte deformability impairing but not the hematological alterations after limb ischemia-reperfusion in rats. *J. Investig. Surg.* **2006**, *19*, 47–56. [CrossRef]
29. Nemeth, N.; Kiss, F.; Hever, T.; Brath, E.; Sajtos, E.; Furka, I.; Miko, I. Hemorheological consequences of hind limb ischemia-reperfusion differs in normal and gonadectomized male and female rats. *Clin. Hemorheol. Microcirc.* **2012**, *50*, 197–211. [CrossRef]
30. Rosero, O.; Nemeth, K.; Turoczi, Z.; Fulop, A.; Garbaisz, D.; Gyorffy, A.; Szuak, A.; Dorogi, B.; Kiss, M.; Nemeskeri, A.; et al. Collateral circulation of the rat lower limb and its significance in ischemia–reperfusion studies. *Surg. Today* **2014**, *44*, 2345–2353. [CrossRef]
31. Nemeth, N.; Deak, A.; Szentkereszty, Z.; Peto, K. Effects and influencing factors on hemorheological variables taken into consideration in surgical pathophysiology research. *Clin. Hemorheol. Microcirc.* **2018**, *69*, 133–140. [CrossRef] [PubMed]
32. Kocman, E.A.; Ozatik, O.; Sahin, A.; Guney, T.; Kose, A.A.; Dag, I.; Alatas, O.; Cetin, C. Effects of ischemic preconditioning protocols on skeletal muscle ischemia-reperfusion injury. *J. Surg. Res.* **2015**, *193*, 942–952. [CrossRef] [PubMed]
33. Johnson, R.M. pH effects on red blood cell deformability. *Blood Cells.* **1985**, *11*, 317–321. [PubMed]
34. Koppensteiner, R. Blood rheology in emergency medicine. *Semin. Thromb. Hemost.* **1996**, *22*, 89–91. [CrossRef]
35. Reinhart, W.H.; Gaudenz, R.; Walter, R. Acidosis induced by lactate, pyruvate, or HCl increases blood viscosity. *Crit. Care* **2002**, *17*, 38–42. [CrossRef] [PubMed]
36. Cicha, I.; Suzuki, Y.; Tateishi, N.; Maeda, N. Changes of RBC aggregation in oxygenation-deoxygenation: pH dependency and cell morphology. *Am. J. Physiol. Heart Circ. Physiol.* **2003**, *284*, H2335–H2342. [CrossRef]

37. Uyuklu, M.; Meiselman, H.J.; Baskurt, O.K. Effect of hemoglobin oxygenation level on red blood cell deformability and aggregation parameters. *Clin. Hemorheol. Microcirc.* **2009**, *41*, 179–188. [CrossRef]
38. Lipowsky, H.H. Microvascular rheology and hemodynamics. *Microcirculation* **2005**, *12*, 5–15. [CrossRef]
39. Pries, A.R.; Secomb, T.W. Blood flow in microvascular networks. In *Handbook of Physiology, Microcirculation*, 2nd ed.; Tuma, R.F., Duran, W.N., Ley, K., Eds.; Elsevier Academic Press: Amsterdam, The Netherlands, 2008; pp. 3–36.
40. Chandran, K.B.; Rittgers, S.E.; Yoganathan, A.P. Rheology of blood and vascular mechanics. In *Biofluid Mechanics*; Chandran, K.B., Rittgers, S.E., Yoganathan, A.P., Eds.; CRC Press: Boca Raton, FL, USA, 2012; pp. 109–154.
41. Baskurt, O.K. In vivo correlates of altered blood rheology. *Biorheology* **2008**, *45*, 629–638. [CrossRef]
42. Schoen, M.; Rotter, R.; Gierer, P.; Gradl, G.; Strauss, U.; Jonas, L.; Mittlmeier, T.; Vollmar, B. Ischemic preconditioning prevents skeletal muscle tissue injury, but not nerve lesion upon tourniquet-induced ischemia. *J. Trauma* **2007**, *63*, 788–797. [CrossRef]
43. Turchanyi, B.; Korei, C.; Somogyi, V.; Kiss, F.; Peto, K.; Nemeth, N. Beneficial postoperative micro-rheological effects of intraoperative administration of diclophenac or ischemic preconditioning in patients with lower extremity operations—Preliminary data. *Clin. Hemorheol. Microcirc.* 2021; Epub ahead of print. [CrossRef]
44. Hardeman, M.; Goedhart, P.; Shin, S. Methods in hemorheology. In *Handbook of Hemorheology and Hemodynamics*; Baskurt, O.K., Hardeman, M.R., Rampling, M.W., Meiselman., H.J., Eds.; IOS Press: Amsterdam, The Netherlands, 2007; pp. 242–266.
45. Baskurt, O.K.; Boynard, M.; Cokelet, G.C.; Connes, P.; Cooke, B.M.; Forconi, S.; Liao, F.; Hardeman, M.R.; Jung, F.; Meiselman, H.J.; et al. International Expert Panel for Standardization of Hemorheological Methods. New guidelines for hemorheological laboratory techniques. *Clin. Hemorheol. Microcirc.* **2009**, *42*, 75–97. [CrossRef] [PubMed]
46. Baskurt, O.K.; Hardeman, M.R.; Uyuklu, M.; Ulker, P.; Cengiz, M.; Nemeth, N.; Shin, S.; Alexy, T.; Meiselman, H.J. Parameterization of red blood cell elongation index—Shear stress curves obtained by ektacytometry. *Scand. J. Clin. Lab. Investig.* **2009**, *69*, 777–788. [CrossRef] [PubMed]

MDPI
St. Alban-Anlage 66
4052 Basel
Switzerland
Tel. +41 61 683 77 34
Fax +41 61 302 89 18
www.mdpi.com

Metabolites Editorial Office
E-mail: metabolites@mdpi.com
www.mdpi.com/journal/metabolites